CONTROL AND DYNAMIC SYSTEMS

Advances in Theory and Applications

Volume 42

CONTRIBUTORS TO THIS VOLUME

ROSS BALDICK
ANJAN BOSE
G. S. CHRISTENSEN
MARIESA L. CROW
M. J. DAMBORG
J. ENDRENYI
MOTOHISA FUNABASHI
ANIL K. JAMPALA
P. KUNDUR
KWANG Y. LEE
Y. MANSOUR
TAKUSHI NISHIYA
YOUNG MOON PARK
S. A. SOLIMAN
DANIEL J. TYLAVSKY
S. S. VENKATA
LU WANG

CONTROL AND DYNAMIC SYSTEMS

ADVANCES IN THEORY AND APPLICATIONS

Edited by
C. T. LEONDES

Department of Electrical Engineering
University of Washington
Seattle, Washington
and
School of Engineering and Applied Science
University of California, Los Angeles
Los Angeles, California

VOLUME 42: ANALYSIS AND CONTROL SYSTEM TECHNIQUES FOR ELECTRIC POWER SYSTEMS
Part 2 of 4

ACADEMIC PRESS, INC.
Harcourt Brace Jovanovich, Publishers
San Diego   New York   Boston
London   Sydney   Tokyo   Toronto

Academic Press Rapid Manuscript Reproduction

Academic Press, Inc.
San Diego, California 92101

United Kingdom Edition published by
ACADEMIC PRESS LIMITED
24-28 Oval Road, London NW1 7DX

Library of Congress Catalog Card Number: 64-8027

ISBN   0-12-012742-3   (alk. paper)

PRINTED IN THE UNITED STATES OF AMERICA
91 92 93 94   9 8 7 6 5 4 3 2 1

CONTENTS

CONTRIBUTORS vii
PREFACE ix

Concurrent Processing in Power System Analysis 1

Mariesa L. Crow, Daniel J. Tylavsky, and Anjan Bose

Power System Protection: Software Issues 57

S. S. Venkata, M. J. Damborg, and Anil K. Jampala

Voltage Collapse: Industry Practices 111

Y. Mansour and P. Kundur

Reliability Techniques in Large Electric Power Systems 163

Lu Wang and J. Endrenyi

Coordination of Distribution System Capacitors and Regulators:
An Application of Integer Quadratic Optimization 245

Ross Baldick

Optimal Operational Planning: A Unified Approach to Real
and Reactive Power Dispatches 293

Kwang Y. Lee and Young Moon Park

Multistage Linear Programming Methods for Optimal Energy
Plant Operation 341

    Takushi Nishiya and Motohisa Funabashi

Optimization Techniques in Hydroelectric Systems 371

    G. S. Christensen and S. A. Soliman

INDEX 473

CONTRIBUTORS

Numbers in parentheses indicate the pages on which the authors' contributions begin.

Ross Baldick (245), *Department of Electrical Engineering and Computer Sciences, University of California, Berkeley, California 94720*

Anjan Bose (1), *Department of Electrical Engineering, Arizona State University, Tempe, Arizona 85287*

G. S. Christensen (371), *Department of Electrical Engineering, University of Alberta, Edmonton, Alberta, Canada T6G 2G7*

Mariesa L. Crow (1), *Department of Electrical Engineering, Arizona State University, Tempe, Arizona 85287*

M. J. Damborg (57), *Department of Electrical Engineering, University of Washington, Seattle, Washington 98195*

J. Endrenyi (163), *Ontario Hydro Research Division, Toronto, Ontario M8Z 5S4, Canada*

Motohisa Funabashi (341), *Systems Development Laboratory, Hitachi, Ltd., Kawasaki, 215 Japan*

Anil K. Jampala (57), *ESCA Corporation, Bellevue, Washington 98004*

P. Kundur (111), *Ontario Hydro Research Division, Toronto, Ontario M8Z 5S4, Canada*

Kwang Y. Lee (293), *Department of Electrical and Computer Engineering, Pennsylvania State University, University Park, Pennsylvania 16802*

Y. Mansour (111), *PowerTech Laboratories, Inc., Surrey, British Columbia V3W 7R7, Canada*

Young Moon Park (293), *Department of Electrical Engineering, Seoul National University, Seoul 151, Korea*

S. A. Soliman (371), *Electrical Power and Machines Department, Ain Shams University, Abbassia, Cairo, Egypt*

Daniel J. Tylavsky (1), *Department of Electrical Engineering, Arizona State University, Tempe, Arizona 85287*

S. S. Venkata (57), *Department of Electrical Engineering, University of Washington, Seattle, Washington 98195*

Lu Wang (163), *Ontario Hydro Research Division, Toronto, Ontario M8Z 5S4, Canada*

PREFACE

Research and development in electric power systems analysis and control techniques has been an area of significant activity for decades. However, because of increasingly powerful advances in techniques and technology, the activity in electric power systems analysis and control techniques has increased significantly over the past decade and continues to do so at an expanding rate because of the great economic significance of this field. Major centers of research and development in electrical power systems continue to grow and expand because of the great complexity, challenges, and significance of this field. These centers have become focal points for the brilliant research efforts of many academicians and industrial professionals and the exchange of ideas between these individuals. As a result, this is a particularly appropriate time to treat advances in the many issues and modern techniques involved in electric power systems in this international series. Thus, this is the second volume of a four volume sequence in this series devoted to the significant theme of "Analysis and Control System Techniques for Electric Power Systems." The broad topics involved include transmission line and transformer modeling. Since the issues in these two fields are rather well in hand, although advances continue to be made, this four volume sequence will focus on advances in areas including power flow analysis, economic operation of power systems, generator modeling, power system stability, voltage and power control techniques, and system protection, among others.

The first contribution to this volume, "Concurrent Processing in Power System Analysis," by Mariesa L. Crow, Daniel J. Tylavksy, and Anjan Bose, deals with the application of parallel processing to power system analysis as motivated by the requirement for faster computation. This is due to interconnected generation and transmission systems that are inherently very large and that result in problem formulations tending to have thousands of equations. The most common analysis problem, the power flow problem, requires the solution of a large set of nonlinear algebraic equations, approximately two for each mode. Other important problems of very substantial computational complexity include the optimal power flow problem, transient stability. In the case of transient stability problems, a 2,000 bus power network with 300 machines can require on the order of 3,000 differential equations and 4,000 (nonlinear) algebraic equations. Other application areas include short circuit calculations, steady-state stability analysis, reliability calculations, production costing, and other applications. This contribution focuses on techniques for the application of parallel computer methods to these large-scale power system problems which require such methods. Other contributions to this four volume sequence that treat the large-scale power system present methods and algorithms that are potentially applicable to parallel computers.

The next contribution, "Power System Protection: Software Issues," by S.S. Venkata, M. J. Damborg, and Anil K. Jampala, provides a rather comprehensive review and analysis of the past, present, and future of power system protection from a software point of view. Next generation power systems and beyond will operate with minimal spinning margins, and energy transportation will take place at critical levels due to environmental and economic constraints. These factors and others dictate that power systems be protected with optimum sensitivity, selectivity, and time of operation in order to assure maximum reliability and security at minimal costs. Naturally, one of the keys to all this and more will be the associated software issues, as treated in this contribution.

The voltage stability phenomenon has emerged as a major problem currently being experienced by the electric utility industry. The next contribution, "Voltage Collapse: Industry Practices," by Y. Mansour and P. Kundur, presents a rather comprehensive review and analysis of this problem of voltage stability. Major outages attributed to this problem have been experienced on a worldwide basis, and two in-depth surveys of this phenomenon have been conducted on the international scene. Consequently, major challenges in establishing sound analytical procedures and quantitative measures of proximity to voltage are issues facing the industry. This contribution will be an invaluable source reference for researchers and practicing engineers working in this problem area of major significance.

In the next chapter, "Reliability Techniques in Large Electric Power Systems," by Lu Wang and J. Endrenyi, an overview is given of the techniques used in the reliability evaluation of large electric power systems. Particular attention is paid to the reliability assessment of bulk power systems which are the composite of generation and high-voltage transmission (hence often called composite systems). Reliability modeling and solution methods used in these systems are unusually complex. This is partly because of the sheer size of bulk power systems, which usually consist of hundreds, possibly thousands, of components, and partly because of the many ways these systems can fail and the multiplicity of causes for the failures. At an EPRI-sponsored conference in 1978, the observation was made that while reliability methods for other parts of the power system were reasonably well developed, the methods for transmission and composite systems were still in an embryonic stage. The reasons were the same difficulties as those mentioned above. Impressive efforts have been made since then to close the gap, and this review attempts to reflect this development. In fact, this chapter can be considered an update of the relevant chapters in the 1978 book by the second author, *Reliability Modeling in Electric Power Systems*, published by John Wiley & Sons.

External forces such as higher fuel costs, deregulation, and increasing consumer awareness are changing the role of electric utilities and putting pressure on them to become more "efficient." Until recently, increases in efficiency were mostly due to improving generation technology; however, the potential for such improvements has been almost completely exploited. Efficiency improvements are increasingly due to nongeneration technologies such as distribution automation systems, which increase the options for real-time computation, communication, and control. This technology will prompt enormous changes in many aspects of electric power system operation. The next contribution, "Coordination of Distribution System Capacitors and Regulators: An Application of Integer Quadratic Optimization," by Ross Baldick, investigates the potential of such technology to improve efficiency in a radial electric distribution system through the coordination of switched capacitors and regulators. System performance criteria and constraints are, of course, examined in depth, and the relationship between the capacitor and regulation expansion design problem and the coordination problem is carefully considered.

The optimal operation of a power system requires judicious planning for use of available resources and facilities to their maximum potential before investing in additional facilities. This leads to the operational planning problem. The purpose of the operational planning problem is to minimize the fuel costs, system losses, or some other appropriate objective functions while maintaining an acceptable system performance in terms of voltage profile, contingencies, or system security. The operational planning problem was first formulated as an optimal power flow problem by selecting the fuel cost as the objective function and the network or load-flow equations as constraints. The problem was solved for an optimal allocation of real power generation to units, resulting in an economic dispatch. Recently, voltage stability or voltage collapse has been an increasingly important issue to utility as the power system is approaching its limit of operation due to economical and environmental constraints. Generators alone can no longer supply the reactive power that is needed to maintain the voltage profile within the allowed range throughout the power system. Additional reactive power or var sources need to be introduced and coordinated with generators. This has motivated many researchers to formulate optimal reactive power problems, wherein the system loss is used as an objective function, resulting in an economic reactive power dispatch. The next contribution, "Optimal Operational Planning: A Unified Approach to Real and Reactive Power Dispatches," by Kwang Y. Lee and Young Moon Park, is an in-depth treatment of these issues which are substantially complicated as a result of the large system scale nature of these problems.

The development of optimization methods has a long history. However, algorithmic innovation is still required particularly for operating large-scale dynamic plants, which are characteristic of electric power systems. Because of the high dimensionality of the plants, technological bases for the problem in operating such large plants are usually found in the area of linear programming. In applying linear programming for optimal dynamic-plant operation, the constraint appears in the form of a staircase structure. In exploiting this structure, several attempts at devising efficient algorithms have been made. However, when applied to the large-scale system problem area of operational scheduling in electric power systems, significantly greater improvements in speed are required. The next contribution, "Multistage Linear Programming Methods for Optimal Energy Plant Operation," by Takushi Nishiya and Motohisa Funabashi, presents techniques for achieving these requisite speed improvements, which are so essential to electric power systems.

The hydro optimization problem involves planning the use of a limited resource over a period of time. The resource is the water available for hydro generation. Most hydroelectric plants are multipurpose. In such cases, it is necessary to meet certain obligations other than power generation. These may include a maximum forebay elevation not to be exceeded because of the danger of flooding and a minimum plant discharge and spillage to meet irrigational and navigational commitments. Thus, the optimum operation of the hydro system depends upon the conditions that exist over the entire optimization interval. Other distinctions among power systems are the number of hydro stations, their location, and special operating characteristics. The problem of determining the optimal long-term operation of multireservoir power systems has been the subject of numerous publications over the past forty years, and yet no completely satisfactory solution has been obtained, since in every publication the problem has been simplified in order to be solved. The next contribution, "Optimization Techniques in Hydroelectric Systems," by G.S. Christensen and S.A. Soliman, presents an in-depth treatment of issues on effective techniques in this broadly complex area.

This volume is a particularly appropriate one as the second of a companion set of four volumes on analysis and control techniques in electric power systems. The authors are all to be commended for their superb contributions, which will provide a significant reference source for workers on the international scene for years to come.

Concurrent Processing in Power System Analysis

Mariesa L. Crow, Daniel J. Tylavsky, and Anjan Bose

Electrical Engineering Department
Arizona State University
Tempe, Arizona 85287-5706

I   Introduction

The application of parallel processing to power systems analysis is motivated by the desire for faster computation. Except for those analytical procedures that require repeat solutions, like contingency analysis, there are few obvious parallelisms inherent in the mathematical structure of power system problems. Thus, for a particular problem a parallel (or near-parallel) formulation has to be found that is amenable to formulation as a parallel algorithm. This solution has then to be implemented on a particular parallel machine keeping in mind that computational efficiency is dependent on the suitability of the parallel architecture to the parallel algorithm.

The interconnected generation and transmission system is inherently large and any problem formulation tends to have thousands of equations. The most common analysis, the power flow problem, requires the solution of a large set of nonlinear algebraic equations approximately two for each node. The traditional method of using successive linearized solutions (Newton's method) exploits the extreme sparsity of the underlying network connectivity to gain speed and conserve storage. Parallel algorithms for handling dense matrices are not competitive with sequential sparse matrix methods, and,

CONTROL AND DYNAMIC SYSTEMS, VOL. 42

since the pattern of sparsity is irregular, parallel sparse matrix methods have been difficult to develop. The power flow problem defines the steady state condition of the power network and thus, the formulation (or some variation) is a subset of several other important problems like the optimal power flow or transient stability. Hence an effective parallelization of the power flow problem is essential if these other problems are to be handled efficiently.

The transient stability program is used extensively for off-line studies but has been too slow for on-line use. A significant speed up by parallel processing, in addition to the usual efficiencies, will allow on-line transient stability analysis, a prospect that has spurred research in this area. The transient stability problem requires the solution of differential equations that represent the dynamics of the rotating machines together with the algebraic equations that represent the connecting network. This set of differential algebraic equations (DAE) have various nonlinearities and some sort of numerical method is usually used to obtain a step-by-step time domain solution. Each machine may be represented by two to twenty differential equations, and so a 2000 bus power network with 300 machines may require 3000 differential equations and 4000 algebraic equations. In terms of structure, the differential equations can be looked upon as block diagonal (one block for each machine) and the sparse algebraic equations as also providing the interconnection between the machine blocks.

This block diagonal structure has made the transient stability problem more amenable to parallel processing than the power flow problem. Research results to date seem to bear this out. Other power system analysis problems are slowly being subjected to parallel processing by various researchers. Short circuit calculations require the same kind of matrix handling as the power flow and the calculation of electromagnetic transients is mathematically similar to the transient stability solution although the models can be more complicated. Steady-state stability (or small disturbance stability) analysis requires the calculation of eigenvalues for very large matrices. The optimal power flow (OPF) optimizes some cost function using the various limitations of the power system as inequality constraints and the power flow equations as equality constraints. Usually the OPF refers to optimization for one operating condition while unit commitment and hydro-thermal coordination requires optimization over time. This optimization problem, especially if there are many overlapping water and fuel constraints, can be extremely large even without the power flow constraints. Reliability calculations, especially when considering generation and transmission together, can be quite extensive and may require Monte Carlo techniques. Production costing is another large example.

It is the size of these above problems and the consequent solution times that encourages the search for parallel processing approaches. Even before parallel computers became a potential solution, the concept of decomposing a large problem to address the time and storage problems in sequential computers has been applied to many of these power system problems. In fact, there is a rich literature of decomposition/aggregation methods, some more successful than others, that have been specifically developed for these problems. The use of parallel computers can take advantage of these decomposition/aggregation techniques but usually a certain amount of adaptation is necessary. Much of the research in applying parallel processing to power systems has its roots in this literature. This report, however, is confined to examining the efforts that apply parallel computers to specific power system problems rather than the much larger area of methods and algorithms that are potentially applicable to parallel computers.

II   Classification of Parallel Architectures

Parallel processing is a type of information processing in which two or more processors, together with an interprocessor communication system, work cooperatively on the solution of a problem [1]. This cooperative arrangement typically takes the form of *concurrent processing*, or the performance of operations simultaneously with some transfer of information between processors. Two distinct methods of performing parallel tasks have emerged:

- Pipelining

- Replication

Pipelining is the overlapping of parts of operations in time, while replication implies that more than one functional unit is applied to solving the desired problem. These two methods underlie the development of a variety of diverse computer architectures and software algorithms. Unfortunately, progress often breeds complexity. Traditional serial programs are generally transportable from machine to machine with, at most, only a nominal amount of software alteration. However, the diversity of the parallel techniques that makes parallel processing so attractive, also tends to make the software difficult to transport from machine to machine. Therefore, algorithms are usually developed for a specific architecture, rather than striving for general application. Before describing the algorithms which have been developed, or are in development at the time of this writing, it is instructive to characterize the various types of parallel hardware currently available. Most of the parallel processors which are commercially available at this time fall into one of three architectural categories: *vector computers*, which exploit pipelining, *processor arrays*, and *multiprocessors*, both of which utilize replication by containing up to a thousand or more interconnected processing units. The difference between these latter processors lies in the means by which the processing units are managed.

II.A   Vector (Pipelined) Computers

Pipelining is essentially the overlapping of subtasks in time. An often used analogy to pipeline processing is that of assembly lines in an industrial plant [1]. Pipelining is achieved if the input task may be divided into a sequence of subtasks, each of which can be executed by a specialized hardware stage that operates concurrently with other stages in the pipeline. Ideally, all the stages should have equal processing speed, otherwise the stream of tasks will form a bottleneck at the point of the slowest task stage. For a linear sequence of S pipeline stages, the total transversal time for a given job will be S_τ, where τ is the time required for the completion of the slowest task stage. After the pipeline is filled (and before it is drained), the steady-state throughput will be one job per cycle. Given enough jobs of a similar nature, this will represent the average throughput, insensitive, and certainly not proportional to the number of pipeline segments [2]. Whenever a change of operation occurs, the pipeline must be emptied, reconfigured, and then refilled for the new operation. One of the disadvantages of exploiting parallelism by pipelining is that the problem must contain large amounts of identical, repetitive sets of instructions. This requirement stems from the considerable amount of overhead time required to set up, fill, and drain a functional pipeline. One type of problem which is conducive to pipelining, is one in which there exists a high percentage of vector calculations. A specific class of processors, known as vector processors, has been developed to solve these types of problems.

A vector processor is a computer that recognizes instructions involving vector variables as well as scalar variables. It is a natural step to implement vector computers as pipelined processors, since operations on vectors often involve a series of identical, repetitive tasks. It is common for applications to manipulate vectors with 50,000 or more elements [3]. Three of the most common pipelined vector processors are the Cyber-205, built by Control Data

Corporation, the Cray-1, and the Cray X-MP, both by Cray Research, where the X-MP model is an improved version of the Cray-1.

II.B   Processor Arrays

Processor arrays are distinguished by several features unique to their architecture. The processor array is designed to consist of a group of processing modules led by a single centralized control unit. This architecture is especially well suited for applications in which it is desired to perform a series of repetitive calculations on various sets of data, using one set of instructions. All of the processors in an array architecture have identical hardware structure and are programmed to execute the same set of instructions on differing sets of data either in an "enabled" or "disabled" mode of operation. This allows each processing element to respond to conditional statements so that it need not participate in all instruction cycles dictated by the control unit, but otherwise the individual processing units do not have the capability of independent operation. This approach has several practical advantages. By replicating identical models, the design process is simplified because fewer overall designed are needed. The repair problems are reduced by needing only one set of diagnostic tools and replacement parts can service a larger area. Another design advantage is the simplification of control. Since the control unit treats all array modules identically, the task of controlling a large number of modules is logically the same as a small number [4]. Two well known processor arrays are the Burrough's PEPE and the Goodyear Aerospace Massively Parallel Processor (MPP).

II.C   Multiprocessor Systems

In contrast to processor arrays, a multiprocessor system consists of a group of processors which are capable of independent operation, and are synchronized to work in unison on a common problem. The categorical characterization of a multiprocessor system can be described by two primary attributes: first, a multiprocessor is a single computer that includes multiple processors, and second, processors may communicate and cooperate at different levels in solving a given problem [1]. Multiprocessor systems are usually referred to as MIMD (Multiple-Input-Multiple-Data) machines, where "input" refers to the stream of instructions given to the individual processors. Since each processor has the ability to operate independently, the input stream to each processor may be different. The most significant difference between types of multiprocessor systems is the difference in architecture of their storage, communication, and data recovery devices. The two most commonly used categories of interconnection are the *tightly coupled* and *loosely coupled* multiprocessor architectures.

In the general tightly coupled system configuration, the individual processors have access to common memory resources and exchange data among themselves via successive read and write operations performed on designated memory locations. This is commonly known as a *shared memory* multiprocessor. The rate at which data can be communicated from one processor to another is on the order of, and limited by, the bandwidth of the memory. When multiple requests to a shared memory must be served simultaneously, the memory is typically partitioned into modules in a hierarchical fashion so that simultaneous conflicting requests to a single module are rare and are handled in a predetermined manner. Access conflicts are manipulated by appropriately configuring the packet-switch paths to memory [5]. However, if there is a large amount of conflict, this conflict management procedure can result in long access delays. This problem may be partially alleviated by supplying each processor with a small individual memory, or cache, which is controlled by the processor only. Examples of shared memory systems are Sequent's Balance 8000, the Alliant FX/8, Denelcor's HEP, Carnegie-Mellon's C.mmp, and the Cray X-MP (note that this is multiprocessor comprised of a set of vector processors.)

In contrast to shared main memory, loosely coupled multiprocessors utilize a main memory which is comprised of the combination of the local memories of the processors. When no shared memory is used, variables that are required by several processors must be accessed through a message passing system that allows one processor to request some other, possibly distant, processor to modify or transmit the requested data. As a result, this type of multiprocessor is often called a *message passing* processor. These systems do not generally encounter the degree of memory conflicts experienced by shared memory system [1], but the length of time needed to access a particular memory location will depend on the distance of the requesting processor to the memory location. A disadvantage of this type of system is that in order to move data from the memory of one processor into the memory of another processor, both the transmitting and receiving processors must be involved for the duration of the transfer. The time required to pass information among elements that are not adjacent to each other is therefore of concern in selecting a specific topology . The *hypercube* is one approach to minimizing the "maximum distance" between processing elements in large networks [6]. Other interconnection topologies include the *ring, butterfly, hypertrees*, and *hypernets..*

II.D   Multicomputer Systems (Distributed Processing)

Parallel processing and distributed processing are closely related. The primary difference between these two classifications is that distributed processing is achieved by a multiple computer system with several autonomous computers which may or may not communicate with each other [1]. Each autonomous computer has its own memory, therefore there is no global memory referencing, and all communication and synchronization of processes between individual processors is achieved via message passing, or through memory shared between pairs of processors. In the most general case, each processor/computer will have its own operating system and may execute programs written in a different computer language from some or all of the other processors/computers in the distributed system.

Distributed processing is an attractive alternative to centralized computing systems fueled by the decreased cost of standalone computing facilities, improved reliability, and high computing speeds for relatively independent, parallel tasks [7]. A disadvantage to distributed processing is the difficult nature of analysis and control of these systems as compared to centrally controlled parallel processing systems.

Although there has been progress in applying vector processors to power system simulation [8], as well as processor arrays [9], [10], and distributed processing [11], the majority of research efforts have been in the application of multiprocessors for power system simulation. The remainder of this chapter will concentrate on algorithm development for these specific types of machine architectures.

III   Parallel Algorithms for Power System Analysis

There are several methods for designing a parallel algorithm to solve a problem. One possibility is to determine and exploit any inherent parallelism in an existing sequential algorithm. Several compilers exist which can perform simple identifications and may work quite well for some programs. Unfortunately, there exist a significant number of sequential algorithms which have no obvious parallelism. It then becomes necessary to approach the design process from a new direction.

III.A   Factors Which Impact Algorithm Efficiency

Parallelism may be exploited at four different levels:

1.   job level

2.   task level

3.   interinstruction level

4.   intrainstruction level

The last two levels require the user to have an intimate knowledge of the system hardware on which the job is to be processed, with the intrainstruction parallelism being directly exploited by the hardware measures. The job level parallelism is typically the type of parallelism utilized in distributed computing. Therefore, this chapter will address only the exploitation of task parallelism.

Two important measures used in determining how well a parallel algorithm performs when implemented on MIMD machines are speedup and efficiency. The *speedup* achieved by a parallel algorithm running on p processors is the ratio between the time taken by that parallel computer executing the fastest serial algorithm and the time taken by the same parallel computer executing the parallel algorithm using p processors. The *efficiency* of a parallel algorithm running on p processors in the speedup divided by p [3], which is a measure of the percentage of time that all processors are operating. It is desired to minimize any "idle" time, i.e., the time any of the processors are forced to wait for communication delays or synchronization. A number of factors contribute to the upper limit of speedup obtainable by a particular parallel algorithm on a specific MIMD machine. In the following sections, several items which affect the speedup and efficiency of parallel algorithms will be discussed.

Many early parallel algorithms were proposed using the assumption that the time required for communication and memory access would be negligible compared to computation time. This assumption has been shown to be erroneous, with communication time often comprising a large percentage of the total run-time. Unfortunately, predicting communication time is probabilistic in nature and can be considered analogous to predicting when a customer will arrive at a store and how much service they will demand. In order for two or more processors to work in unison on a problem, they must be able to communicate and synchronize their combined efforts.

Synchronization refers to the control of deterministic aspects of computation. The objective of synchronization is to guarantee the correctness of parallel computations such that the results obtained from a parallel execution of a program are the same as those of a sequential execution [6]. Synchronization serves two purposes. Firstly, synchronization insures the integrity of the data being shared between processors in a shared memory system. In a system where local or cache memory is used, it is possible for caches to differ. For example, if cache A and cache B contain the same copy of the data in central memory, and processor 1 updates A, then if processor 2 accesses cache B for the same variable, it will retrieve the incorrect value. There are several schemes available for avoiding this problem. These schemes generally utilize some method of "testing" and "flagging" data to ensure that the data being retrieved is the most recent copy. In message passing systems only one processor "owns" a variable, therefore the problem of data inconsistencies is alleviated.

The second purpose of synchronization is to determine the precedence and sequentiality of a set of tasks. In almost all parallel programs there are critical sections that must be performed serially. Only one processor should be executing the section at a time. Similarly, if all or part of a certain task A must precede task B, the processor to which task B is assigned must wait until the critical portion of task A is completed and the updated information is available to B.

Since testing and waiting is expensive in terms of system efficiency, algorithm design should strive to make the length of time between required synchronizations as large as possible. This is often achieved by making the *grain size* of the the algorithm as large as possible.

The *grain size* of a parallel algorithm is the relative number of operations done between synchronizations in a MIMD algorithm [3]. If an algorithm has a fine grain size, then the completion of tasks will often require information from other tasks [12]. It is important that an algorithm of this type be implemented on a computer that requires relatively little interprocessor time. If, on the other hand, the computer has a relatively large communication time, then it is advantageous to be able to divide, or *partition*, the problem into coarse grain tasks which require little interprocessor communication. In most situations, designing a particular algorithm for a specific machine will involve determining how to partition the algorithm into task sizes to match the communication capabilities of the machine. A mismatch in granularity and machine type may result in an unbalanced increase in communication overhead which will subsequently reduce the maximum obtainable speedup. In general, loosely coupled multiprocessors are best suited to fine grain task parallelism. Tightly coupled multiprocessors exhibit a higher communication time to access the main memory and are therefore better suited to coarse grain tasks.

The remainder of this chapter will be devoted to the study of the characteristics of various parallel algorithms which have been proposed for the solution of traditional power system problems and the implementation of these algorithms on the two main types of multiprocessors previously discussed.

III.B   Parallel Processing Techniques for Power Flow Analysis

Nonlinear Equation Solution

The power flow problem might be most generally stated as a set of simultaneous algebraic equations of the form,

$$f(x) = 0, \text{ or } f_k(x_1, \dots , x_n) = 0, \ 1 \le k \le n$$

The Newton algorithm solves nonlinear equations of this form using the iteration scheme,

$$x^{k+1} = x^k - Df(x^k)^{-1} f(x^k)$$

where D is the derivative operator. Of the many methods for solving simultaneous equations Newton's method has been traditionally used for several reasons. First, generally, a sufficiently good initial estimate of the solution is known apriori so that the behavior of the power flow equations in the region encompassing the solution and the initial guess is approximately quadratic. This means that the a solution can usually be found reliably and quickly. (The problem on multiple solutions and nonexistent real solutions is of interest in general but will not be addressed here.) Second, and because of this, it is faster (and more reliable) on systems of equations which characterize utility applications (i.e., non-radial networks).   Third, quasi-Newton methods are usually faster, and in certain cases more robust. Quasi - Newton algorithms deal with modifications to the above iteration schemes in the following general way

$$x^{k+1} = x^k - (A^k)^{-1} f(x^k)$$

where A^k are approximations to the Jacobian matrices $Df(x^k)$. These algorithms are usually implemented via the update rules,

$$A^k \Delta x^k = - f(x^k), \quad x^{k+1} = x^k + \Delta x^k.$$

Using standard notation the power flow equations may be written for the power injections $S_k = P_k + jQ_k$, into a network with n buses in the form

$$P_k(\theta, V) = \sum_m V_k V_m (g_{km} \cos \theta_{km} + b_{km} \sin \theta_{km}) = P_k^d$$

$$Q_k(\theta, V) = \sum_m V_k V_m (g_{km} \sin \theta_{km} - b_{km} \cos \theta_{km}) = Q_k^d$$

where $Y = G + jB$ is the network admittance matrix, with entries $G_{km} = g_{km}$ and $B_{km} = b_{km}$, the bus voltages have magnitude V_k and phase q_k ($q_{km} = q_k - q_m$) and $P_k^d + jQ_k^d = S_k^d$ are the desired values of the power flows. The real and reactive power mismatches equations are given by,

$$\Delta P_k (\theta, V) = P_k^d - P_k (\theta, V)$$
$$\Delta Q_k (\theta, V) = Q_k^d - Q_k(\theta, V)$$

Newton and Quasi Newton methods are used to solve for zero real and reactive power mismatches. The Newton update rule for the power flow equations can be expressed in the form,

$$\begin{bmatrix} \Delta P(\theta, V) \\ \Delta Q(\theta, V) \end{bmatrix} = \begin{bmatrix} J_{P\theta} & J_{PV} \\ J_{Q\theta} & J_{QV} \end{bmatrix} \begin{bmatrix} \Delta \theta \\ \Delta V/V \end{bmatrix}$$

where the sub matrices J_{xx} are obtained from the Jacobian matrix of the mismatch vector. Figure 1 shows a block diagram of an implementation of Newton's algorithm. Neglecting input/output routines, effective parallelization requires that the mismatch equations, Jacobian entries, LU factorization of the Jacobian, and forward/backward substitution be evaluated in parallel during every iteration. Of these routines, the mismatch and

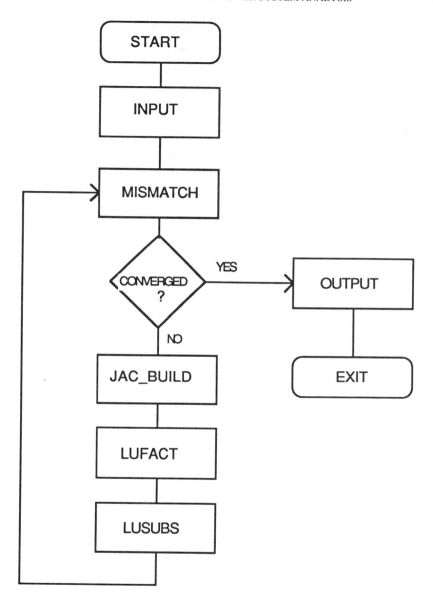

Figure 1. The Newton Power Flow Algorithm

Jacobian entries are easily parallelized while sparse matrix operations (i.e., factorization and substitution discussed later) are difficult to parallelize. Thus effective implementation of a power flow algorithm is usually dictated by how effectively factorization and substitution can be parallelized.

One way of minimizing the deleterious effects of the highly serial sparse matrix operations is to use the quasi-Newton methods. Two quasi-Newton methods which have been shown to be effective in solving power flow problems are known as the XB and the BX method. The XB algorithm may be expressed in terms of the update rules,

$$[1/V_k] \, \Delta Q(\theta_k, V_k) = B'' \Delta V_k$$
$$[1/V_k] \, \Delta P(\theta_k, V_k + \Delta V_k) = [1/x]' \Delta \theta_k$$

and the BX algorithm expressed in terms of the update rules

$$[1/V_k] \, \Delta P(\theta_k, V_k) = B' \Delta \theta_k$$
$$[1/V_k] \, \Delta Q(\theta_k + \Delta \theta_k, V_k) = [1/x]'' \Delta V_k$$

where the B' and B" matrices are given by,

XB Method

$$\left[\frac{1}{X}\right]'_{ij} = B'_{ij} = \frac{-1}{X_{ij}} \ \forall \ i \neq j$$

$$\left[\frac{1}{X}\right]''_{ii} = B'_{ii} = \sum_{j \neq i} \frac{-1}{X_{ij}}$$

$$B'' = [B]$$

Note that the B' diagonal elements include no effect from any shunt elements which includes lines charging, capacitive and reactive shunts, and transformer losses due to off-nominal tap ratios. The B'' matrix is exactly the B portion of the Y matrix. The BX requires re-definition of these matrices in the form,

BX Method

$$B'_{ij} = B_{ij} \ \forall \ i \neq j \ \text{(excluding effect of off-nominal taps)}$$

$$B'_{ii} = \sum_{j \neq i} B_{ii} \ \text{(excludes all shunt effects)}$$

$$\left[\frac{1}{X}\right]''_{ij} = B'_{ij} = \frac{1}{X_{ij}} a_{ij}$$

$$\left[\frac{1}{X}\right]''_{ii} = B'_{ii} + B_{shunt} + \sum_{j \neq i} \frac{1}{X_{ij}} a'_{ij} + \frac{S_{ij}}{2}$$

where:

a_{ij} = the inverse of the tap ratio (1.0 for transmission lines)

$a' = a^2$ if i is the tap side

= 1.0 if i is the impedance side.

S_{ij} = the charging susceptance of the branch

Both the classical (shown in Figure 2) and the successive iteration schemes are employed. Note that this algorithm requires Jacobian matrix construction and factorization only once, reducing the inefficiency if matrix operations can not be parallelized/vectorized efficiency. Since forward and backward substitution must still be carried out at every iteration, high gain parallelization requires that repeat matrix equation solutions be parallelized.

There are two ways of avoiding parallelizing these matrix solutions. First, when a power flow like problem is encountered as a subproblem, for instance in a dynamic simulation, it is often possible to configure the algorithm such that *multiple* matrix solutions are required simultaneously. This often allows individual matrix solution procedures to be assigned to each processor.

When multiple problems cannot be constructed and the parallel machine of interest makes parallelization of the matrix equation problem prohibitively inefficient, then either an alternate method to Newton's method must be found or a highly parallel method for solving Newton linearized equations must be discovered. An alternate approach to using Newton's methods to obtain the solution to a set of nonlinear equations, is to iterate directly on the nonlinear equations f(x)=0. This approach has a rich algorithmic history in power flow literature with the Gauss-Seidel method being the most often cited algorithm. The Gauss-Seidel approach in its basic form (i.e. excluding block Gauss-Seidel) is not suitable for parallel processing because of the inherent sequentiality. Gauss-Jacobi algorithms do not suffer from this sequentiality but may require a large number of iterations to converge. Such algorithms perform better on the algebraic equations characteristic of the transient stability problem than the power flow model because of different methods used in modeling the impact of the generator on the network. In traditional transient stability algorithms, loads are often modeled as varying with voltage magnitude to the 1.5 power as compared to (highly nonlinear) constant power loads (typical of power flow solutions) which have no voltage dependence or linear load models which use 2.0 as the exponent. The authors are not aware of any parallel implementation available for this approach to the nonlinear equation problem.

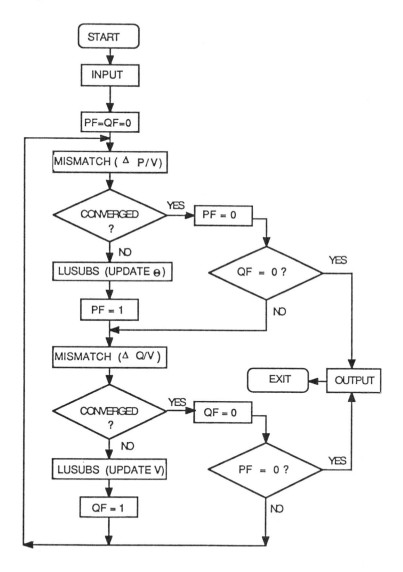

Figure 2. The Fast-decoupled Classic Iteration Scheme.

Algebraic Equations

Many power system problems have large portions which can be easily parallelized. However, solution of most power system problems (including power flow, transient stability, optimal power flow, state estimation, etc.) requires the solution of the linear algebraic problem in the form,

$$A x = b \tag{1}$$

where A is large with random sparsity, is typically incidence symmetric and is often numerically symmetric. Also, x and b may or may not be sparse. There are a large number of direct and indirect algorithms for solving this problem. The most effective method on serial processors for power system application to date is the use of triangular factorization along with forward/backward substitution defined by,

$$L D U = A \tag{2 (a)}$$
$$L y = b \tag{2 (b)}$$
$$D U x = y \tag{2 (c)}$$

The two distinct phases to this problem are the factorization phase, Eq. (2a), and the substitution phase, Eqs. (2b,c). The algorithms in which such solutions are required will dictate whether both phases can be processed simultaneously. For example, in a full Newton power flow the Jacobian and mismatches are recreated on each iteration so Eq. (2) must be solved repeatedly. The fast decoupled power flow requires that Eq. (2a) be solved once and Eqs. (2b,c) be repeatedly solved on each iteration. Thus there exists a need for parallelizing Eqs. (2), (2a), and (2b,c). Much work has been done on algorithms for parallel triangular factorization [12-19] and/or forward and backward substitution [12-18, 20, 21]. Many of these algorithms have attempted to take the serial factorization/substitution problem and exploit available parallelism through reordering/partitioning of the A matrix. This has been effective in reducing the number of precedence relationships which is governed by the maximum factor path length. (The length of the longest factor path in the elimination tree seems to represent a fundamental limit in minimizing the number of precedence relationships during factorization [22].) While algorithm development has yielded good theoretical results, little software has been developed for parallel machines to date. Results contained in [12, 23] show parallel factorization and substitution results on the iPSC which are unimpressive. Full factorization can be accomplished with speed gains on the order of 2 and with parallel gains of about 10 when (partial) factorization is halted before the densest portion of the matrix is encountered. Partial factorization is shown schematically in Figure 3. The factor path (elimination tree) of the IEEE 118 bus system is shown in Figure 4. It is easy to see that a great deal of fine grain parallelism exists in the beginning of the factorization stage. However, after several parallel elimination steps, the amount of available fine grain parallelism decreases significantly. Coarse grain parallelism can be achieved by grouping nodes so that precedence relationships for each group are wholly contained within the group as shown in Figure 4. However, the amount of available coarse grain parallelism vanishes after the first step leaving a fine grain not-so-parallel process.

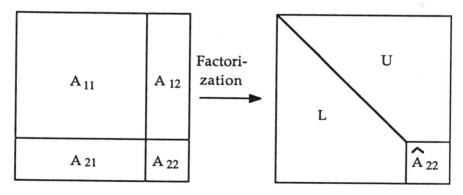

Figure. 3 Partial Triangular Factorization

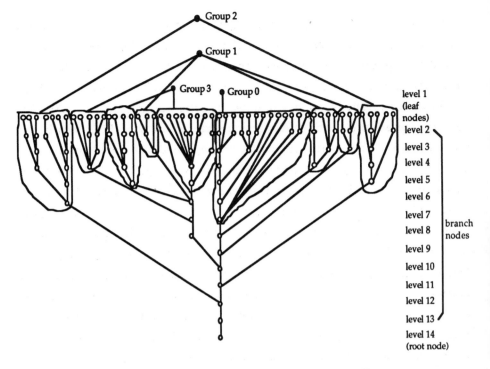

Figure 4: Factorization Path Tree of IEEE 118 Bus System

It is interesting that often on the order of 30% of the total serial execution time of factorization is involved in solving the dense factorization problem (i.e., after the coarse grain parallelism in Figure 4 is exhausted). Amdahls law gives a quick approximation of the gain obtained for parallelizing an algorithm as,

$$G = \frac{1}{f_s + f_p/N} \leq \frac{1}{f_s}$$

where f_s and f_p are the serial and parallelizable fractions of the serial code and N is the number of processors available. The upper bound for this case says that *at best* a speed up of 3.0 might be expected for the factorization problem.

Much work has been done on parallelizing the substitution problem under the assumption that parallel factorization exists or is unnecessary. (A large number of repeat solution, such as in the fast decoupled power flow, may dominate the number of calculations making parallelizing the factorization unnecessary.) Fundamentally new algorithms attempting to minimize the precedence relationships with forward and backward substitution problems include the multiple factorization scheme [24] and the use of sparse inverse factors [21, 25, 26]. Multiple factorization has yet to implemented on a parallel machine (but it may be implemented soon). A method for highly parallel substitution coupled with multiple factorization could make this approach attractive.

It is well know that Eqs. (2b,c) may solved with using the inverses of the L factors without additional computation or sacrificing sparsity as shown in Eq. (3).

$$y = L_p^{-1} \ \dots\dots \ L_2^{-1} L_1^{-1} \, b \tag{3a}$$

$$x = U_1^{-1} U_2^{-1} \ \dots\dots \ U_p^{-1} \, y \tag{3b}$$

This is one end of the factorization spectrum which saves no precedence relationships and hence is of no practical interest. The other end of the spectrum involves multiplying the inverse factor to get L^{-1} which reduces the number of precedence relationships to one. This is not cost effective because the number of nonzeros in L^{-1} will be substantially more than contained in L as shown in Table 1. The principle idea behind sparse inverse factors is that, like L, L^{-1} may be represented as the inverse of the factors, greater than 1 and fewer than the matrix rank (or the maximum factor path length),with little or no increase in sparsity. A simple scheme for grouping rows so that relatively little fill-in occurs can yield a significant reduction in the number of precedence relationships (involved in completing the inner product operation associated with one inverse factor before preceding to the next factor) as shown in Table 2. The sparse inverse factors schemes are promising; however an algorithm for parallel construction of the inverse factors has yet to be developed. Further no parallel implementation using sparse inverse factors in place of substitution has as yet appeared leaving in question its ascendancy over traditional substitution. This approach cuts down significantly on the number of precedence relationships involved but it still leaves a fine grain problem for which easily programmed processors are not available.

Table 1. The Number of Nonzeros in L and L^{-1} Factors

system	118	494	662
non-zeros in L	390	1443	2652
non-zeros in L^{-1}	1089	8497	18810
$\dfrac{\text{non-zeros in } L^{-1}}{\text{non-zeros in } L}$	2.79	5.89	7.09

Table 2. The Number of Partitions of #118, #494, and #662 systems

System Size (Buses)	Extremum of Set of Factor Path Lengths	Number of L/U Inverse Factors
#118	11	8
#494	25	11
#662	36	13

The parallel algorithms discussed thus far have mostly been fine grained algorithms designed for multiprocessor systems. The work in [27] has produced experimental results for solving Eq. (2) using a vector processor. These results show promise of being able to take advantage of a portion of the capability of the widely available vector machines. These experimental results are helping to distinguish some of the real issues in parallel processing from non-issues.

Indirect methods have been revisited in an attempt to minimize the number of precedence relationships in solving Eq. (2) [28]. While this fine grained algorithm can be shown to require about the same number of precedence relationships as traditional substitution, its implementation on a coarse grain parallel processor has shown results no more optimistic than those from traditional substitution implementations.

Summary

The results concerning parallelizing traditional power flow algorithms indicates that the modus of parallelizing the power flow problem is the solution of the linear matrix equations. Much theoretical research is available for the linear matrix equation problem however implementation of these algorithms on available coarse grain machines show gains which are modest at best. New algorithms which are capable of using coarse grain machines are much in need. Fine grain machines currently available are not particularly suitable for the power system problem because the random sparsity pattern of the associated matrices prevents facile implementations of the existing algorithms.

III.C   Parallel Processing Techniques for Transient Stability Simulation

Power systems in general are considered to be large scale, often involving several hundred equations to describe the behavior of an interconnected power system during and following a fault on the system. Due to the formidable size of the problem, most of the simulations are performed off-line. As power system operations become increasingly complex, it becomes desirable to be able to perform on-line analyses of voltage conditions and system stability. Because such calculations are computationally involved, present utility computers may take minutes to simulate events which in actuality occur in a few seconds, thus precluding real-time analysis [29]. However, recent advances in computer technology have made it feasible to re-evaluate the existing simulation algorithms and introduce new methods of computer simulation utilizing this technology.

In order to create an effective parallel-based simulator, it is necessary to understand the operation of customary circuit simulators. The generalized model of an interconnected power system, consists of a set of nonlinear differential equations coupled to a set of algebraic equations

$$\dot{x}(t) = f(x(t), y(t)) \tag{4}$$

$$0 = g(x(t), y(t)) \tag{5}$$

where $x \in R^n$ corresponds to the state variables of the generating units, $y \in R^m$ to the network variables and the loads are modeled statically. Equation (1) describes the machine dynamics and Eq. (2), the static behavior of the network .

There are two basic traditional approaches to solving this system of mixed differential and algebraic equations. The first approach is to solve Eq. (4) by an integration method (usually an explicit method such as a fourth-order

Runge-Kutta) and at every time interval using the previous values for the static variables y. Equation (5) is then solved for updated values of y using the new x values just obtained. This process repeats, alternating between solving Eq. (4) and Eq. (5) until convergence is achieved. A second approach to solving this system of equations, called the simultaneous implicit approach, is to discretize Eq. (1) using a backwards difference method (trapezoidal, Gear's methods, ...) and then solve both equations simultaneously at each time step using a Newton method to solve the nonlinear algebraic equations. The simultaneous implicit approach has been the method most used in the first attempts at parallelizing the transient stability problem [30].

The standard simultaneous implicit method to solving the system of differential/ algebraic equations is commonly referred to as a direct method and is used in most standard circuit simulators (such as SPICE2). This approach is composed of three steps [31]:

- *An implicit integration method is used to convert the differential equations into a sequence of systems of nonlinear algebraic equations.*

- *A damped Newton method is used to convert the nonlinear equations into linear equations.*

- *The resulting sparse linear equations are solved using Gaussian elimination or LU decomposition.*

Parallelism of this algorithm may be exploited at several levels. Much attention has been given to parallel methods applied to the last stage: the solution of a system of sparse linear equations. In general, this parallelism can be described as fine-grained. The simultaneous implicit method is analogous to the methods presented in the previous section on power flow parallel methods. The dynamic problem offers a wider range of parallelism possibilities.

In addition to the direct method approach, applying relaxation techniques have been found to be effective in increasing medium- and coarse-grain parallelism. In relaxation, the jth equation in a set of equations is solved for the jth variable while holding all other variables to either the previous iteration value (Gauss-Jacobi) or the present iteration values (Gauss-Seidel). The process is repeated until convergence is achieved. The Gauss-Jacobi relaxation is inherently parallel. Each equation may be solved independently, and concurrently, with all other equations. This observation tacitly assumes that all necessary information about any other variable previous iteration value is readily available. The idea in relaxation-based methods is to obtain the solution of the system without directly solving the large linear system of equations, thereby, drastically reducing the simulation run time. Another advantage of the relaxation-based methods is that they may be used to directly exploit inherent latent behavior easily as compared to direct methods [31].

Relaxation methods may be applied at any one of the three stages of the solution process. For example, the Gaussian elimination portion of the third stage may be replaced by a linear relaxation scheme. In addition to using relaxation to find the solution of linear equations, it may also be effectively applied at one of the earlier stages in the solution process [31]. Note that the first two steps are relatively independent; hence, they can be performed in either order. An overview of the resultant set of algorithms is shown in Figure 5 with their common names. There are many hybrids of these methods which have received significant attention and in many cases have been quite effective which are not included in this figure in an attempt to simplify the presentation of the basic algorithms.

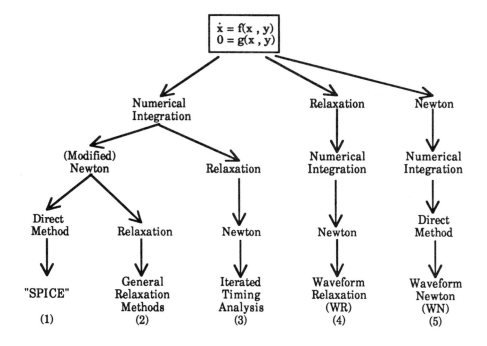

Figure 5: Overview of Algorithms

The algorithm on the far left is the traditional direct method for solving Eqs. (4) and (5). Any parallelism is fine-grained and is exploited either at stage three in the solution of the linearized equations or in the parallel evaluation of the components of the Jacobian matrix and the "right-hand-side" function vector. There are several approaches to iteratively solving the nonlinear equations which arise as result of the numerical integration process. Consider the following system of equations

$$F(Z)=0 \qquad (6)$$

where $F : R^{n+m} \rightarrow R^{n+m}$, and $Z = [X\ Y]^T \rightarrow R^{n=m}$ and z_i may represent either a dynamic or static variable. The general iterative method is to find some matrix A such that

$$Z^{k+1} = Z^k - \left[A^k\right]^{-1} F\left(Z^k\right) \quad k = 0, 1, \tag{7}$$

The well-known Newton's method takes $A^k = F'(Z^k)$ where $F'(Z^k)$ is the Jacobian of $F(Z)$ evaluated at Z^k. Many modifications to this method exist and have been successfully implemented on various types of problems.

One "modified" Newton method which is commonly used to solve the nonlinear system of Eq. (6) is a "very dishonest Newton (VDHN)." The VDHN iteration is given as

$$Z^{k+1} = Z^k - \left[F'\left(Z^0\right)\right]^{-1} F\left(Z^k\right) \quad k = 0, 1, \ldots \tag{8}$$

where A^k in the previous formulation is now the constant Jacobian matrix $\left[F'\left(Z^0\right)\right]$ which is not updated unless the convergence is deemed to be progressing too slowly.

The VDHN has been applied to the power system transient stability problem in [32]. The linearized equations of Eq. (6) at time t are

$$-\begin{bmatrix} J_{1,t} & J_{2,t} \\ J_{3,t} & J_{4,t} \end{bmatrix} \begin{bmatrix} \Delta X_t^k \\ \Delta Y_t^k \end{bmatrix} = \begin{bmatrix} F_{1,t}^{k-1} \\ F_{2,t}^{k-1} \end{bmatrix} \tag{9}$$

where

$$J_{1,t} = \frac{\partial F_{1,t}}{\partial X_t} \qquad J_{2,t} = \frac{\partial F_{1,t}}{\partial Y_t}$$

$$J_{3,t} = \frac{\partial F_{2,t}}{\partial X_t} \qquad J_{4,t} = \frac{\partial F_{2,t}}{\partial Y_t}$$

and $\Delta X_t^k = X_t^k - X_t^{k-1}$ and $\Delta Y_t^k = Y_t^k - Y_t^{k-1}$. After performing Gaussian elimination of Eq. (9), the following relationships are obtained [32]

$$\widehat{F}_2 = F_2 - J_3 J_1^{-1} F_1 \tag{10}$$

$$\widehat{J}_4 = J_4 - J_3 J_1^{-1} J_2 \tag{11}$$

A relaxation scheme is used to solve the linearized set of equations resulting in update equations

$$\Delta Y_t^k = \left[\widehat{J}_4^{(0)}\right]^{-1} \widehat{F}_2\left(X_t^{k-1}, X_{t-1}^{k-1}, Y_t^{k-1}, Y_{t-1}^{k-1}\right) \tag{12}$$

$$\Delta X_t^k = - \left[\widehat{J}_{1,t}^{k-1}\right]^{-1} \left(F_{1,t}^{k-1} + J_{2,t}^{k-1} \Delta Y_t^k\right) \tag{13}$$

The constant matrix $\left[\widehat{J}_4^{(0)}\right]$ is computed only when convergence is proceeding excessively slowly or a system change occurs. Note that in addition to being held constant for the Newton iteration, it is also held constant over time as well. Parallelism in this method occurs as a result of the block-diagonal structure of the sub-Jacobian $\left[J_{1,t}\right]$ (each block corresponds to an individual generator). The solution of each block of equations may be processed concurrently. The solution of Eq. (12) at each iteration must be executed sequentially, but may still be relatively inexpensive in terms of total run-time if the LU factors of $\left[\widehat{J}_4^{(0)}\right]$ are stored in memory and are not computed explicitly at every iteration.

An alternate modified Newton method which has been used in power system applications is the *Picard iteration*. This iterative method arises when the nonlinear system F(Z) in Eq. (6) may be decomposed into the special form

$$F(Z) = AZ - G(Z) \tag{14}$$

where $G : R^{n+m} \rightarrow R^{n+m}$ is nonlinear and A is a nonsingular matrix (possibly the identity matrix *I*). This leads to the following iterative process

$$Z^{k+1} = A^{-1} G\left(Z^k\right) \quad k = 0, 1, \dots \tag{15}$$

In power systems, this iterative method has sometimes also been called the "Gauss-Seidel" method. This process was first proposed as an algorithm well suited for parallel implementation in [34] and implementation results of this method on a hypercube computer have been reported in [35]. These two algorithms would be in the category of algorithm (2) in Figure 5.

In addition to using relaxation methods to solve the set of linearized equations, parallelism may be obtained by interjecting relaxation at stage two of the solution process. The discretized equations are decomposed into blocks of equations which are then solved iteratively using one Newton iteration. The equations are solved repeatedly until convergence to a consistent solution is achieved. The solution is then advanced to the next time point where the process is repeated. Since the blocks of equations are decoupled, any block which has not changed significantly with respect to time can be skipped or neglected at that time point, thereby, exploiting any inherent system latency. This method is referred to as iterated timing analysis (ITA) and is shown in Figure 5 as algorithm (3).

A similar method has been applied to the transient stability problem in power systems [32]. This method, called one-step SOR-Newton, uses scalar partitioning rather than blocks of equations. The $(k+1)$ updated iterative value of any variable z_i at time t is given by

$$z_{i,t}^{k+1} = z_{i,t}^{k} - \omega_i \frac{f_i\left(X_t^k, Y_t^k\right)}{\partial_i f_i\left(X_t^k, Y_t^k\right)} \tag{16}$$

where z may represent either a dynamic variable or a static variable. The function $\partial_i f_i\left(X_t^k, Y_t^k\right)$ is the diagonal element of the Jacobian of the nonlinear system of equations. Thus, only the diagonal partial derivatives are required in this method. The factor ω_i is the relaxation factor for the ith element of Z. It has been experimentally determined that convergence is most rapid when the dynamic variables X are overrelaxed ($\omega_i > 1$) and the static variables Y are underrelaxed ($\omega_i < 1$) [32].

A method which applies the relaxation directly to the original differential/ algebraic system, rather than to the algebraic equations after numerical integration, is known as waveform relaxation (WR). In the WR method, the system is broken into subsystems which are solved independently, each subsystem using the previous iterate waveforms as "guesses" about the behavior of the state and algebraic variables in other subsystems. Waveforms are then exchanged between subsystems, and the subsystems are resolved with updated information about the other subsystems. This process is repeated until convergence is achieved. The subsystems are solved independently and may therefore be discretized independently. This trait is especially useful if the system contains states which are varying at significantly different rates. The WR method is shown as algorithm (4) in Figure 5 and is summarized below in Algorithm 1.

Algorithm 1 - The Gauss-Jacobi WR Algorithm.

$k \leftarrow 0$;
Guess some $x^{k+1}(t)$ such that $x^{k+1}(t_0) = x(t_0)$;
Guess some $y^{k+1}(t)$ such that $y^{k+1}(t_0) = y(t_0)$;
repeat{
   $k \leftarrow k + 1$
    foreach ($i \in \{1, ..., r\}$) solve on $[t_0, T]$

$$x_i^{k+1} = F_i\left(x_1^k, ..., x_i^{k+1}, ..., x_r^k, y_1^k, ..., y_i^{k+1}, y_r^k, t\right) \quad x_i^{k+1}(t_0) = x_i(t_0)$$

$$O = G_i\left(x_1^k, ..., x_i^{k+1}, ..., x_r^k, y_1^k, ..., y_i^{k+1}, y_r^k, t\right) \quad y_i^{k+1}(t_0) = y_i(t_0)$$

   }**until** $\left(\iota x^{k+1} - x^k \iota \le \varepsilon_x \text{ and } \iota y^{k+1} - y^k \iota \le \varepsilon_y\right)$;
where ε_x and ε_y are small positive values.

Successful implementation of the WR method depends on the convergence rate of the function space iterates. The convergence is directly dependent on the choice of partitioning of the system and the length of the time interval over which the waveforms are relaxed. These two aspects of the WR algorithm are discussed in relation to power system simulation in [36] and [37]. Although the WR algorithm will converge for any chosen partitioning [38], the choice of partitioning greatly influences the rate of convergence. If tightly coupled nodes are not grouped together or if too many nodes are placed in the same subsystem, the rate of convergence may be greatly decreased and the WR method is no longer efficient. One approach to partitioning is to exploit the natural decoupling of the physical system [36]. Specifically for power systems, the *coherency* characteristics inherent in power system behavior can be exploited. This method has been experimentally shown to be effective for partitioning the system for transient stability simulation [37].

The convergence speed of the WR algorithm also depends on the size of the interval over which the waveforms are computed. To increase the rate of convergence, the usual practice is to break the simulation interval [t_0, T] into subintervals, or windows, [t_0, T_1], [T_1,T_2], ..., [T_{n-1}, T] and compute the waveforms over each window [38]. Again, though the WR method will converge for any finite length window, if the window size is too large, convergence may be very slow. Conversely, if the window size is too small, the advantages of the waveform relaxation may be lost if too many extra time points are unnecessarily added at the window boundaries. Thus, for the WR algorithm to be efficient, it is important to pick the largest windows over which the iterations actually contract uniformly.

Algorithm (5) in Figure 5 is the waveform-Newton (WN) algorithm. This method is a function-space extension of the classical Newton's method in the same manner that waveform relaxation is a function-space extension of classical Gauss-Jacobi and Gauss-Seidel relaxation methods. In the WN algorithm, each nonlinear differential equation is linearized about an initial guess waveform (on the interval $[t_0, T]$) whose value at t_0 matches the given initial condition for the differential equation. The "guess" waveform is then updated by solving the resulting linearized differential equation. The original differential equation is then relinearized about the updated guess, and the process is repeated until convergence is achieved [31]. The WN algorithm has not been applied to power system transient stability simulation. This method does give rise to an interesting hybrid algorithm known as the waveform-relaxation-Newton (WRN) algorithm. In the WRN algorithm, the WR iteration equations are solved approximately by performing only one step of the waveform-Newton method for each waveform relaxation iteration. This method can be considered to be the function-space extension of the one-step SOR-Newton method described earlier.

The above algorithms can be categorized as "parallelism-in-space" methods. Each algorithm consists of a spacial partitioning step where subsystems are determined and then solved independently. The solution procedures still approach the time dependence of the integration portion of the algorithm in a sequential manner, i.e., it is assumed that $z(t_{n-1})$ must be computed before $z(t_n)$. It has been suggested that if "time" could be parallelized, this could add another dimension to the parallel implementation of circuit simulation.

A method which has been proposed in conjunction with the waveform relaxation algorithm is *time-point-pipelining* (TPP). This method was introduced to preserve the Gauss-Seidel ordering, which generally converges more rapidly than a Gauss-Jacobi ordering, but is, by nature, more sequentially oriented, while increasing the amount of computation which can be performed in parallel. A simple example of TPP is a computer which has two processors, where processor 1 is computing the waveform for $z_1(t)$ and processor 2 is computing the waveform for $z_2(t)$. In the Gauss-Seidel WR, the computation of $z_2(t)$ cannot begin before processor 1 finishes the computation of

$z_1(t)$. In TPP, however, processor 2 can begin computing $z_2(t_n)$ as soon as $z_1(t_{n-1})$ is known.

Time-point-pipelining can also be used across iterations to generate even more parallelism. Consider again the above example. When processor 2 computes the first time point for $x_2^k(t)$, another processor can also be concurrently computing $x_1^{k+1}(t)$. This method will not only generate more parallelism for the Gauss-Seidel algorithm, but can also be applied to the Gauss-Jacobi algorithm to generate further parallelism. This method has been successfully applied to the simulation of VLSI circuits [31], and has been proposed for power system implementations [39].

Another means for exploiting "parallelism-in-time" is the use of *block implicit* integration methods [32], [34], [40]. Consider the linear set of equations

$$\dot{Z} = AZ + f(t)$$

to which the trapezoidal integration method has been applied

$$Z_{n+1} = Z_n + \frac{h}{2} A (Z_{n+1} + Z_n) \tag{17}$$

where $h = t_{n+1} - t_n$ is a constant, $A \in R^{n+m} \times R^{n+m}$, and $z \in R^{n+m}$. If the integration is carried out for N time steps, then a set of successive nonlinear functions arise

$$z_1 = z_0 + \frac{h}{2} A (z_1 + z_0)$$

$$z_2 = z_1 + \frac{h}{2} A (z_2 + z_1)$$

$$\vdots$$

$$z_N = z_{N-1} + \frac{h}{2} A (z_N + z_{N-1})$$

which can be written succinctly

$$A \, \widetilde{Z} = b \qquad (18)$$

where

$$a_{ij} = \begin{cases} I - \dfrac{h}{2} A & i = j \\[2mm] -\left(I + \dfrac{h}{2} A\right) & i = j + 1 \\[2mm] 0 & \text{else} \end{cases}$$

$$b_i = \begin{cases} \left(I + \dfrac{h}{2} A\right) z_i + \dfrac{h}{2}(f_i + f_{i-1}) & i = 1 \\[2mm] \dfrac{h}{2}(f_i + f_{i-1}) & i \neq 1 \end{cases}$$

and $\widetilde{Z} = [z_1, z_2, ..., z_n]^T \in R^{(n+m)N}$. The equation (18), when solved, will now yield all of the n+m values of $z = [X \; Y]^T$ for all N time steps! The proposed parallel-in-time methods suggest that Eq. (18) not only be relaxed in space, but also in time. If the system of equations is partitioned into p partitions in space, which p processors can compute in parallel, and the N time steps can be solved concurrently by N processors, thus, requiring $p \times N$ processors, then all N time steps can be be computed simultaneously and a large degree of parallelism has been achieved.

Although the parallel-in-time method was described by an example using a linear system of equations, this methodology can be extended directly to the resultant linearized equations of the modified Newton step of algorithm (1) in Figure 5.

The parallel-in-time methods have improved the speed of computation, but the convergence of the time steps is sequential, i.e. later steps require the converged results from previous steps before converging themselves. This means that processors which are calculating earlier time steps are idle during the time it takes the latter steps to converge, and thus introduce inefficiencies. The compromise is to use a few steps in parallel at a time, or a window of time, and once the time points in this window have converged, the next window of steps is initialized and processed.

The parallelism in space and in time with the windowing technique has been quite effective in speedup improvement as compared with pure parallelism in space. The processors which have completed the convergence for the initial time steps of a window may just sit idle or do the redundant calculations while processors in later time steps are busy with computation. To reduce the processor idling, a travelling window technique or the *Toroidal* method is used [41]. If one time step reaches convergence, it will send a signal to the groups of processors both preceding and following the current time step processor.   Then the "converged" processor will immediately move to simulate the next unassigned time step, which is actually at the end of a new window.   This new window is one time step foreword, and hence the name travelling window, or toroidal for the circular configuration of the processor communication.   The result of this procedure is that all processors are busy until the last window of the study interval is converged and the last time step CPU exits.

III.D   Implementation Issues

For a parallel algorithm to be efficient, there are several aspects of implementation   which must be considered.   The algorithm and its implementation are no longer separate entities. Many theoretically attractive algorithms have not produced significant gains because of resultant bottlenecks when the suitability of the algorithm to the machine architectures was not properly addressed. The factors hindering parallelism are very often massive communication costs, poor implementation strategies, unavoidable sequentiality, and the slow convergence of parallel relaxation methods.

Successful implementations are those which match the communication requirements and granularity of the algorithm to the communication architecture, communication speed, and granularity of the processor. To better understand the general differences between parallel code for a message passing architecture versus a shared memory architecture, two representative machines will be compared and contrasted for efficiency in performing a parallel transient stability simulation. The Intel iPSC/2 hypercube will be the representative for message passing machines, and the Alliant FX/8 will represent shared memory computers.

To illustrate the differences between algorithms and architectures, two algorithms, the VDHN and SOR-Newton, were implemented on both the iPSC and Alliant, with two power systems examples, the IEEE 118 and a US Midwestern 662 bus network [32]. These particular algorithms are chosen to be representative of the classes of parallelized direct methods (VDHN) and parallelized relaxation methods (SOR-Newton). These programs were checked against the EPRI production grade program [42] and the results from all three were almost identical.

In these simulations, a fourth order generator models with typical values are used. Nonlinear loads are assumed with voltage exponents equal to 1.5 for both active and reactive power. A three phase to ground fault at a generator bus is simulated, using a small ground impedance of 0.001 pu. For all tested cases, a time step of 0.02, a fault clearing time of 0.06 and a solution time of 1.12 seconds for 118 bus and 0.48 for 662 bus systems, are adopted. A fast exciter model (IEEE AC4) is used for all generators. A convergence tolerance of 0.001 on both the voltage and state variables is enforced.

The sequential execution times on one node of both the Alliant and iPSC for the two test systems are listed in Table 3. The execution speed in the Alliant is several times faster than in the iPSC hypercube mainly because of the difference in the microprocessors' CPU capabilities. The vectorizing capability of the Alliant is not used here to focus the comparisons on the parallelization. The solution time between the serial VDHN and SOR-Newton methods are quite close for the case shown in the table. In general, the SOR-Newton iterations are sensitive to the severity of the fault and the execution time can be similar to VDHN for moderate faults and twice as slow for severe ones.

system time (s)	662 bus system		118 bus system	
	VDHN	SOR-N	VDHN	SOR-N
iPSC	119.612	216.056	33.823	34.365
Alliant	51.347	59.290	6.174	6.332

Table 3. Sequential run time of VDHN and SOR-Newton Algorithm on iPSC and Alliant

The following few figures and tables are all based on the 662 bus system. The speedup factors are defined as:

$$SP_1 = \frac{\text{Serial SOR-Newton run time on 1 processor}}{\text{Parallel SOR-Newton run time using N processors}}$$

$$SP_2 = \frac{\text{Serial VDHN run time on 1 processor}}{\text{Parallel VDHN run time using N processors}}$$

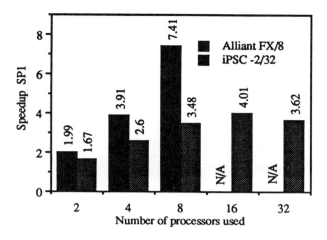

Figure 6.   Speedup of SOR-Newton vs. # of Processors

The speedup vs. the number of processors on both the iPSC and the Alliant for SOR-Newton is shown in Figure 6.   There is a significant difference in speedups between these two types of parallel computers, and it increases with the number of processors.   For example, with 8 processors, the speedup on the Alliant is 7.41 (efficiency of 93%), but on the iPSC the speedup is only 3.48 (44% efficiency).   Even on 4 processors, this difference is noticeable (efficiency of 65% vs 98%).   As there is no sequential overhead (like the forward and backward substitution in VDHN) in the SOR-Newton , the difference is mainly caused by the communication overhead. The communication speed on the iPSC is much slower (2.8 Mbyte/sec vs. 376 Mbyte/sec) than that on the Alliant.   Communication overhead always exists on the iPSC, while on the Alliant, the communication overhead is negligible if there is no bus contention.   This is usually the case when the number of processors is small and the amount of data transfer is moderate.   As a result, the speedup on the iPSC tends to saturate quite rapidly as the number of processors increases.   The efficiency falls below 50% for 8 processors and the speedup actually drops between 16 and 32 nodes.   For the shared-memory Alliant, the speedup saturation is not apparent up to 8 processors. Although the Alliant in this study had only 8 processors, the same program was run on the 24 processor Sequent Balance, which has a architecture similar to the Alliant, with an efficiency greater than 50%. The transient dynamics of the 118 bus system was simulated for 1.12 seconds with a execution run-time of approximately 1.0 second using the SOR-Newton algorithm. This means that even on existing parallel computers, it is quite possible to achieve the real time transient dynamic simulation for moderate sized systems [32].

The comparison of speedups of the parallel VDHN method on both shared and local memory machines are shown in Figure 7. The speedup increases as more processors are used but starts saturating when using more than 8 processors. Only around 50% efficiency on 8 processors can be reached and for 32 processors on the iPSC, the speedup is only 5.61 (efficiency < 20%). The algorithm requires frequent variable data exchange after each iteration, and sequential calculations like LU factorizations. This overhead becomes a significant part of the overall solution time as the number of processors increases. Also, it can be seen that the speedups obtained in the Alliant is less than that in the iPSC and appear to be saturating faster. This is because the compiler in the Alliant cannot recognize and achieve concurrentization of some do-loops partly due to the way sparsity coding is implemented. This enlarged sequential portion considerably slows down the speedup gains and in some way offsets the relative advantage of the fast communication speed over the message passing machine.

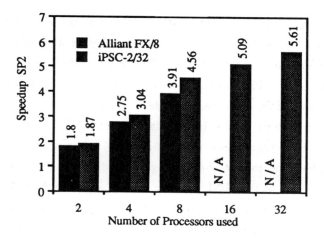

Figure 7. Speedup of VDHN vs. Number of Processors

Generally, the iPSC is good for complex and computationally intensive algorithms, while the Alliant is better suited for inherently parallel algorithms [32]. This is a result of the granularity of the machine type appropriately matched to a like grained algorithm. Coarse grain machines, such as the Alliant, generally tend to have comparatively few processors relative to fine grain machines. Similarly coarse grain parallel algorithms are those with parallel processes that allow a large number of computations for every word of intercommunication required while fine grain algorithms require large amounts of communication per few computations. A mismatch between algorithm and architecture can lead to significant degradation of performance [12],[32], while a good match can lead to significant gains.

In a message passing machine, such as the IPSC, when a message needs to be broadcast to all nodes, the efficient way to accomplish this is to have the instigating processor send the message to its nearest neighbors and have them resend it to their nearest neighbors without duplicating any destination. This is known as a pseudo-binary tree strategy. The complete process needs only d

steps (d is the dimension of the cube) instead of 2^d-1 steps. As d grows, time savings become very significant. The same strategy is also adopted for a message transfer from one node to a few other nodes or in sub-cube communication [32]. The overall communication overhead for various algorithms varies widely since required data exchange after each iteration and between time steps of the integration process differs extensively as different parallel-in-space and parallel-in-time schemes are used and iterations required for convergence change accordingly. Thus, a message passing machine, such as the iSPC, is better suited for a direct iterative method like VDHN than a relaxation algorithm. The computational intensity for each iteration in a relaxation method is comparatively less than a direct iterative method. Thus, the communication overhead tends to be heavier as a percentage of the execution time during each iteration period even though only one data exchange is needed.

The iPSC provides considerable flexibility and programming manipulation. The iPSC programmer can decide which part of the code is run in parallel, and all memory space is local, thus the programmer does not need to worry about data integrity or bus contention. This aspect may also be disadvantageous however. Although the programmer has complete control over synchronization of processes and data communication, the programmer must have an intimate knowledge of the hardware structure of the computer and software development is currently very time intensive. In general, the hypercube requires more programming effort to explicitly recognize the different cube configurations that may be used. The Alliant FX/8 is transparent, but the maximum advantage can only be obtained if the code is written in a proper and parallel fashion.

Shared memory architectures also exhibit several advantageous characteristics. First, there are several programming tools available, such as Parallel Directives, Fortran macros, library subroutines, as well as some UNIX system calls to help identify the loops or processes to be executed in parallel, classifying the data types as shared, private, or locked, synchronizing the processes during critical sections, etc. This leads to a great deal of transparency. This important advantage makes porting code to other shared-memory multiprocessors much easier than porting code to the distributed-memory multiprocessors such as the iPSC. The original code run on a sequential machine can be executed directly on these parallel computers with a certain number of processors by inserting a concurrent directive at the beginning of the main routine or by selecting the appropriate compiler options.

High efficiency and gains can be achieved by taking advantage of several features that shared-memory processors possess. The cache of each processor can provide faster access and also reduce the effective access time and the traffic on the common bus if a repeatedly used vector element is redefined as a local data element and stored in the cache. For example, in the portion of the transient stability code which is used to update generator variables, sinusoidal functions are evaluated frequently at each iteration. These can be put in the local memory to reduce the access time [32].

The excessive interprocessor communication overhead on local memory machines like the hypercube cause a fast saturation in speedup gains when more CPU's are used, while the bus contention and variable-locking are the most serious limiting factors in shared memory multiprocessors. These potential bottlenecks in each multiprocessor type must be recognized and minimized by the wise choice of type of parallel algorithm to be implemented on a particular architecture.

IV   Future Research Directions

In the last decade, a significant amount of research has been conducted in parallel processing of power system problems. The majority of this work has been in developing algorithms suitable for the parallel solution of power system problems, but only recently has actual testing on multiprocessor architectures been done. Some of the most recent results at the time of this writing are included in the section on implementation issues. At this juncture, there are three main areas in which further research may be focused [7].

● *Processor architecture* - includes aspects of design, performance, and control of new or existing hardware technologies.

● *Software development* - includes reformulating existing software to optimize transparency, portability, task scheduling, vectorization, and performance evaluation.

● *Algorithm development* - includes the design and analysis of new numerical and symbolic methods to match existing or new architectures.

The main thrust of the research so far has been in the latter two categories, with the majority of the emphasis on the third issue. Algorithm development has been pursued with a specific power system problem in mind. However, the search for better algorithms has not yet fully run its course and the application of parallel algorithms to other power system problems has to be continued. In addition to matching hardware architecture with algorithm, a third leg of dependency has been suggested and examined. This dependency of algorithm performance to system characteristics. The previous section mentioned the dependence of the SOR-Newton method performance to the length of fault. The longer the fault was applied to the system, the more ill-behaved the SOR-Newton method became, whereas the VDHN method was relatively impervious to fault length. This dependency has been further supported in using system coherency characteristics as a basis for partitioning in block relaxation methods [37]. In addition, real-time simulation has been achieved on a hypercube computer by imitating the network interconnections [35]. The processors in the hypercube are interconnected along the same pattern as the network itself where each bus is assigned to an individual processor.

As mentioned earlier, the majority of current research has centered on the problem of transient stability, because there is a perceived need for on-line analysis to determine the security of the power system in (better than) real-time. The power flow problem has also received considerable attention. This is due in part to the attention in many fields of solving linear and nonlinear systems in parallel. Many varied algorithms for LDU decomposition, Gaussian-elimination, forward and backward substitution has been proposed. Other areas within power systems have been touched, but not explored to the extent that the power flow and transient stability problems have been. Most control center functions are computationally intensive and are therefore generally too slow for interactive use. Efficiency would be greatly increased by faster computation, thus any breakthroughs in parallelization would be welcomed. The most time consuming function currently performed with regularity in the control center is static contingency analysis. At this time, these studies are performed off-line, but it is perceivable that this is one computation which might be done dynamically with the aid of parallel processors. Once dynamic contingency analysis has been achieved, attention will be turned towards the dynamic determination of corrective and protective control. This will involve security constrained scheduling and economic dispatching of units over time taking into account fuel constraints and load forecast uncertainties. These economic calculations (except for generator dispatching), as well as system planning, reliability, and production costing are not performed on-line, but are computationally intensive and would benefit even in the off-line performance of parallel processing.

Another area of research which would gain considerably from further progress in parallel processing is artificial neural network technology. Neural nets have been proposed for several power system problems including such critical problems as short-term load forecasting [43] and control center alarm processing. Artificial neural network computing is a series of independent fine grained processes which are amenable to parallelization and implementation on fine grained SIMD processors. Although only preliminary results are available at the time of this writing, neural networks have performed with promising results on problems of small dimension, where the input variables have number less than 100. This is another area where the power industry may benefit from the experiences of researchers in other fields.

V Conclusion

As power systems become increasingly interconnected and the traditional methods for analysis of their steady-state condition as well as dynamic behavior become similarly complex and compuationally intensive, new and faster methods for performing the necessary calculations will be required. The use of vector and array processors to speed up certain calculations and the design of distributed processors for system monitoring and control are already ubiquitous in the power industry. The advent of parallel processing has opened up many new arenas for fast and efficient processing. This has resulted in a great deal of interest in the development of algorithms and software for use with these processors. Experience has shown however, that it is foolhardy to propose a parallel algorithm without thoroughly considering the type of hardware architecture for which is it best suited. The most efficient algorithms to date have not been those algorithms which are attractive on paper, but those algorithms which are amenable to implementation. This chapter has discussed several types of processors and the types of algorithms which have been determined to be an effective match. This chapter has attempted to educate the reader in the pertinent issues of parallel processing.

There still remain an abundance of problems in power systems which have not yet been tapped for possible parallel applications, as well as parallel architectures which have not been examined. There still exists an open area for the development of algorithms for the solution of optimization problems, as well as further development of differential, linear, and nonlinear equation solution methods. The challenge of matching architectures with algorithms with system characteristics remains a primary research issue, while introducing the effects of system characteristics is growing in interest. The biggest challenge is to learn to look at problems with new eyes and to not be persuaded by "traditional" methods of parallelizing a given problem.

References

[1]   Kai Hwang and Faye A. Briggs, *Computer Architecture and Parallel Processing.* New York, New York: McGraw-Hill Book Company, 1984.

[2]   T. C. Chen, "Parallelism, pipelining, and computer efficiency," *Computer Design*, pp. 69-74, January 1971.

[3]   Michael J. Quinn, *Designing Efficient Algorithms for Parallel Computers.* New York, New York: McGraw-Hill Book Company, 1987.

[4]   D. K. Stevenson, "Numerical algorithms for parallel computers," *Proceedings of the National Computer Conference*, AFIPS Press, vol. 49, 1980.

[5]   D. J. Kuck, E. S. Davidson, D. H. Lawrie, and A. H. Sameh, "Parallel supercomputing today and the Cedar approach," *Science*, vol. 231, pp. 967-974, February 1986.

[6]   Kai Hwang and Douglas DeGroot, editors, *Parallel Processing for Supercomputers and Artificial Intelligence*, New York, New York: McGraw-Hill Book Company, 1989.

[7]   D. J. Tylavsky and A. Bose, Co-Chairmen, IEEE Computer and Analytical Methods Subcommittee, "Parallel Processing in Power Systems Computation," submitted for publication in the *IEEE Transactions on Power Systems*, 1991.

[8]   M. Takatoo, S. Abe, T. Bando, K. Hirasawa, M. Goto, T. Kato, and T. Kanke, "Floating vector processor for power system simulation," *IEEE Transactions on Power Apparatus and Systems*, vol. PAS-104, pp. 3361-3366, December 1985.

[9]   H. Taoka, S. Abe, and S. Takeda, "Fast transient stability solution using an array processor," *PICA 1983*, Houston, Texas, May 1983.

[10] A. O. M. Saleh and M. A. Laughton, "Cluster analysis of power-system networks for array processing solutions," *IEE Proceedings*, vol. 132, Pt. C. No. 4, pp. 172-178, July 1985.

[11] S. N. Talukdar and E. Cardozo, "An environment for rule-based blackboards and distributed problem solving", *Readings in Distributed Artificial Intelligence*, Morgan Daufmann Publishers, San Nateo, CA, 1988.

[12] K. Lau, D. J. Tylavsky, and A. Bose, "Coarse grain scheduling in parallel triangular factorization and solution of power system matrices," *IEEE Transactions on Power Systems*, paper # 90 SM 485-3 PWRS, to appear.

[13] D.E. Barry, C. Pottle and K. Wirgau, A Technology Assessment Study of Near Term Computer Capabilities and Their Impact on Power Flow and Stability Simulation Program, EPRI-TPS-77-749 Final Report (1978)

[14] F.M. Brasch, J.E. Van Ness and S.C. Kang, Evaluation of Multiprocessor Algorithms for Transient Stability Problems, EPRI-EL-947, (1978)

[15] F.M. Brasch, J.E. Van Ness and S.C. Kang, Design of Multiprocessor Structures for Simulation of Power System Dynamics, EPRI-EL-1756, (1981)

[16] A.M. Erisman, "Decomposition and Sparsity with Application to Distribution Computing," Exploring Applications of Parallel Processing

to Power System Analysis Problems, ed. P.M. Anderson, EPRI-EL-566-SR, pp. 183-208, (1977)

[17] A.M. Erisman, K.W. Neves, M.H. Dewarakanath, eds., Electric Power Problems: The Mathematical Challenge, SIAM, Philadelphia, (1980)

[18] J. Fong and C. Pottle, "Parallel Processing of Power System Analysis Problems via Simple Parallel Microcomputer Structures," Exploring Applications of Parallel Processing to Power System Analysis Problems, Electric Power Research Institute, Palo Alto, pp. 265-286, (1977).

[19] D.J. Tylavsky, "Quadrant Interlocking Factorization: A Form of Block L-U Factorization," Proc. IEEE, pp. 232-233, (1986)

[20] Abur A., "A Parallel Scheme for the Forward/Backward Substitutions in Solving Sparse Linear Equations," *IEEE Trans. on PWRS-3*, pp. 1471-1478, 1988.

[21] F.L. Alvarado, D.C. Yu, Ramon Betancourt,"Partitioned Sparse A^{-1} Methods," IEEE PES 1989 Summer Meeting, Paper # 89 SM 679-2 PWRS.

[22] W.F. Tinney, V Brandwajn, S.M. Chan, "Sparse Vector Methods," *IEEE Trans. on Power Apparatus and Systems,* vol. PAS 104, no. 2, pp. 295-301, February, 1985.

[23] S.Y. Lee, H.D. Chiang, K.G. Lee and B.Y. Ku, "Parallel Power System Transient Stability Analysis on Hypercube Multiprocessors," IEEE Power Industry Computer Applications Conference, Seattle, WA, May 1989.

[24] J.E. Van Ness and G. Molina, "The Use of Multiple Factoring in the Parallel Solution of Algebraic Equations," Power Industry Computer Applications Conference, (1983)

[25] R.Betancourt, F.L.Alvarado, "Parallel Inversion of Sparse Matrices," *IEEE Trans. on Power Systems*, Vol. PWRS-1, No.1, pp.74-81, Feb. (1986).

[26] M.K. Enns, W.F. Tinney, F.L. Alvarado, "Sparse Matrix Inverse Factors,"IEEE PES 1988 Summer Meeting, Paper # 88 SM 728-8.

[27] A. Gomez, and R. Betancourt, "Implementation of the Fast Decoupled Load Flow on a Vector Computer," IEEE PES 1990 Winter Power Meeting.

[28] D.J. Tylavsky and B. Gopalakrishnan, "Precedence Relationship Performance of an Indirect Matrix Solver," PES 89 Summer Meeting and IEEE Trans. on Power Systems (accepted) (1989)

[29] J. Douglas, " Supercomputers for the utility future," *Electric Power Research Institute Journal*, pp. 6-15, October/November 1988.

[30] H. H. Happ, C. Pottle, and K. A. Wirgau, "Future computer technology for large power system simulation," *Automatica*, vol. 15, no. 6, pp. 621-629, November 1979.

[31] R. A. Saleh, K. A. Gallivan, M-C Chang, I. N. Hajj, D. Smart, and T. N. Trick, "Parallel circuit simulation on supercomputers," *Proceedings of the IEEE*, vo. 77, no. 12, December 1989.

[32] N. Zhu, J. S. Chai, A. Bose, and D. J. Tylavsky, "Parallel Newton type methods for power system stability analysis using local and shared memory multiprocessors," accepted for publication in *IEEE*

Transactions on Power Systems, IEEE Winter PES Meeting, New York, New York, 1991.

[33] J. M. Ortega and W. C. Rheinboldt, *Iterative Solution of Nonlinear Equations in Several Variables*. New York, New York: Academic Press, Inc., 1970.

[34] M. La Scala, A. Bose, D. J. Tylavsky, and J. S. Chai, "A highly parallel method for transient stability analysis," *IEEE Proceedings of the PICA Conference*, Seattle, Washington, pp. 380-386, May 1989.

[35] H. Taoka, et. al., "Real-time digital simulator for power system analysis on a hypercube computer," accepted for publication in *IEEE Transactions on Power Systems*, IEEE Winter PES Meeting, New York, New York, 1991.

[36] M. Ilic-Spong, M. L. Crow, and M. A. Pai, "Transient stability simulation by waveform relaxation methods," *IEEE Transactions on Power Systems*, vol. PSWR-2, no. 4, pp. 943-953, November 1987.

[37] M. L. Crow and M. Ilic, "The parallel implementation of the waveform relaxation method for transient stability simulation," accepted for publication in *IEEE Transactions on Power Systems*, IEEE Winter PES Meeting, paper # 91 WM 158-6 PWRS, 1990.

[38] J. K. White and A. Sangiovanni-Vincentelli, *Relaxation Techniques for the Simulation of VLSI Circuits*. Boston, MA: Kluwer Academic Publishers, 1987.

[39] M. La Scala, R. Sbrizzai, and F. Torelli, "A Pipelined-in-time parallel algorithm for transient stability analysis," accepted for publication in *IEEE Transactions on Power Systems*, IEEE Summer PES Meeting, 1990.

[40] F. L. Alvarado, "Parallel solution of transient problems by trapezoidal integration," *IEEE Transactions on Power Apparatus and Systems*, vol. PAS-98, May/June, pp. 1080-1090, 1979.

[41] J. S. Chai and A. Bose, "Bottlenecks in parallel algorithms for power system stability analysis," submitted to the *IEEE Transactions on Power Systems*, summer meeting, San Diego, CA, 1991.

[42] Extended Transient-Midterm Stability Package: Technical Guide for the Stability Program, EPRI EL-2000-CCM-Project 1208, Jan. 1987.

[43] K. Y. Lee and J. H. Park, "Short-term load forecasting using an artificial neural network," accepted for publication in the *IEEE Transactions on Power Systems,* IEEE PES Winter Meeting, paper # 91 WM 199-0 PWRS, New York, New York, 1991.

POWER SYSTEM PROTECTION:

SOFTWARE ISSUES

S. S. Venkata
M. J. Damborg
University of Washington
Seattle, WA 98195

Anil K. Jampala
ESCA Corporation
Bellevue, WA 98004

I.   INTRODUCTION

Power systems of the 21st century will be more modern, and complex, utilizing the latest available technologies. At the same time, generating plants will have to operate with minimal spinning margins and energy transportation has to take place at critical levels due to environmental and economic constraints. These factors dictate that the power systems be protected with optimum sensitivity, selectivity and time of operation to assure maximum reliability, and security at minimal cost.

With an increasing role played by digital computers in every aspect of protection, it is important to take a critical and fresh look at the art and science of relaying and protection. The main objective of this chapter is to review the past, present and future of power system

CONTROL AND DYNAMIC SYSTEMS, VOL. 42

protection from a software point of view. A companion chapter by Phadke and Thorp addresses some of the hardware issues of relaying. The authors sincerely hope that this chapter will arouse the curiosity of the readers and draw more researchers to work in this difficult and needed area of power system protection.

The organization of this chapter is as follows. First, a historic view of the increased role of computers in protection is provided. Following this review, the state-of-the-art in computer aided approaches to transmission relay coordination is discussed in detail. Then newer concepts such as subsystem coordination and adaptive protection are discussed qualitatively. With a strong belief that both transmission and distribution systems should be viewed as a single entity, rather than as two separate systems in planning and operation of power systems, the authors propose ideas for integrating the role of computers in the overall protection and coordination of power systems. The chapter concludes with thought-provoking ideas for protecting future power systems "intelligently".

A. Historical Perspective

A protective relay, whether computer-based or otherwise (electro-mechanical, solid-state), senses abnormal events in a power system such as short circuits, and signals the associated protective device (circuit breaker) to operate. The relay operation depends on its characteristics, settings and the currents and the voltages sensed.

According to Alvarado, et al [1], *"the short-circuit or fault problem was first solved by a digital computer program as early as 1956"*. A few years later, Albrecht, et al [2], reported a relay coordination program using a "batch" off-line approach. A decade later, Rockefeller [3] was the first to propose comprehensive substation protection based on a digital computer.

During the 1980s, researchers at the University of Washington (UW), under the sponsorship of the Electric Power Research Institute

(EPRI) and technical support from Puget Sound Power and Light Company (PSPL), have developed a computer-aided protection software package [4,5] culminating in a tool called RELAY. Both mainframe and personal computer versions are available.

All the above pioneering work broke new ground at that time, using the then available technology, and paved the way for the technology as it exists now.

B. Computer-Aided Coordination System

A Working Group of the IEEE Power System Relaying Committee (PSRC) investigated the application of computers to the task of setting and verifying the settings of line protective devices. The functional requirements of a computer-aided coordination system are [6]:

a)   System database containing short circuit data, protective relay data and protective equipment data.

b)   Software including DataBase Management System (DBMS), short circuit software and coordination software.

c)   Presentation of results including tabular outputs, graphical displays and interactive color graphics.

d)   Guidelines for computer system requirements.

C. Scope of this Chapter

Section II presents the state of the art in transmission protection software. Section III presents two advanced topics: subsystem coordination and adaptive protection.

Due to the nature and scope of this chapter, short-circuit software and digital relays will not be covered here. The reader should refer to [5] for additional details.

II.   STATE OF THE ART TRANSMISSION PROTECTION SOFTWARE

RELAY is a computer-aided protection system developed at the University of Washington [4,5] and is a precursor to the work reported by IEEE PSRC Working Group [6]. RELAY is used only as a representative example, mainly because of the authors' familiarity with it.

RELAY software has been designed to be "modular". Each program acts as a separate module and the interaction between modules is achieved through the central database. There is no direct data communication between any two modules. This concept facilitates adding a user's own fault study software, or additional modules such as power flow or stability analysis to the software at a future date. An overview of the structure of RELAY is shown in Figure 1.

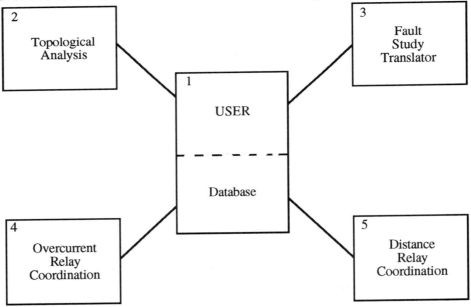

Figure 1. RELAY, Computer-Aided Engineering Protection Software Organization.

A. Database Structure

As Figure 1 suggests, a DataBase Management System (DBMS)

is an integral part of the RELAY software system. A DBMS is extremely useful when developing a package such as RELAY because it provides a stable data repository which serves to integrate different software applications programs or "modules". More importantly, the DBMS serves the user by providing an organization for the data and a manipulation capability which permits the user to understand and modify the data easily once the DBMS interface is mastered. These comments would be true for most engineering tools but are especially valid when the volume of data becomes large, as it does in relay coordination [4].

A database management software package called Relational Information Management (RIM) is used for RELAY. The details about the capabilities and guidelines for users can be found in the RIM User Guide [7] for the mainframe version and R:Base 5000 User's Manual [8] for the IBM-compatible Personal Computer version.

The basic concepts involved in using a DBMS with an application package like RELAY are illustrated in Figure 2. As this figure shows, the DBMS stands between the user (including the applications software) and the disk containing the actual data. A detailed discussion of database concepts is covered in Chapter 4 of [4]. Only an abbreviated summary is given here.

The DBMS can be thought of as a series of models or mappings of the data from one form into another. At the machine level is the "internal model" which is the interface to the actual storage on the disk. The user generally does not concern himself or herself with this model.

At the center of the DBMS is the "user model" (or conceptual model). This is probably the most important concept in a DBMS. The user model is the basic organization tool and consists of structures containing all the data. In relational DBMS's such as RIM, the user model is a set of relations or tables. Each table consists of columns labeled by data item or "attribute" and rows containing the occurrences of the data items. An example of such a relation is labeled LINES in

Table I. This relation contains the data for test system number 1 (TS 1) and will be described in detail later. For now it is enough to observe that each row indicates which bus and breaker occurs on the ends of some line. In cases of parallel lines, a circuit number distinguishes between them. Hence, this relation provides the basic structural information on the system.

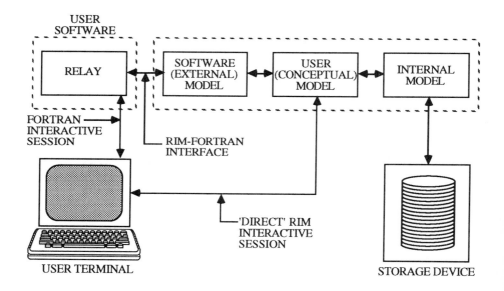

Figure 2. Overview of RIM Organization and Links to RELAY and User

Table I Data in LINES Relation for TS 1

LINENO	FROMBUS	TOBUS	FROMBKR	TOBKR	CKTNO
D	1	2	4	5	0
E	6	3	12	9	0
G	1	3	6	7	2
H	4	3	14	10	0
I	5	4	17	15	0
K	6	5	13	16	0
L	6	2	11	8	0

The complete user model consists of 27 such relations. All data items (attributes) used by RELAY must be assigned to a column in at least one such relation. In fact, the organization of the relations (the assignment of attributes to columns) must satisfy basic principles which make the relations easy to maintain (modify, update) and easy for the user to understand while conveying the necessary information [4,9].

The "software models" (or external models) also consist of 9 relations and contain a subset of the attributes in the user model. These relations are organized to enhance the performance of the application programs which depend upon them. Normally more than one software model exists with each designed for the specific needs of a particular software program.

Reference 5 provides basic information on RIM, specifically on distinguishing the direct interactive mode from access through application programs. It also documents how the different code modules of RELAY depend on specific relations in the database.

B. Topological Analysis

1 Introduction

The process of coordinating a system of directional relays (either overcurrent or distance) involves setting relays one by one so that at each stage the relay being set performs a backup operation which coordinates with all its primary relays, i.e., those relays located in the downstream direction. When this process is carried out on a transmission system with loops, it is apparent that an iterative scheme is required. As explained in [4], the system-wide coordination may necessitate a large number of iterations through all the relays or, at times, may even be impossible to achieve if the relays were considered in an arbitrary order. An approach called topological analysis which yields an efficient sequence for setting relays has been developed. The coordination process of a system of directional relays will converge rapidly to the final solution if the relays are set using this sequence.

The details of this topological analysis are initially presented using the full matrix methods [4,10] and more recently with the application of sparse techniques [11]. This section describes the topological analysis process.

The main problem faced by the protection engineer in the coordination process of a system is to determine the starting points for this activity and the order for setting all other relays. These starting points are revisited in the coordination process, due to the loops present in the system. Hence, it is very important that such starting points are minimized to ensure the fast convergence of this iterative process. The topological analysis process finds a minimum number of starting point locations called the break point (BP) set. This set spans all the loops in the network; that is, the removal of all the relays in the BP set will destroy all the loops. Then an efficient relay sequence called the relative sequence vector (RSV) to set all other relays is obtained. This process ensures that whenever any relay beyond the BP relays is set, all its primary relays have already been set in the previous steps and, hence, the subject relay can be set to coordinate with all its primary relays. Setting relays in this sequence also ensures that each relay is visited only once during an iteration through all the relays. The absence of such a sequence may necessitate visiting each relay many times during each iteration. Finally, since the relay coordination is centered around the primary/backup (P/B) relay pairs concept, this algorithm enumerates the set of sequential pairs (SSP), which is the ordered sequence of all P/B pairs in the system. The order is such that the backup relays appear according to the RSV.

It has been verified that significant amounts of memory and computation time are saved by adopting sparsity techniques [11] to this topological analysis when compared to the full matrix methods used earlier [4,10].

2 Sparse Data Scheme

A data structure based on the LINKNET scheme [12] is used

to represent the system topology. LINKNET was developed as a general purpose structure for representing networks in a computer. It uses a linked-list type of data structure. High storage efficiency is achieved by using vectors containing pointers which are used to describe the connectivity of the given network.

An algorithm based on the Depth-First Search (DFS) and Back-Tracking (BT) technique [13] is developed for the enumeration of all the loops in the system. These loops are then used to compute the Break-Point (BP) set which contains a minimum number of starting relays. This set is further used in enumerating all the Primary/Backup relay pairs in an "optimally" ordered sequence. The Topological Analysis Program (TAP) carries out the topological analysis of the given multi-loop system and arrives at an "optimal" sequence of relays. A flow chart indicating the major steps involved in Topological Analysis is shown in Figure 3.

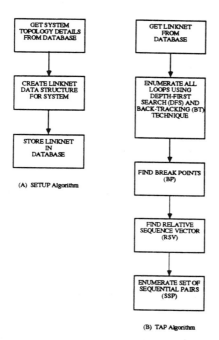

Figure 3. Algorithms for Topological Analysis

3. Algorithm for Loop Enumeration

A prerequisite for determining the BP set is the enumeration of all the loops in the system. It is known from graph theory concepts that the number of fundamental loops in a network of V buses and E lines is given by the number of tree links L given by,

$$L = E - V + 1 \tag{1}$$

All the loops of the network are obtained by taking all possible combinations of the fundamental loops. This gives rise to a maximum possible of $2^L - 1$ loops. However, it is necessary to exclude from this collection all the multiple loops which are the union of two or more edge-disjoint fundamental loops. Hence in general, the total number of valid loops NL is bounded as follows:

$$2^L - 1 > NL > L \tag{2}$$

Note that since the systems analyzed consist of directional relays at both ends of each line, one needs to find loops in both directions. For each physical loop, there are two directional loops, one by traversing the loop in the clockwise direction and the other in the counter-clockwise direction.

A detailed flow chart indicating the loop enumeration procedure using DFS/BT is shown in Figure 3.6 of reference [5]. This algorithm ensures a systematic and efficient procedure of enumerating each loop only once.

4 Enumeration of BP, RSV and SSP

Once the loops of the system are found, the next step is to determine the break point set, or the minimum set of relays which will open all these loops. This problem belongs to the general category of problems known as "set covering problems" in graph theory [14,15]. The set covering problem is known to be NP-complete, that is, no efficient polynomial time algorithms for solving this problem are known and the best available methods for the exact solution require exponential time. It is then the normal practice to adopt a suitable heuristic method and accept a solution which may be less than optimum. A heuristic approach suggested in reference [16] is made use of. In this approach, each loop is assigned a weighting factor equal to the number of relays participating in that loop. Each relay is also assigned a weighting factor equal to the sum of all weights of the loops in which the relay participates. These loop and relay weights are used to identify the BP relays one by one based on the criterion that the relay selected at each step will have the greatest impact on opening the loops of the system. Then proceed to determine the Relative Sequence Vector (RSV), which is an ordered sequence of all the relays in the system. It can be verified that beyond the BP set, whenever a relay is considered, all its primary relays should have preceded it in this sequence.

Since relay coordination is centered around the P/B relay pairs concept, one must enumerate the Set of Sequential Pairs (SSP), which is the ordered sequence of P/B pairs of the system. The order is such that the backup relays appear according to RSV. The enumeration must be carried out by taking each relay from the RSV as a backup and obtaining all its primary relays from the topological information contained in the LINKNET structure.

5 Use of SSP in Relay Coordination

Initial settings are assigned to all the relays (overcurrent and distance) in the break point set. The first pass begins with the P/B relay pair from the SSP that is just beyond the BP set. The P/B relay pairs are taken one at a time from the SSP and the backup relay is set to coordinate with the primary relay for all the fault currents (impedances) for overcurrent (distance) relays considered. Upon exhausting the SSP, move to the top of the SSP and check the backup relays corresponding to the BP set for proper coordination with their primary relays. If these settings have not converged to the final solution, take a second pass through other relay pairs. This process must be repeated until satisfactory coordination is obtained for all the relay pairs.

5.1 An Expression for the Number of P/B Relay Pairs

The number of primary/backup (P/B) relay pairs grows quite rapidly with the system size. Consider a bus I with $N2_I$ two terminal lines and $N3_I$ three terminal lines incident at this bus. Let the total number of lines incident at bus I be N_I, that is,

$$N_I = N2_I + N3_I \tag{3}$$

The total number of P/B relay pairs for a system with NB buses is given by,

$$PB = \sum_{I=1}^{NB} (N_I - 1) \cdot (N2_I + 2.N3_I) \tag{4}$$

One can generalize the above expression for systems consisting of multi-terminal lines with more than three terminals [5].

6 Further Enhancements

The algorithm for the Topological Analysis Program (TAP), which uses depth-first search and back-tracking, was presented earlier. When parallel lines are present in the system, they increase the number of loops dramatically which in turn increases the execution time. In order to significantly reduce the execution time of TAP, certain properties of parallel lines are made use of, in enumerating break points [5,25].

Further enhancements in this area are reported by other researchers recently [17,18]. Instead of relying on heuristic methods for enumerating break-point relays, they applied graph-theoretic concepts, which tend to give minimal break-point sets.

C. Overcurrent Relay Coordination

1 Coordination of Overcurrent Relays

The coordination process of a system of directional overcurrent relays involves determining the three operating parameters normally associated with a typical relay. These parameters are the instantaneous tap, the time delay pickup tap and the time dial tap.

Suitable setting values for these three taps are to be obtained to satisfy the coordination criteria described in earlier work [4]. The instantaneous and the time delay pickup taps are set based on the relay's primary operation and hence are not constrained by the other relays in the system. The computations involved in obtaining these settings are therefore relatively straight forward. But the time dial setting of a relay should not only ensure the fastest possible operation, but also provide coordinated action while performing the backup operation to a number of its primary relays on adjacent lines. The presence of a large number of loops in a typical transmission network requires many iterative calculations for the time dial settings. This problem is tackled by utilizing the set of sequential pairs (SSP) of primary/backup relay pairs, which have been described in Section II B. The SSP drastically reduces the number of iterations and facilitates the rapid convergence of the coordination process.

2 Relay Settings

The overcurrent relay coordination program sets and coordinates all the overcurrent relays in the system. The basic issues in selecting each relay parameter are discussed below. The background is covered more thoroughly in [4].

2.1 Relay Models

In order to represent the inverse-time overcurrent relay characteristics, several models are proposed [4] and utilized. Currently a Working Group of IEEE PSRC is investigating a simplified model that can fit electro-mechanical, solid-state and digital relays. The model has the following general form:

$$t = TD^a [b/(M^c - d) + e], \text{ where}$$

t is the time of operation (s), TD is the time delay, M is the current expressed in multiples of pick-up, a, b, c, d and e are constants that depend on the relay. For example, the CO-9 very inverse overcurrent relay has the following parameters:

$$a = 1.0, b = 5.616; c = 2.0; d = 1.0; e = 0.026.$$

2.2 Instantaneous Tap Setting

As indicated in [4], the instantaneous unit of the directional overcurrent relay should be set so that it provides instantaneous protection for as much of the main line as possible without extending beyond the remote bus. This setting is computed based on the maximum current through the relay for a fault at the remote bus. Since the instantaneous unit performs only primary protection, there are no coordination issues associated with this unit. Hence for each relay a proper instantaneous setting can be individually assigned.

2.3 Contingency of Instantaneous Unit Failure

An additional user option is included in the coordination algorithm by which the user can study the effect of considering the failure of the instantaneous unit of the overcurrent relays. The algorithm then ignores the instantaneous unit during the coordination process and all relays are allowed to act only in the time-delay mode.

2.4 Time Delay Pickup Tap Setting

The pickup tap of the time delay unit of an overcurrent relay is set so that the relay operates both "selectively" and "sensitively". The selectivity is ensured by computing a lower bound based on the maximum load current through the relay, whereas the sensitivity is obtained by determining an upper bound based on the minimum current through the relay for any fault in its primary zone of protection. The protection engineer can then choose the pickup tap setting of each relay to lie anywhere in between these two bounds since he/she can make a better judgement on the pickup setting value based on his/her experience with the system. The remaining relays are automatically determined by the algorithm.

2.5 Inclusion of Radial Lines

Provisions have been made to analyze systems consisting of a mixture of loops and radial lines. The directional relays located on radial lines pose added constraints to be met by the relays in the loops. These constraints arise due to the fact that the settings of some relays on the radial lines may depend only on the system external to the system being studied. The external system can be a transmission, sub-transmission or distribution system. Hence these relays should have fixed settings based on the external system. The user enters a time dial setting for those relays which will not appear as a backup relay in the SSP and these will not be modified by the program. The coordination algorithm is modified to set the relays in the study system to satisfy the constraints posed by the relays in the external system.

2.6 Time Dial Tap Setting

The selection of the time dial setting is the most involved part of the coordination process. Each primary/backup relay pair in the system is checked for proper coordinated action for all the faults that affect the current through these relays. The basic coordination criterion to be satisfied by each relay pair is that the operating time of the backup relay should exceed that of the primary relay by a Coordination Time Interval (CTI) for every fault (typical value of 0.3 s). The faults considered encompass both normal and first contingency, that is, one line out conditions, for various strategic locations in the system. The number of such faults for which the relays are to be coordinated grows rapidly with the system size and it is this large number of fault current pairs that make the coordination problem computationally very intensive.

As mentioned earlier, the computation of the time dial settings is an iterative process due to the loops present in the system. Each iteration is to be carried out considering all the current pairs for coordination. The set of sequential pairs described in Section II B is employed to achieve fast convergence of this iterative coordination process. A detailed flow chart indicating this iterative process of obtaining the time dial settings is presented in [4,10]. Basically, all the relays in the system are initially assigned a user entered value, which is typically the minimum tap available on the relays. Then, following the order of the SSP, each relay is considered for its backup operation and coordinated with all its primary relays for all the relevant fault current pairs. Setting the time dials of the relays in this sequence ensures that whenever a relay is set, all its primary relays are already set in the previous steps. Therefore a suitable setting for the backup operation of the subject relay can be assigned to satisfy the coordination constraints imposed by all the primary relays. Once a pass through all the relay pairs of the SSP is completed, the initial relay pairs are revisited. One can recall from Section II B that the break point relays appear as backup relays in this initial part of the SSP. New settings for the break point relays are computed if necessary, and an iterative process is carried out through all the other relay pairs until convergence is achieved. Due to the efficient relay sequence and the inverse nature of the overcurrent relay characteristics, this iterative process converges rapidly [4].

One can observe that the concept of the SSP has been central to the coordination process by ensuring that each relay is "visited" only once per iteration, that is, when its backup operation is coordinated. It can also be claimed that the coordination algorithm is not changed for different applications of these relays, such as for phase or ground protection. These functions are achieved by simply considering the corresponding fault data. Additional fault types and network contingencies can be included by augmenting the fault data with the corresponding data.

The fault current pairs grow dramatically with the increase in system size. Naturally, faster results can be achieved if some of these pairs are not required for computation during the entire coordination process.

Two enhancements are presented: For a given fault current pair, if the primary relay operates in the instantaneous mode and the backup relay's time dial is adjusted during the first iteration of coordination process, this pair need not be considered further.

The second enhancement is based on practice. For any fault

current pair, if the backup relay's operating time is more than 10.0 s, then that pair is not considered beyond first iteration.

4 Man-Machine Dialog in Overcurrent Relay Coordination

The program is designed to be interactive in nature and makes use of queries, and prompts. Detailed outputs are displayed on the screen as well as written to a file. An experienced user can also run the program in batch mode.

Graphics appear at three different phases in the graphics version of overcurrent relay coordination program. During the initial phase, if the user elects to choose the pick-up settings of the relays, then the system diagram with all the relays appears and one can select the desired relay in a "Query" mode. The information provided includes the relay number, the range and available pickup taps. If there are any problems, such as inadequate pickup taps, messages also appear.

The second phase of the dialog starts with the selection of the instantaneous setting of each relay in the system. Interactive messages are displayed for those relays which do not satisfy the stipulated criteria [4,10]. The process, then, proceeds to determine the time-dial setting of the relays. If there are any problems while coordinating primary/backup relay pairs such as slow operation of the primary relay or miscoordination, the appropriate details will appear on the graphical display. Color is used to "coordinate" this information but does not appear in this figure. In one window, the system diagram with appropriate relays appears. In the second window, the time-dial characteristics and fault currents of primary and backup relays appear. The times of operation and line-out conditions, if any, are also provided. To accommodate large systems, windowing features are used thus showing only the required part of the system.

During the output phase, one can examine the final settings in a "Sequential" or in a "Query" mode. In "Sequential" mode, one relay after the other is shown. In "Query" mode, as displayed in Figure 4, one selects the required breaker. Provisions are made to toggle between these two options.

5 Coordination Review and System Performance Evaluation

The protection engineer is quite often interested in reviewing and evaluating the performance of the existing relay settings rather than designing new settings. Hence a feature has been included in the algorithm by which the user can check the existing relay settings for proper relay action for all faults of interest. This process brings out situations where undesired relay operations or miscoordinations exist in the system. It also provides some overall statistical figures of the operating times of the relays and identifies slowly operating and miscoordinated relays. Another option is also provided by which two different sets of settings (for example, one obtained from manual calculations and the other from the computer) can be compared. The results of the comparison can be used to evaluate alternate settings. This feature is used to validate the settings against manually obtained parameters.

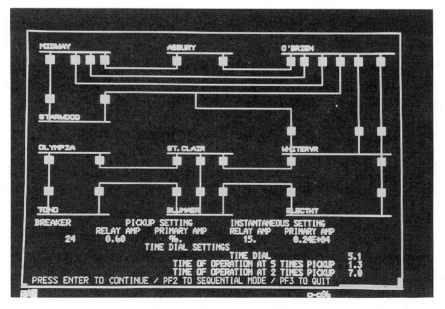

Figure 4. Final Settings of Overcurrent Relay (Output Phase)

D. Distance Relay Coordination

1 Coordination of Distance Relays

Directional distance relays with three zones of operation are considered. The setting and coordination process of a system of distance relays involves determining the impedance tap settings for all the three zones and the time delays associated with zones 2 and 3. Zone 1 is intended for instantaneous primary protection for any fault on the line where the relay is located. Since this zone does not perform any backup protection, it is not constrained by any other relay in the system. Zone 1 for all the relays in the system can therefore be determined directly. On the other hand, zones 2 and 3 both perform the backup operation. Zone 2 also provides primary protection to a portion of the line on which the relay is located. These zones need to be properly coordinated with a number of primary relays for faults on adjacent lines: the second or third zones of any P/B relay pair must never intersect, or else the time delay of the backup relay must exceed that of the primary relay by a coordination time interval, TDMC. A typical value for TDMC is 0.3 s. The determination of the settings for zones 2 and 3 is complicated due to the presence of adjacent lines of greatly different lengths and infeeds in the system. The coordination process is also iterative due to the loops in the system. Similar to the coordination of the overcurrent relays, the SSP discussed in Section II B is utilized to ensure fast convergence of the distance relay coordination process.

In this section, the following procedures which ensure the complete coordination of all the three zones of distance relays for both two and three terminal lines are reported [5, 10, 11]:

1. Procedure for setting zone 1 for three terminal lines.
2. A Successive Zone Coordination Algorithm (SZCA) which performs complete coordination of zones 2 and 3 on both two and three-terminal lines.

This algorithm also ensures that maximum backup coverage for all the lines are accorded without sacrificing coordination.

2 Relay Models

Distance relays of the Mho type [4] are modeled in this work. Specifically, models for the circular characteristics of Westinghouse KD relays have been developed. The impedance setting is specified in terms of a set of impedance taps. For example, these taps for the KD-10 relay are termed as S, M and T taps. The S and T taps provide coarse adjustment while the M tap provides the fine adjustment needed to get the desired impedance setting. All these taps have been suitably modeled so that the final setting of the distance relay can be given in terms of these taps. A similar procedure can be applied to other types of distance relays.

3 Zone 1 Setting

As described earlier, zone 1 is set to provide instantaneous protection for any fault on the "primary line". It is set short of the line length to ensure that it does not overreach the remote bus. Since zone 1 is intended only for primary protection, there are no coordination issues associated with it. Hence relays can be considered in any arbitrary order and assigned proper zone 1 settings. It may be noted that for a relay on a two-terminal line, the zone 1 setting is purely based on the ohmic impedance of the line. However, in the case of a relay on a three-terminal line, the setting depends on the fault current profiles on the line segments of the three-terminal line in order to account for the infeed and/or outfeed at the junction of the three-terminal line [11].

4 Successive Zone Coordination Algorithm

Zones 2 and 3 are employed to provide backup protection for faults on transmission lines. These zones of a relay should be set so that maximum backup coverage is provided for all the adjacent lines and also to ensure that the subject relay coordinates with all of its primary relays for faults on these adjacent lines. These two objectives are not fully met by the previously existing methodologies [4, 19-21] of coordinating a system of distance relays. A new approach termed the

Successive Zone Coordination Algorithm (SZCA) to ensure maximum backup coverage and complete coordination of zones 2 and 3 of all the distance relays is presented here.

The Successive Zone Coordination Algorithm (SZCA) has been developed to ensure complete coordination of all the zones of distance relays. In this algorithm, the limiting values for the zone 2 setting of any backup relay is computed using the zone 1 settings of all its primary relays. Similarly, the limiting zone 3 setting value of any backup relay is found from the zone 2 settings of all its primary relays. Thus the zones of the distance relays are set successively to ensure proper coordination as well as provide maximum coverage of the lines being protected. A general analysis procedure to compute these limiting impedance values for zones 2 and 3 is comprehensively described in reference [11].

5 Man-Machine Dialog in Distance Relay Coordination

The program, just like the overcurrent coordination program, is designed to be interactive and makes use of queries, and prompts. Detailed outputs are displayed on the screen as well as written to a file. An experienced user can also run the program in batch mode.

The graphics option is available in the graphics version of the distance relay coordination program. The determination of zone 1 settings for relays on two-terminal lines is a straight forward process. For relays on three-terminal lines, due to infeed and/or outfeed conditions, the zone 1 reach may not cover the desired portion of the line. A graphics display will appear when such a situation occurs. The line is drawn to scale (the length of line drawn from each relay to the tap point is proportional to the line impedance from that relay to the tap) and color coding is used to display the zone reach of the relay under consideration. Such diagrams will help the engineer to quickly comprehend the situation.

During the coordination of zones 2 and 3, if there are any problems with the coordination process, appropriate messages will appear. All the necessary details will be provided and the user will be asked to select one of the options available such as a reduced zone reach or delayed operation to overcome the miscoordination problem. The output phase is similar to that of the overcurrent relay coordination program.

III. ADVANCED TOPICS

In this section, two advanced topics, namely Subsystem Coordination and Adaptive Transmission are discussed.

A. Subsystem Coordination

1 Definition of Problem [22,23]

The protection engineer, while working with an existing system, often encounters a situation where he/she would like to study a portion of the large system for proper coordination of the protective relays. One reason may be that this subsystem was coordinated earlier, but may not be responding properly due to present system changes. More specifically, the engineer may also like to determine the changes in relay settings needed in response to structural changes like addition or removal of lines, generators, or significant changes in system loading conditions. The changes in system topology may be due to temporary maintenance activities, permanent network reconfiguration, or the result of switching actions after a fault. For a large transmission system, it may not be desirable and, at times, may even be impossible due to the system size, to carry out a complete system coordination process for every one of these modifications. Coordination of the full system as described in the previous sections may be prohibitively expensive in terms of the manpower and computer effort involved in obtaining new operating parameters for the protective relays. A more efficient way to solve this problem is to adopt a subsystem coordination approach where only that part of the network which has been affected by the change is considered for recoordination.

This section presents a new algorithm called the Subsystem

Coordination Algorithm (SCA) which automatically identifies a subsystem of the original network as a "window" containing the region of change(s). This window is identified by studying the sensitivity of the relay setting parameters with respect to the new network conditions. Only those relays within this window will require new settings due to the system change. The settings of the relays outside the window will be unchanged. This subsystem is then coordinated for proper relay actions considering the new system configuration. This coordination process also accounts for the "boundary" conditions which are the coordination constraints imposed by the relays just outside the window on the relays within the window.

Consider the system shown in Figure 5 which is a part of a large system. Assume that the entire system has been coordinated earlier. Consider the problem of removing the line M and studying the effect of this line removal on all the relays in the system. The removal of this line can be viewed as a "structural disturbance" at the end buses of this line, namely buses 1 and 2. This disturbance will affect the fault current profiles on the neighboring lines and hence affect the coordination of the relays under the existing settings. However, it is known that normally the changes in fault current profiles diminish as one moves farther from these disturbed buses. Hence, it can be reasonably expected that relays which are far away from the disturbed zone may still remain coordinated even with this system change. The subsystem coordination identifies those relays in and around the disturbed zone which are affected by the system change and obtains new parameters for those relays which will ensure coordination under the present system conditions.

1.1 Terminology

1. An area called the "**window**" is automatically identified around the disturbed zone. This window contains all the relays which are to be *reset* to obtain coordination for the new system conditions. This window is identified by studying which relays need

S. S. VENKATA, M. J. DAMBORG, AND ANIL K. JAMPALA

"significant" changes in their parameters for proper coordination. The criteria for these significant changes will be explained later.

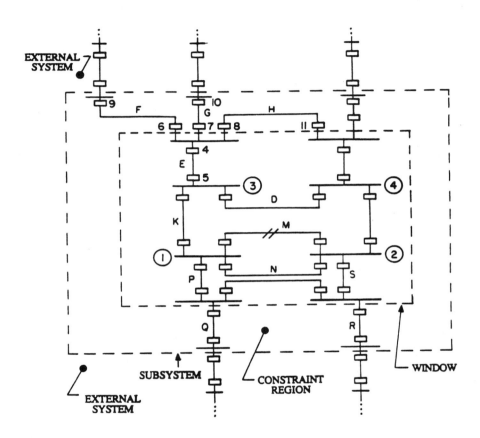

Figure 5. An Example System for Definitions of Subsystem
Terminology

2.   Considering the window shown in Figure 5, one can observe that
the relays within the window at the boundary are constrained in
their operation by the relays which are just outside the window.
Since only *one* level of backup is considered for coordination
purposes, the relays outside the window which impose the
boundary constraints are located on lines which are within a
radius of *one* bus from each of the boundary buses. Hence the
"**constraint region**" consists of all the lines which are one bus
away from the boundary buses of the window.

The significance of this constraint region is illustrated with
examples later in this section. Note that by definition, these
relays in the constraint region themselves have fixed settings,
that is the changes in the system structure have not affected their
settings and they remain same as before. The relays in this
constraint region impose constraints on both the primary and the
backup operation of the relays in the window. Therefore these
constraints can be classified into two categories, "**primary
constraints**" and "**backup constraints**". The primary constraints
are those imposed by the relays in the constraint region on the
primary operation of the relays in the window and the backup
constraints are those put forth by the relays in the constraint
region on the *backup* operation of the relays in the window.
Again referring to Figure 5, relays 9, 10 and 11 in the constraint
region backup relay 4 in the window. Hence, relay 4 is to be set
for its *primary* operation to coordinate for the existing, fixed
settings of relays 9, 10 and 11 for faults on line E and bus 3.
These relays 9, 10 and 11 are therefore said to impose primary
constraints on relay 4. Similarly, relay 5 in the window backs up
relays 6, 7 and 8 in the constraint region. Hence, relay 5 is to be
set for its *backup* operation to coordinate for the existing fixed
settings of relays 6, 7 and 8 for faults on lines F, G and H
respectively. Relays 6, 7 and 8 are therefore said to impose

backup constraints on relay 5. Thus, in addition to coordinating all the relays in the window, the subsystem coordination procedure must also include these primary and backup constraints imposed by the relays in the constraint region.

3.   The combination of the "window" and the "constraint region" is called the "**subsystem**". Thus the subsystem consists of all those relays which are to be considered in the subsystem coordination process though only a portion of this subsystem, namely the "window", contains relays which will have changed settings.

4.   The remaining portion outside the subsystem is referred to as the "**external system**". The settings of relays in this external system remain unaltered.

2   Preliminary Investigation of Relay Response to System Changes

The structural changes (addition or removal of lines) generate significant changes in the parameters of the overcurrent relays in the neighborhood of the disturbed region; but the distance relays have negligible changes in their parameters. Changing load levels had minimal effect on the overcurrent relays but have significant effect on the third zone settings of the phase distance relays. It can therefore be expected that coordinating the overcurrent and the distance relays may require different subsystems.

3 Subsystem Coordination Procedure

The steps for the Subsystem Coordination Procedure are as follows:

1.   Modify the existing database to reflect changes after determination about changes is made. The engineer may add or delete transmission lines or relays from the database relations describing the system topology or relay specifications. Normally the modifications needed will be minor.

2.   Build the LINKNET data structure for the modified system.

3.   Obtain the fault current data for the modified system. This study

must be executed on the <u>entire</u> network so that all changes in fault currents can be identified.

4.   Obtain software model relations corresponding to the fault data.

5.   Use the subsystem coordination program, which is explained next, to carry out the identification and coordination of the subsystem.

4 Algorithms for Subsystem Identification and Coordination

The process of identifying and coordinating a subsystem is carried out through the following three main steps:

4.1. Subsystem identification

Once the buses involved with the system changes (the "disturbed" buses) are specified, a radial search starting from these buses is carried out. Those relays whose settings need to be changed are identified. The window containing all the lines on which these relays are located is obtained. The constraint region is identified (as explained later), and the subsystem is formed.

4.2. Subsystem Database Setup

All the data corresponding to the topology, relays and fault information for the subsystem are extracted from the full database and these are loaded into the relations pertaining to the subsystem. The subsystem coordination program accesses only these relations to carry out the coordination process.

4.3. Subsystem Coordination

The LINKNET data structure for the subsystem is first formed. The topological analysis process is carried out on the subsystem to identify the break points and the set of sequential pairs (SSP) for the subsystem. Next, a complete loop coordination is carried out on the subsystem. This coordination process includes all the constraints imposed by the relays in the constraint region on those in the window.

A detailed flow chart indicating various steps involved in the subsystem identification and coordination process, carried out by the Subsystem Coordination Program, is shown in Figures 2 and 3 in reference [23].

5 Test System and Results

The subsystem identification and coordination program has been tested using the complete 115-kV transmission network of PSPL. A portion of this system is shown in Figure 6. Consider the case of removal of the line OB-MI1 (the line #1 from O'Brien substation to Midway substation). Since this line is an electrically short one (low impedance), it is expected to significantly affect the fault profiles on adjacent lines and thus disturb the coordination of the relays in the neighborhood. The subsystem coordination procedure is applied to obtain new setting values for the modified system.

The results obtained from the subsystem approach are validated by comparing them with those obtained when the full system is coordinated after this line is removed. Identical precision values for checking the convergence of the relay settings are used in both the full and subsystem coordination procedures. The results of the full system coordination confirmed that the window obtained through the subsystem approach is indeed valid. Table II compares the settings obtained for the relays in the window from both the full system approach and the subsystem approach. It can be observed that the settings obtained through the subsystem method are in very good agreement with those obtained through the full system method. The time dial settings obtained through the subsystem method are shown in two steps, one after the window identification process (column indicated as "WI") and the other after the loop coordination of the window (column indicated as "WC"). For the test system, it can be observed that for most of the relays the final settings are obtained at the end of the window identification process itself.

Relays 162 and 152 are the only relays which changed their settings during the window coordination process to reach their final settings.

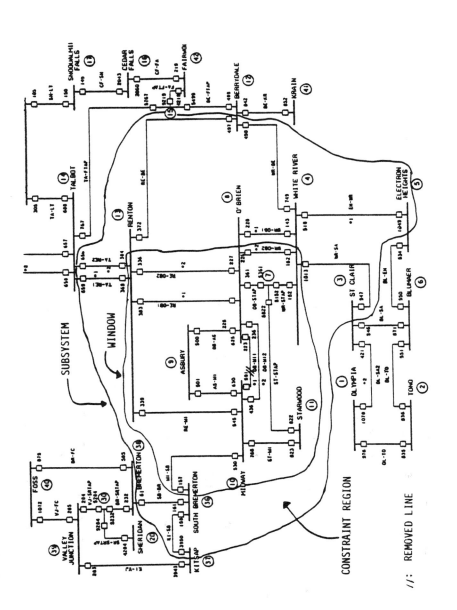

Figure 6   Subsystem Identified by the Identification Algorithm

Table II Comparison of Relay Settings by
Full and Subsystem Methods

RLY NO.	PICKUP SETTING			INSTANTANEOUS SETTING			TIME DIAL SETTING			
	OLD	NEW FSC	NEW SSC	OLD	NEW FSC	NEW SSC	OLD	FSC	NEW SSC WI	SSC WC
162	1.5	1.5	1.5	20	20	20	2.04	2.19	2.18	2.19
226	0.6	0.6	0.6	20	20	20	2.78	2.78	2.78	2.78
152	0.6	0.5	0.5	24	24	24	5.67	6.21	5.77	6.21
361	0.5	0.5	0.5	15	15	15	4.10	4.22	4.22	4.22
822	0.6	0.6	0.6	15	15	15	5.53	5.56	5.53	5.53
823	1.0	1.0	1.0	24	30	30	5.52	5.15	5.15	5.15
766	1.0	1.0	1.0	24	24	24	2.72	2.68	2.72	2.72
236	0.8	1.0	1.0	20	30	30	3.59	4.95	4.95	4.95
436	0.5	0.6	0.6	20	20	20	2.60	2.66	2.60	2.60
625	1.5	1.5	1.5	40	40	40	5.11	4.75	4.75	4.75

LEGEND

OLD: System before change; TS83   SSC: Subsystem Coordination Method

NEW: System after change; TS84   WI: After Window Identification

FSC: Full System Coordination Method   WC: After Window Coordination

86

Table II (continued)

RLY NO.	PICKUP SETTING			INSTANTANEOUS SETTING			TIME DIAL SETTING			
	OLD	NEW		OLD	NEW		OLD		NEW	
		FSC	SSC		FSC	SSC		FSC	SSC WI	SSC WC
500	0.8	1.0	1.0	24	30	30	4.30	2.82	2.82	2.82
339	0.6	1.0	1.0	24	24	24	7.06	4.45	4.45	4.45
545	1.0	0.8	0.8	16	16	16	4.47	3.91	3.91	3.94
501	0.8	0.8	0.8	16	16	16	3.24	2.74	2.74	2.74
.630	2.5	2.5	2.5	60	48	48	2.51	2.37	2.37	2.37
383	0.8	0.8	0.8	20	24	24	4.35	2.86	2.86	2.86
225	1.0	1.0	1.0	30	30	30	4.31	4.58	4.59	4.59
336	0.6	0.8	0.8	20	24	24	4.88	2.84	2.84	2.84
227	1.0	1.0	1.0	40	40	40	4.62	4.84	4.84	4.84
530	0.5	0.5	0.5	10	10	10	6.25	6.16	6.18	6.18
157	0.6	0.6	0.6	8	8	8	3.63	4.04	4.04	4.04

The subsystem method reduced the amount of CPU time by a factor of about 3.5 when compared with the full system method.

B. Adaptive Transmission Protection

1 Definition of Adaptive Protection [24,25]

The term "adaptive protection" refers to the ability of the protection system to automatically alter its operating parameters in response to changing network conditions in order to maintain optimal performance. Optimal performance implies best relay settings which result in fastest relay operation, while satisfying the basic coordination criteria. These criteria should satisfy protection engineers if they needed to consider only the present network conditions. It is implied that recomputing and transmitting the new operating parameters is to be done in real-time such that the time interval between the initiating event and the modification of parameters is of little practical significance.

In this section, the adaptive protection concept is defined and described. The computational issue of this concept is investigated further, since it is an essential part of an adaptive protection scheme.

2 Why Adaptive Protection ?

Any protective scheme possesses three important characteristics: sensitivity, selectivity, and speed. Any, or all, of the three can be improved by adopting an adaptive scheme. For example, consider the settings of overcurrent relays. Each relay has three parameters: pickup setting, instantaneous setting, and time-dial setting. At present, the protection engineer normally takes all possible single line out contingencies into account while computing these settings. Since the relays are set manually, changing their settings according to changing network conditions is extremely difficult. With the advent of digital relays, if the relays can be set to respond to "existing" network conditions (that vary from time to time), they can be made to operate faster and can be made more sensitive to various types of fault conditions without losing their selectivity. This intuitive idea is illustrated using a simple study conducted by the investigators.

Two sets of settings (for overcurrent relays), with and without line out conditions, are obtained using the RELAY software described in Section II. These settings are reviewed for their performance and the results, documented in Table III, are summarized below. Without the line-out conditions:

- mean operating time of primary relays is 18.4 % faster.
- mean operating time of backup relays is 14 % faster.
- individual primary relays as well as backup relays consistently operate faster.
- the instances in which primary relays operated slowly are eliminated.

These results demonstrate that setting the relays for existing network conditions on an on-line and real-time basis results in speedier operation of the relays. However, the relay settings need to be computed more often because of possible changes that may occur in the state of the network. Then, a natural question that arises is: Under what circumstances, and how often, do the relay settings need to be changed ? The answer is: either at an operator's request, or on a periodic basis when the relay settings are reviewed with the forecast network state. The relay settings will be changed at least twice, once for peak conditions and the other time for off-peak conditions, during a day. In practice, it may be done more often: once in a few hours.

The next question is: what if a contingency such as a permanent fault occurs ? Or, if there is a localized change in generation or load ? The "subsystem coordination" approach for dealing with such changes from planning point-of-view is described earlier. In this approach, a "window" is identified around the disturbed region. The relays in this region need to be re-coordinated whereas the relays outside the window retain their existing settings. This is an appropriate approach, even in real-time, if the entire process can be achieved within a few minutes in an automated mode. The entire process means fault detection and identification, on-line computation of new settings, communication and decision times. However, if the window identified is too large to fit real-time needs, then further speeding up of the computational process is warranted.

Table III: Performance Analysis of Overcurrent Relay Settings

Performance Index	Normal and Contingencies	Normal Only
# of Fault Current Pairs	1376 [*]	1376
Primary Relays		
Operating Time (S)		
Mean	0.473	0.386
Standard Deviation	0.571	0.445
# Pairs Operating Slowly (> 1.0 S)	32	0
# Pairs Relatively Faster	0	588
Backup Relays		
# Pairs operating < 10.0 S	958	965
Operating Time (S)		
Mean	1.696	1.457
Standard Deviation	2.548	2.320
# Pairs Relatively Faster	0	710

[*]When contingencies are also included, there are 5,024 fault current pairs for coordination of the relays. However, the comparison is based on the same 1,376 pairs which result for normal operations.

In the case of slowly varying network conditions, all the relays in the network need to be re-coordinated. Based on the authors' experience, existing techniques cannot handle large networks in real-time. In order to improve existing tools, enhanced algorithms described earlier are developed. Also inherent parallelism in the problem is exploited so that multi-processing techniques can be utilized. The details of this latter development are explained later in this section.

Adaptive protection can also offer new benefits; one is enhancing the reliability of relays. When a relay operates, all of its backup relays will be asked to check whether they have sensed a fault current, and, if so, how long would each of them take to operate. By doing this off-line simulation, one can ensure that the relays are operating properly. Perhaps, some of the proposed Artificial Intelligence (AI) tools, to be discussed later, may be used for this purpose.

Having discussed how an adaptive protection scheme can be of value under various network operating conditions, some of the benefits offered to utilities by such a scheme are mentioned below:

1.   Better transmission line loading due to significantly improved speed of relaying,

2.   Improved life expectancy of transmission lines and switchgear because they are exposed to extreme fault conditions for shorter times,

3.   Better relays due to their self-checking capabilities which increase system availability and, thus, are cost-effective,

4.   Permissive over-loading of transmission lines by increasing the pickup setting of the overcurrent relays to prevent tripping when service is restored following an outage,

5.   Verification of data obtained from multiple sources to detect errors, thus permitting prompt action to be taken,

The existing protection schemes are not fully adaptive in nature and the system performance is not optimum (for example, the various components are exposed to a fault condition longer than required, the loading of the transmission lines cannot be fully exploited). After identifying the various components required to realize the adaptive protection concept in the next section, an example is provided to illustrate how it works.

3 Description of Adaptive Protection Concept

In order to realize the adaptive protection concept the following four main components need to be addressed:

1) Hardware
2) Communication and Control
3) Software
4) Human factors

3.1 Hardware

The term hardware, in the adaptive context, refers to relay hardware and is identified in Appendix-C in reference [5]. Several possible hardware architectures are:

i)   A single computer performing all the relaying functions in a substation.

ii)   A set of relays that implement a specified protection function such as line protection, or

iii)   A relay with multiple processors to realize a single protection scheme.

The first alternative was proposed by Rockefeller [3]. However, this would be in sharp contrast with the existing protection schemes in which there is a separate and independent relay for each protective function. Some experts do not prefer the first alternative (see discussion offered by Lewis in reference [26]). It will be difficult to convince protection engineers to use the first alternative unless it has been implemented for a sufficiently long time.

In the second alternative, a given protection scheme may have several relays working on different principles. For example, the protection of a single transmission line may consist of overcurrent, distance and travelling-wave based relays. This would be in close conformation with the existing protection practices and may gain acceptance much more easily than the first alternative.

In the third alternative, a given relay may be realized by multiple processors. Several examples of this type are mentioned in Appendix-C in [5]. To cite one such example, Girgis and Brown discussed how a distance relay could be implemented with multiple microprocessors (see the closure to discussions of reference [27]).

A combination of the above alternatives is an attractive proposition. Consider the following: each protective function will have more than one relay (each of these relays may be implemented by using more than one processor). A station computer oversees the operation of these relays and may provide backup operation in case any relay malfunctions.

3.2 Communication and Control

A power system is a clear example of a totally distributed system. The generating units and the loads are usually far apart and the transmission system interconnects them. However, control is centralized and many control decisions require system-wide knowledge. Each substation has complete information on its current status, but not on any other substation. Data is gathered from major substations and is sent to a central control center for processing and further action. At a given time, only the control center has the up-to-date performance data of the system and the computing power to process it. It is believed that this trend in control centers will continue despite the fact that cheaper and powerful computers may start appearing at each substation.

Control and communications issues are inseparable. In order to take control decisions, data needs to be communicated from substations to the control center. On the other hand, the control action is to be communicated to the respective devices. Traditionally, utilities use telephone, microwave and carrier communications schemes between substations and the control center. In the future, these will coexist with fiber optic systems. Communication between substations is used for protective relaying purposes. In the future, this communication may be completely based on fiber-optic systems, particularly if ground or static wires are present.

Communications within a substation will change dramatically in the future. For example, computer based relays acquire information every 1 ms (16 samples in a cycle based on 60 Hz system frequency) which could be used for data acquisition purposes. In other words, the same data will be compressed and sent to the control center. Further, some protection schemes, such as bus differential protection, need sharing of data among processors.

Computer communications is becoming increasingly important. The future developments in utilities will be incorporating the International Standards Organization (ISO) Open Systems Interconnection (OSI) model [28] between the central computer and substation computers. At each substation, a local area network will be commonplace.

3.3 Software

The software issue is addressed in the context of the proposed concept. The software may be a part of the Energy Management System (EMS) software. It should have two major components: on-line relay coordination programs and protection-related dispatcher aids.

On-line relay coordination programs: These should be able to encompass scenarios such as those described in subsection 2 on **Why Adaptive Protection**. The software should be able to

- review existing settings,
- re-compute the settings in case of a contingency or of localized changes in generation or load,
- re-compute the settings to accommodate the slowly varying network conditions, and
- check the backup relay operation for improving reliability.

The computation should be extremely fast so that it can be done "almost instantaneously" in response to a change in network condition.

The principal tool that is used is the RELAY software. Later in this section, various issues involved in making this an on-line tool are addressed.

Dispatcher aids: The aim of these tools is to help the dispatcher make intelligent operational decisions, including those concerning protection. These tools are likely to include an expert system for restoration of service to customers when contingencies occur [29] and an expert system for analyzing the historical data. These tools will help in analyzing the type and location of a fault, the action to be taken, and whether a similar fault has occurred in recent history, so that appropriate remedial action can be taken.

The two expert system aids mentioned in [30] can be used in conjunction with the fourth feature mentioned in subsection 3.3. For example, consider the disturbance in the form of removal of a line. The events that actually take place should match with the events that are predicted by the Discrete Event Simulator [30] for the given disturbance. The operator can make sure that this indeed is the case. Similarly, the Diagnostician program can be invoked when the network configuration changes due to a fault. The operator can verify whether the event has happened or not based on the hypothesis predicted by the Diagnostician.

3.4 Human Factors

The issue of human factors is one of the important issues which is not addressed in detail in this chapter. Besides the man-machine interaction provided by the EMS software, the tools mentioned in subsection 3.3 will be beneficial to the operator. In the future, the operator may also be a protection engineer in the control room. For example, Horowitz expressed doubts about the dispatcher's ability to make decisions about protection (see discussion offered by Horowitz in reference [31]). All these points are mentioned to underscore the importance of an experienced operator to achieve the proposed adaptive protection concept.

3.5   Illustration of the Concept

In order to better understand the adaptive protection concept, consider the following two examples:

1.   The existing relay settings are periodically reviewed. If there are any miscoordinations, the operator can determine whether a complete recoordination is warranted. If the settings are to be computed, then the programs mentioned earlier will be run and the new settings are transmitted to the relays.

2.   If a permanent fault occurs on a transmission line, the fault is identified by the associated relays within a cycle (16.67 ms on a 60 Hz system frequency basis) after the occurrence of the fault. After the line is removed from service, the first action is to determine whether service needs to be restored to customers affected by the line removal. For this purpose, the expert system proposed by Tomsovic et al [29] can be used. Concurrently, a window can be identified around the disturbance to identify the region affected, and, if necessary, the new relay settings can be computed and sent to the affected relays through the communication system. Next, with the help of the Discrete Event Simulator [30], the sequence of events, which took place from the moment the line is removed, can be verified. Also, from the initial and final states of the system, the Diagnostician can hypothesize the cause of the disturbance which needs to be verified by the operator.

In the following sub-sections, the results of different approaches proposed and investigated are presented with the ultimate aim of developing a fast, on-line tool for computing relay settings for the adaptive protection of transmission systems. Two approaches that significantly improve the execution times of relay coordination studies are discussed. These approaches make the adaptive protection concept begin to appear feasible.

4 Parallelism in Relay Coordination

In Section II, efficient modifications to existing algorithms, which make the computation of relay settings very fast, are presented. But, even this speeding up is not sufficient for adaptive protection where settings need to be computed "almost instantaneously". Hence, the possibility of exploiting the natural parallelism in solving the problem is investigated.

Figure 7 shows two systems A and B which are connected by a single transmission line. Relay # 1 acts as backup relay for relays in system B and hence, its settings are determined by the relays in system B. However, the settings of relays in system A will be influenced by those of relay # 1. Relay # 2 is also in a similar position. Instead of solving a composite system, formed by combining the above two systems and the transmission line, a distributed computing approach is proposed here in which each system is solved on a separate processor and the processors communicate to exchange the settings of the boundary relays. The emphasis is on testing whether such a concept works and if so, how to implement it. Two approaches, not necessarily mutually exclusive, are proposed:

- Distributed Computing
- Supercomputing

5 Distributed computing approach

In this sub-section, distributed computing is investigated with the intent to demonstrate the feasibility of the concept. Supercomputing is considered in the next sub-section.

Figure 8 shows the Test System 4, the complete 115-kV transmission network of PSPL, divided into four subsystems, which can be identified with the four operating divisions of PSPL.

(a) System A and B connected by transmission line

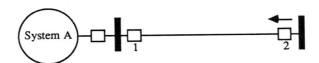

(b) Relay 2 setting determined by its primary relays in System A

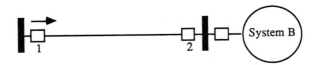

(c) Relay 1 settings determined by its Primary relays in System B

Figure 7 Illustration of Distributed Computing Approach

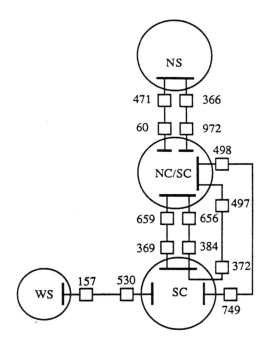

Figure 8 Conceptual View of the Four Subsystems of Test System 4

To demonstrate the distributed computing concept, a shared memory option, available on IBM 4341 mainframe computer through a software package, is chosen. This allows multiple virtual processors to share a segment of memory. In this scheme, each processor has its own memory, disk space, console and access to a shared memory. Each processor has limited read/write access and exclusive write access on request.

The first test case has four processors sharing the memory. The overall speed-up is less than three. The partitioning (based on divisional basis) is not adequate. Another case is tested in which subsystems WS and SC are combined as one. The speed-up in this case is closer to 3, suggesting that the earlier partitioning is not optimal.

6 Supercomputing Approach

Another alternative to achieve better computational times is to utilize faster hardware such as a CRAY type computer. The RELAY software is tested on a CRAY-XMP Supercomputer. The availability of a compatible database management system simplified the testing. The programs and the data (unloaded from the database) are easily transported. No modifications are needed to compile the programs. The data is loaded into the database using the interactive mode of the database software.

The important conclusion is that the systems about the size of 10-buses and 15-lines can be studied in less than a second of CRAY-XMP CPU time. A speed-up greater than 20 is realized. Further speed-up is possible by optimizing the programs.

7. Summary

It has been demonstrated that the distributed computing concept does speed up the relay coordination process time. The implementation details depend on the particular computer system used. Consider for example a CRAY-XMP computer with four processors. The same approach demonstrated by the authors can be implemented without any problem by using the primitives provided on that system. By doing so and, by further optimizing the programs, the time for computation of relay settings can be reduced to the order of a second of CPU time for the TS 4 network. Indeed, such a speed-up can be achieved with the availability of the faster hardware.

C. FUTURE DIRECTIONS OF ADAPTIVE TRANSMISSION PROTECTION

A. Introduction

As one scans the adaptive protection literature, many different concepts emerge. The first is that there is no agreement on the definition of the term "adaptive protection". Some papers [32,33,34] contain a very broad interpretation of the term. However the term is accorded a narrow interpretation in this work. It signifies adapting relay performance as a result of changing network conditions, e.g. changes in system load or topology.

Another concept that emerges is a number of different levels of complexity due to adaptive protection. In this section , five levels of use of computer relaying are identified. This subdivision of the literature is certainly subjective and open for discussion. However, it is useful to think on expanding traditional protection concepts.

Finally, it is clear that all of adaptive protection has been motivated by the appearance of digital relays and solid state or static relays. A simplistic view is that a digital relay performs traditional relaying operations, but is managed by an electronic rather than a mechanical device. However, as explained later, traditional relaying operations are not a part of some of the more advance concepts of adaptive protection.

B. Levels of Complexity in Adaptive Protection

Five hierarchical levels of complexity are proposed for this new concept.

Level 1:

A primary advantage of digital relays is that the settings are very easy to change. It is an easier, more reliable process to alter relay action by changing data in a memory unit than to alter a setting on a mechanical device. As a result, engineers can consider changing the settings more frequently than is now the case even though field personnel is required. For example, changing relay settings could be a standard feature of major maintenance operations or response to seasonal load changes.

Level 2:

Digital relay settings can be changed remotely since only transmission of data is required. With a communication link, the engineer can think of changing settings more often, perhaps in response to weekly load changes or standard maintenance changes.

Level 3:

If all relays in a system can be changed remotely from a single central site, then the time between changes could become much shorter still. In fact, this time is limited only by the time required to calculate new settings. Since computer software exists for coordinating all relays in a system, one can think of arbitrarily frequent changes if the software will operate fast enough. A logical goal in speed would be a few seconds which would be fast enough to respond to even unexpected switching actions with new settings. The concept is elaborated further in the next subsection.

Level 4:

The traditional role of relays may not serve the desired goals at this level. Here, the idea is that a substation could have a unified relaying subsystem in which a single substation computer controls all switches, i.e. a central decision process for the substation rather than local decisions at each relay. If each pair of adjacent substation computers communicate they could function comparably to pilot relays. If not, a traditional concept of coordination must be retained where local decisions depend upon stored data which was determined from a "total system" perspective assuring that individual substation actions will be best for the entire system. Of course, this coordination would not use traditional relay parameters but it would be guided by the traditional concepts that relays should disturb as little of the network as possible while switching as rapidly as possible.

Level 5:

In the final stage of computer control, one central computer could conceivably manage all switching operations by gathering all relevant data from the system. This concept is fundamentally appealing since central management normally seems desirable in control problems. Of course, the heavy burden on communications is a major concern since a complete communication system is increasingly difficult to sustain as the physical scale of the power system becomes large. In power systems, size approaches the extreme. Leaving these problems aside, having a single computer detect all system faults, and initiate appropriate switching actions in response is an appealing concept.

C. Proposed Approach to Adaptive Protection

The proposed approach starts at level 3 and requires the use of the RELAY code developed as an off-line, planning tool [4-5]. This code has been tested and verified on a variety of actual systems chosen from the Puget Sound Power and Light Company's (PSPL) 115 kV transmission network. This code has met the PSPL requirements and is being used by them for system planning studies.

A goal of any protection system is to switch as rapidly as possible in response to faults. Another is to switch selectively to disturb as little of the network as possible. The goals of rapid and selective switching are in conflict, however, when they are based upon a single collection of settings that accommodate a wide variety of network conditions (contingencies). That is, if the relay settings are to respond appropriately for many different cases of fault location, for single-line-out contingencies or for changes in system loading, this appropriate switching is managed at the expense of speed.

Then, an appropriate approach to adaptive protection is to reduce the number of contingencies that settings are designed for, and achieve faster switching. When a contingency occurs, calling for new settings, they are rapidly changed before another switching action is needed. The fastest switching action would be achieved when the settings are set for only the existing network conditions (loading and configuration or topology).

Two experimental investigations are carried out to test this concept. One determined what gains in switching speed could be realized if relay settings are determined for only existing network conditions rather than for all single line out contingencies. The details are presented in Section III. B and no doubt depend on the network studied. However, not only did the average response time of the primary and backup relays decline significantly, but also the number of relay operations classified as "slow" is greatly reduced.

The other experiment is to see how fast a full collection of relay settings could be computed. The coordinated settings were computed for all overcurrent relays in the PSPL system in a few seconds on a CRAY X-MP computer. Since today's CRAY is probably representative of future reasonably priced hardware, the concept is feasible.

A modest investigation of this adaptive protection concept is in order so that its potential as an on-line operational tool is understood better. The following three steps might form the basis for such an investigation in the future.

1.   For faults that are actually recorded on a system, determine the response of the system relays as set conventionally and compare it to the response if the relays were set to the existing network conditions only. This study could be done "off-line" using utility data and does not require special relay equipment. A realistic assessment is to evaluate performance improvements that can be

achieved for faults that are actually recorded.

2.   Refine the coordination code to speed up the computation of settings. Two refinements are desirable: (i) increase the basic computation speed through changes in the algorithms and the database management system and (ii) implement the ability to identify the smallest subsystem of the network whose relays must be reset so the coordination problem remains as small as possible.

3.   Develop some simple on-line tests to begin the process of phasing into the operating environment. Alarm data could be acquired from a SCADA computer and fed to a computer which would compute the necessary new settings to test the computational response in an actual setting. New settings could be transmitted to a few computer controlled relays on a trial basis to further test the concept.

III   IDEAS FOR FUTURE INVESTIGATION

A. Combined Protection of Transmission & Distribution

1.   During the entire 20th century, utility executives, power system planners and, operators have treated transmission and distribution as two separate entities. If both belong to the same power system, why should they be treated independently? There may be sound and obvious technical, operational and corporate reasons to keep them disparate. However, it is not desirable to keep these two functions apart in the future because of deregulation, the emergence of Independent Power Producers, etc. Distribution systems are no longer power carriers to loads alone. Cogenerators springing up at various locations pose challenging problems in protecting these systems. If these generating facilities were close enough to a distribution substation, then the protection problems might even permeate the nearby subtransmission, or even, to the bulk transmission system itself.

One can now envision the challenges posed by the protection of a distribution substation with two way power flow between the transmission and distribution sides of it. Only future developments can dictate other possible scenarios for combining these two parts of the system. This will be particularly significant when automation becomes commonplace, and a centralized control center operator has to deal with both parts of the utility system. If protection has to be achieved adaptively and on an on-line basis, dealing with both parts of the utility system becomes more challenging.

B. Intelligent Approaches to Protection

There are several challenging functions in transmission & distribution protection such as fault diagnosis, location and restoration. Improved dynamic security will seem to assume paramount importance. Then it is the authors' contention that the application of AI methods are bound to enhance the reliability of the power system.

Acknowledgements

The authors gratefully acknowledge   the contributions Dr. R. Ramaswami of Electrocon International, Mr. John Postforoosh of Puget Sound Power & Light Company, Mr. James V. Mitsche of Power Technologies Inc., and Mr. Dominic Maratukulam and Mark Lauby of Electric Power Research Institute.   Finally the authors thank Mrs. Padma S. Venkata for editing and correcting this chapter and many other related articles.

References

[1]   F.L. Alvarado, S.K. Mong, M.K. Enns, "A Fault Program with Macros, Monitors and Direct Compensation in Mutual Groups", IEEE Trans. on Power Appratus and Systems, Vol. PAS-104, No. 5, May 1985, pp 1109-1120 (referring to L. W. Coombe, D. G. Lewis, "Digital Calculation of Short-circuit Currents in Large Complex-Impedance Networks", AIEE Trans, Vol. 75, Part III,

pp 1394-97, Feb 1957.

[2]   R.E. Albrecht, et al, "Digital Computer Protective Device Coordination Program - I, General Program Description", <u>IEEE Trans. on Power Apparatus and Systems</u>, Vol. 83, No.4, April 1964, pp 402-410.

[3]   G.D. Rockefeller, "Fault Protection with a Digital Computer", <u>IEEE Trans. on Power Apparatus and Systems</u>, Vol. PAS-88, pp 438-64, Apr 1969.

[4]   M.J. Damborg, S.S. Venkata, <u>Specification of Computer-Aided Design of Transmission Protection Systems</u>, Final Report EL-3337, RP 1764-6, EPRI, Jan 1984.

[5]   S.S. Venkata, et al, <u>Computer-Aided Relay Protection Coordination</u>, Final Report EL-6145, RP 2444-2, EPRI, Dec 1988.

[6]   IEEE Committee Report, "Computer Aided Coordination of Line Protection Schemes", <u>IEEE Trans. on Power Delivery</u>, April 1991, Vol. 6, No. 2, pp 575-583.

[7].   <u>User Guide: RIM 5.0</u>, Boeing Commercial Airplane Co., Seattle, WA, 1982.

[8].   <u>R:Base Series 5000 User's Manual & Tutorial</u>, Microrim Inc., Redmond, WA., 1985.

[9].   M.J. Damborg , R. Ramaswami , et al., "Application of Relational Database to Computer-Aided-Engineering of Transmission Protection Systems", <u>IEEE Trans. on Power Systems</u>, Vol. PWRS-1, No. 2, May 1986, pp. 187-93.

[10]   M.J. Damborg,et al.,"Computer Aided Transmission Protection System Design, Part I: Algorithms", <u>IEEE Trans. on PAS</u>, Vol. PAS-103, January 1984, pp. 5-57.

[11]   R. Ramaswami , M.J. Damborg , et al., "Enhanced Algorithms for Transmission Protective Relay Coordination", <u>IEEE Trans. on Power Delivery</u>, Vol. PWRD-1, Jan. 1986, pp. 280-7.

[12]   R. Podmore and A. Germond, <u>Development of Dynamic Equivalents for Transient Stability Studies</u>, Final   Report PL-456, Volume 2, RP 763, EPRI, November 1977.

[13]   R. Tarjan, "Depth-First Search and Linear Graph Applications", <u>SIAM Journal on Computing</u> , June   1972,

pp.146-160.

[14]   M. R. Gray and D. S. Johnson, Computers and Intractibility, - A Guide to the Theory of NP-Completeness, W.H. Freeman, San Francisco, 1979.

[15]   R. V. Slyke, "Redundant Set-Covering in Telecommunications Network", IEEE 1982 Large Scale Systems Symposium, Virginia Beach, Virginia, Oct. 11-13 1982, pp. 217-222.

[16]   M.H. Dwarakanath and L. Nowitz, "An Application of Linear Graph Theory for Coordination of Directional Overcurrent Relays ", Electrical Power Problems-The Mathematical Challenge, SIAM ,1980, pp.104-114.

[17]   V.V. Bapeswara Rao and K. Sankara Rao , "Computer Aided Coordination of Directional Relays: Determination of Break Points", IEEE Trans. on Power Delivery, Vol. PWRD-3, Apr. 1988, pp. 545-8.

[18]   V. C. Prasad, K.S. Prakasa Rao, A. Subba Rao, "Coordination of Directional Relays Without Generating All Circuits", IEEE Trans. on Power Delivery, April 1991, Vol. 6, No. 2, pp 584-590.

[19]   D.E. Schultz and S.S. Waters, "Computer-Aided Pro-tective Device Coordination- A Case Study", IEEE Trans. on PAS, Vol. PAS-103, pp. 3296-301, November 1984.

[20]   R.B. Gastineau, et.al., "Using the Computer to Set Transmission Line Phase Distance and Ground Backup Relays", IEEE Trans. on PAS, Vol. PAS-96, No. 2, March/April 1977, pp. 478-484.

[21]   W.M. Thorn, et.al., "A Computer Program for Setting Transmission Line Relays", Proceedings of American Power Conference, Vol. 35, Chicago, 8-10 May 1973, pp. 1025-1034.

[22]   R. Ramaswami, "Transmission Protective Relay Coordination - A Computer-Aided-Engineering Approach for Subsystems and Full Systems", Ph. D. Dissertation, University of Washington, Seattle, Wa, 1986.

[23]   R. Ramaswami, M. J. Damborg, S.S. Venkata, "Coordination of Directional Overcurrent Relays in Transmission Systems - A Subsystem Approach", IEEE Trans. on Power Delivery, Vol. 5, No. 1, Jan 1990, pp 64-71.

[24]   A. K. Jampala, "Adaptive Transmission Protection - Concepts and Computational Issues", Ph. D. Dissertation, University of

Washington, Seattle, Wa, 1986.

[25]   A. K. Jampala S. S. Venkata, M. J. Damborg, "Adaptive Transmission Protection - Concepts and Computational Issues", IEEE Trans. on Power Delivery, Vol. 4, No. 1, Jan 1989, pp 177-185.

[26]   A. G. Phadke, T. Hlibka, et al, "A Digital Computer System for EHV Substations: Analysis and Field Tests", IEEE Trans. on PAS, Vol. PAS-95, pp. 291-301, Jan./Feb. 1976.

[27]   A. A. Girgis, R. G. Brown, "Application of Kalman Filtering in Computer Relaying", IEEE Trans. on PAS, Vol. PAS-100, pp. 3387-97, July 1981.

[28]   J. D. Day, H. Zimmermann, "The OSI Reference Model", Proc. of IEEE, Vol. 71, No. 12, December 1983, pp 1334-40.

[29]   K. Tomsovic, C.C. Liu, et al, "An Expert System as a Dispatcher's Aid for the Isolation of Line Section Faults", IEEE Trans. on Power Delivery, July 1987, pp 736-43.

[30]   S. N. Talukdar, et al, "The Operator's Assistant - An Intelligent Expandable Program for Power System Trouble Analysis", IEEE Trans. on Power Systems, Aug 1986, pp 182-7.

[31]   T. E. Dy Liacco, "The Adaptive Reliability Control System", IEEE Trans. on PAS, Vol. PAS-86, pp. 517-31, Feb. 1967.

[32]   S. H. Horowitz, A. G. Phadke, et al, "Adaptive Transmission System Relaying", IEEE Trans. on Power Delivery, Vol 3, No 4, Oct 1988, pp 1436-45.

[33]   G. D. Rockefeller, C. L. Wagner, et al, "Adaptive Transmission Relaying Concepts for Improved Performance", IEEE Trans. on Power Delivery, Vol. 3, No. 4, Oct 1988, pp 1446-58.

[34]   J. Zaborszky, M. Ilic-Spong, et al, "Computer Control of the Large Power System During for Faults for Inherently Adaptive Selective Protection", IEEE Trans. on PWRS, Vol. 2, No. 2, May 1987, pp. 494-504.

VOLTAGE COLLAPSE: INDUSTRY PRACTICES

Y. Mansour
PowerTech Labs, Inc.
Surrey, BC
Canada

P. Kundur
Ontario Hydro
Toronto, Ontario
Canada

I.   INTRODUCTION

The problem of voltage stability is not new to power system practising engineers. The phenomenon was well recognized in radial distribution systems and was explained in its basic form in a number of text books [1,2]. However, it is only in recent years that the problem has been experienced at bulk transmission system levels in well-established and extensively networked systems. Most of the early developments of the major HV and EHV transmission networks and interties faced the classical machine angle stability limitation. Innovations in analytical techniques and stabilizing measures made it possible to maximize the power transfer capabilities of the transmission systems. The result was increased power transfers over long distances of transmission. As a consequence, many utilities have been experiencing voltage related problems, even when angle stability is not a limiting factor. Some, however, recognized the problem after being exposed to catastrophic system disturbances ranging from post contingency operation under low voltage profile to total voltage collapse.

Major outages, attributed to this problem, have been experienced in the United States, France, Sweden, Belgium, and Japan. There have also been cases of localized voltage collapse (e.g. B.C. Hydro - Canada). Voltage stability has thus emerged as a governing factor in both planning and operation of a number of utilities. Consequently, major challenges in establishing sound analytical procedures and quantitative measures of

CONTROL AND DYNAMIC SYSTEMS, VOL. 42

proximity to voltage instability have been facing the industry for the last few years.

The significance of this phenomenon was emphasized by two surveys in the last decade. One of these surveys, conducted by Electricite de France (EDF) identified world wide 20 major disturbances leading to voltage collapse. Analysis of the characteristics of the affected systems and the disturbances revealed the following [3]:

-       Before the disturbance: The systems were weakened by outages (lines or plants) or temporary operating conditions, due to maintenance, combined with high system loading.

-       The disturbance: In more than half the cases, the loss of only one element was sufficient to initiate the disturbance. In the remaining cases, successive faults lead to loss of parts of the network. In several cases, the disturbance was initiated by a bus fault during substation maintenance. In all the cases, there was at least one event categorized as "should never have occurred" (e.g. human error or equipment malfunction).

-       After the disturbance: Delays in system restoration were usually due to difficulties in matching extreme boundary conditions of various parts of the affected networks.

The second survey was conducted by the IEEE Working Group on Voltage Stability in 1988 to determine the extent of the problem in the utility industry [4]. The following is a summary of the responses received from 116 key operating personnel:

-       About 40% of the respondents indicated that their systems have voltage stability related limitations. The instability could be initiated by a wide variety of contingencies such as loss of a transmission line, reactive support or generation. These contingencies could be within their own systems or in neighbouring systems. Also, some of the contingencies were further aggravated by limitation of voltage control.

-       Only 65% of those who have voltage stability limited systems have related criteria, but almost half of them base their criteria on simple upper and lower operating bus voltage limits.

Thus, the voltage stability phenomenon has emerged as a major problem currently being experienced by the electric utility industry. In this chapter, we have attempted to provide information related to utility experiences and practices with regard to this problem. It is hoped that this chapter would provide researchers and practising engineers working in this area the general background useful in their efforts to deal with the problem.

II.   BASIC CONCEPTS AND DEFINITIONS

In its simple form, voltage stability may be explained by considering a simple radial system of a source feeding a load (P,Q) through a transmission line represented by its series impedance and shunt capacitance.

If the sending end voltage is fixed and the active demand is incremented while maintaining the power factor constant, a curve similar to the one shown in Figure 1 would be obtained by plotting the received active power against the receiving end voltage. If sending end voltage and load active power were fixed while varying load reactive power, one of the family of curves of Figure 2 would be obtained by plotting the received reactive power against the receiving end voltage. If the sending end voltage is allowed to change at constant load, the family of curves in Figure 3 would be obtained.

The right edge of the V_r-P curves in Figure 1 and the bottoms of the Q-Vr curves in Figure 2 and the Vs-Vr curves in Figure 3 are the voltage stability limits of the simple radial network and are interrelated. The top part of the V_r-P and the right parts of the Q-Vr and the Vs-Vr curves characterize the stable operating points of the network. The unstable parts of the curves are characterized by excessive reactive power losses in the network for incremental increase in demand which brings the receiving end voltage further down until complete collapse. While it was possible to obtain the unstable parts of these curves analytically because of the simplicity of the example network, they are not as easy to calculate numerically for a complex network. Special numerical techniques have been suggested in the literature for this purpose.

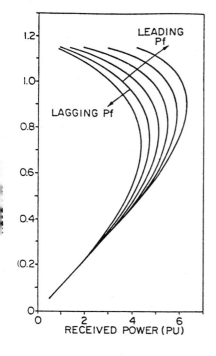

Figure 1: V_r-P Characteristics

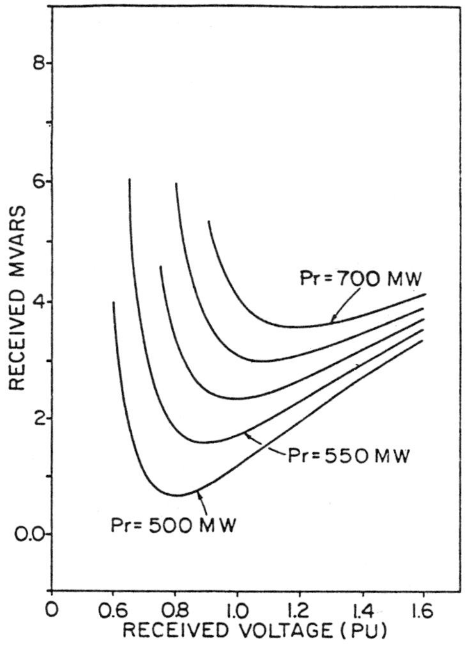

Figure 2: Q-V_r Characteristics

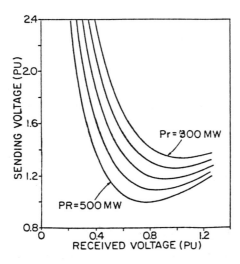

Figure 3: V_s-V_r Characteristics

One can further view the above example as an active dipole feeding a passive load of a certain impedance. The maximum transmitted power is achieved when the load impedance matches the driving point impedance as observed from the source to the load. This maximum power is the voltage stability limit of Figure 1. Any attempt to increase the load demand beyond this limit (by reducing the load impedance) would cause a reduction in voltage and reduction of power.

Based on the foregoing, voltage stability is defined as the ability of a system to maintain voltage such that an increase in load demand is met by an increase in power [4,5]. A load demand here means switching in a load admittance, which may or may not result in an increase in power consumption. As the control and protection devices try to correct the situation (e.g. automatic tap changing, generator excitation limiting, etc.) voltage instability may lead to voltage collapse. The extent of voltage collapse of a given network depends largely on dynamic load characteristics, undervoltage load tripping, and voltage control strategy. For the above reasons we emphasize the difference between voltage instability and voltage collapse.

At this point, it may be useful to explain the dynamic behaviour of some of the relevant elements of a power system [4,6]. Consider a typical mixture of residential, and industrial loads. As the distribution voltage drops, the active and reactive power of the residential load are likely to drop by a factor close to the square of the voltage ratio, thus reduction of the line loadings and reactive losses, while the change in their industrial counterparts would be relatively minor. Moreover, the contribution of shunt capacitors would also be reduced by the square of the voltage ratio which may increase the industrial reactive demand on the system. The changes may nearly balance and tend to stabilize the voltage at a low value.

The action of the voltage regulating devices, primarily the on-load tap changers, would tend to restore the distribution voltage. This would increase the residential load considerably and reduce the reactive demand of the industrial load slightly. The net result is reduction of the primary voltage further which, in turn, reduces the line charging and primary capacitors support to lower the primary voltage further and further unless the tap changers reach their limits. This indicates that under these conditions, it may be advantageous to block the tap changing action as will be seen later.

III.   ANALYTICAL TECHNIQUES

Voltage stability problems normally occur in heavily stressed systems. While the disturbance leading to voltage collapse may be initiated by a variety of causes, the underlying problem is due to an inherent weakness in the power system. In addition to the strength of transmission network and power transfer levels, principal factors contributing to voltage collapse are generator reactive power capability limits, load characteristics, characteristics of reactive compensation devices, and the action voltage control devices such as underload transformer tap changers.

Voltage stability is indeed a dynamic phenomenon and can be studied using time domain stability simulations. However, system dynamics influencing voltage stability are usually slow. Therefore, many aspects of the problem can be effectively analyzed using static methods, which examine the viability of the equilibrium point represented by a specified operating condition of the power system. The static analysis techniques allow examination of a wide range of system conditions and, if appropriately used, can provide much insight into the nature of the problem and identify the key contributing factors. Dynamic analyses, on the other hand, are useful in a detailed study of a specific voltage collapse situations, coordination of protection and controls, and testing remedial measures. Dynamic simulations also examine if and how the steady-state equilibrium point will be reached.

In this section, we will discuss the analytical techniques currently used by the utility industry, the limitations of this approach, and some of the improved techniques that have been found to be attractive for practical application.

A 30 bus test system shown in Figure 4 is used to illustrate application of different analytical techniques considered in this section. The total system load for the base case is 6150 MW. In order to examine conditions near the voltage stability limit, the system load and generation are scaled up uniformly throughout the system. At the highest load level for which a feasible power flow solution can be obtained, the total system active power load is 11,347 MW. All generators, except the two at bus numbers 22 and 23 are assumed to have unlimited reactive power output capability.

A.   **Current Practice**

The electric power utility industry at present depends largely on conventional power flow programs to analyze voltage stability problems. Two characteristics, V-P curves and Q-V curves, are normally used to examine system operating conditions and their proximity to voltage instability. These characteristics were discussed with respect to a simple radial system in Section II, and similar relationships apply to large practical systems.

Figure 5 shows the V-P curve for the test system of Figure 4. Voltage at bus 24, which is a critical bus prone to voltage instability, has been plotted as a function of total system active power load. This curve has been produced using a series of power flow solutions for various system load levels. At the 'knee' of the V-P curve, the voltage drops rapidly with

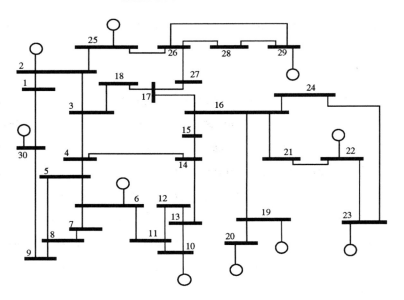

Figure 4: Test System

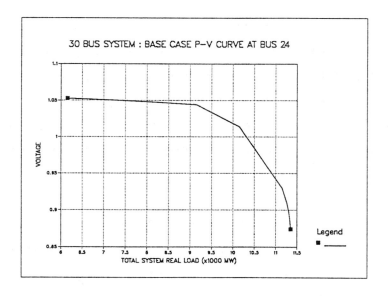

Figure 5: V-P Curve at Bus 24

increase in load power. Power flow solution fails to converge beyond this limit, and this is considered to be indicative of instability. Operation at or close to the stability limit is impractical and a satisfactory operating condition is ensured by allowing sufficient "power margin".

Figure 6 shows the Q-V curves computed at buses 1, 9, 21 and 24 for the base case with a total system load of 6150 MW. Each of these curves has been produced by successive power flow calculations with a variable reactive power source at the selected bus and recording its values required to hold different scheduled bus voltages. The bottom of the Q-V curve, where the derivative dQ/dV is equal to zero, represents the voltage stability limit. For each value of Q above the minimum value, there are two values of V. Since all reactive power control devices are designed to operate satisfactorily when an increase in Q is accompanied by an increase in V, operation on the right side of the Q-V curve is stable and on the left side is unstable. Also, voltage on the left side may be so low that protective devices may be activated. The bottom of the Q-V curve, in addition to identifying stability limit, defines the minimum reactive power requirement for stable operation.

Figure 7 shows the Q-V curves for the four selected buses for the critical case which corresponds to the highest load level (total system load 11,347 MW) for which converged solution could be obtained. It is seen that buses 21 and 24 are on the verge of voltage instability. The bottom of the Q-V curves for buses 1 and 9 could not be established because of power flow convergence problems.

Y. MANSOUR AND P. KUNDUR

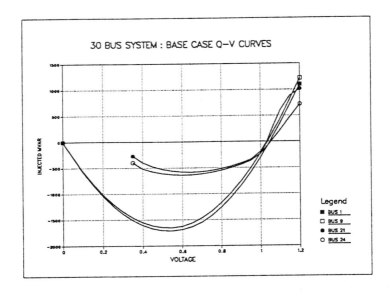

Figure 6

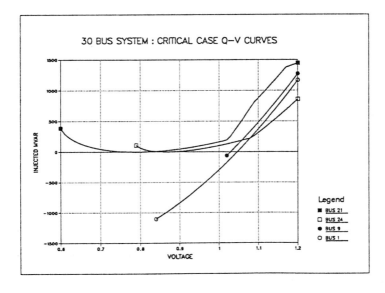

Figure 7

Determination of both V-P and Q-V curves requires execution of a very large number of power flows representing different system and loading conditions. While such procedures can be automated, they are time consuming and do not provide sensitivity information useful in gaining insight into causes of the stability problem. Therefore, as identified in [4], there is a need for analytical tools capable of the following:

- Accurately quantifying voltage stability margins;

- Predicting voltage collapse in complex networks;

- Defining power transfer limits with regard to voltage instability/collapse;

- Identifying voltage-weak points and areas susceptible to voltage instability;

- Determining critical voltage levels and contingencies; and

- Identifying key contributing factors and sensitivities affecting voltage instability/collapse, and providing insight into system characteristics to assist in developing remedial actions.

Further, system modelling used in conventional power flow studies may not be adequate for investigation of voltage stability. The following sections identify the differences in modelling requirements and describe special analytical techniques for voltage stability analysis which have been found to be attractive for practical applications.

B.   Modelling Requirements

The following is a brief description of models for power system components which for voltage stability analysis differ from those for conventional power flow analysis [7].

Generators

For power flow analysis, a synchronous generator is modelled as a P,V bus when operating normally and as a P,Q bus when the reactive power output is at its limit.

For voltage stability analysis, it may be necessary to account for the droop characteristic of the AVR, rather than assume zero droop. If load (line drop) compensation is provided, its effect should be represented. Field current and armature current limits should be represented specifically, rather than as a fixed value of maximum reactive power limit.

Loads

Load characteristics could be critical in voltage stability analysis. Unlike in conventional power flow analysis, expanded subtransmission system in voltage-weak area may be necessary.   This should include underload tap changer (ULTC) action, reactive power compensation and voltage regulators in the subtransmission system.

Voltage and frequency dependence of loads are important. It may also be necessary to model induction motors specifically.

Static VAR compensators (SVC)

When the SVC is operating within the normal voltage control range, it maintains bus voltage with a slight droop characteristic. When operating at the reactive power limits, the SVC becomes a simple capacitor or reactor; this could have a very significant effect on voltage stability. These characteristics of SVC should be represented appropriately in voltage stability studies.

Power flow programs in current use do not in general provide special SVC models. Therefore, SVCs are represented as simple reactive power sources with maximum and minimum output limits. Such a representation when used for voltage stability studies leads to overly optimistic results.

Automatic generation control (AGC)

For contingencies resulting in a significant mismatch between generation and load, the impacts of primary speed control and supplementary tie-line bias frequency control need to be represented.

C.   Special Static Analysis Techniques

A number of special algorithms have been proposed in the literature for voltage stability analysis using the static approach. In this section, we will discuss three of these methods which appear to have certain practical advantages.

V-Q sensitivity

In this method, V-Q sensitivity at each load bus is calculated using the Jacobian matrix associated with the power flow equations.

The linearized steady state system power voltage equations are given by,

$$\begin{bmatrix} \Delta P \\ \Delta Q \end{bmatrix} = \begin{bmatrix} J_{P\Theta} & J_{PV} \\ J_{Q\Theta} & J_{QV} \end{bmatrix} \begin{bmatrix} \Delta \Theta \\ \Delta V \end{bmatrix} \tag{1}$$

where,

ΔP = incremental change in bus real power.

ΔQ = incremental change in bus reactive power injection.

$\Delta \Theta$ = incremental change in bus voltage angle.

ΔV = incremental change in bus voltage magnitude.

If the conventional power flow model is used for voltage stability analysis, the Jacobian matrix in (1) is the same as the Jacobian matrix used when the power flow equations are solved using the Newton-Raphson technique. With enhanced device models included, the elements of the Jacobian matrix in (1) are modified appropriately.

In (1), let $\Delta P = 0$, then,

$$\Delta Q = [J_{QV} - J_{Q\Theta} J_{P\Theta}^{-1} J_{PV}] \Delta V \tag{2}$$
$$= J_R \Delta V$$

and,

$$\Delta V = J_R^{-1} \Delta Q \tag{3}$$

where,

$$J_R = [J_{QV} - J_{Q\theta} J_{P\theta}^{-1} J_{PV}] \tag{4}$$

The matrix J_R^{-1} is the reduced V-Q Jacobian. Its ith diagonal element is the V-Q sensitivity at bus i. For computational efficiency, this matrix is not explicitly formed. The V-Q sensitivity is calculated by solving (2).

A positive V-Q sensitivity is indicative of stable operation; the smaller the sensitivity, the more stable the system. As stability decreases, the magnitude of the sensitivity increases, becoming infinity at the stability limit. Conversely, a negative V-Q sensitivity is indicative of unstable operation. A small negative sensitivity represents a very unstable operation.

For the 30 bus test system of Figure 4, Table I shows the ranking of load buses based on V-Q sensitivities for the base case as well as the critical case. It is seen that, for the critical case, buses 21 and 24 are approaching instability. This is consistent with the results based on Q-V curves shown in Figure 7.

It should be noted that, because of the nonlinear nature of the V-Q relationships, the magnitudes of the sensitivities for different system conditions do not provide a direct measure of the relative degree of stability [7].

Table I. Ranking of Buses Based on V-Q Sensitivities

Base Case			Critical Case		
Rank	Bus #	DV/DQ	Rank	Bus #	DV/DQ
1	12	0.02407	1	21	2.90036
2	24	0.02176	2	24	1.86986
3	9	0.01549	3	16	0.15838
4	1	0.01502	4	15	0.10972
5	27	0.01480	5	17	0.08262

An approach using V-Q sensitivity and piecewise linear power flow analysis is used in [8] and [9] for on-line application of voltage instability analysis.

Modal analysis [7]

Voltage stability characteristics of the system can be identified by computing the eigenvalues and eigenvectors of the reduced Jacobian matrix J_R defined by (4).

Let,

$$J_R = \xi \wedge \eta \qquad\qquad (5)$$

where,

ξ = right eigenvector matrix of J_R

η = left eigenvector matrix of J_R

$\wedge$ = diagonal eigenvalue matrix of J_R

and,

$$J_R^{-1} = \xi \wedge^{-1} \eta \tag{6}$$

From (3) and (6), we have,

$$\Delta V = \xi \wedge^{-1} \eta \Delta Q \tag{7}$$

or,

$$\Delta V = \sum_i \frac{\xi_i \eta_i}{\lambda_i} \Delta Q \tag{8}$$

Where ξ_i is the ith column right eigenvector and η_i the ith row left eigenvector of J_R.

Each eigenvalue λ_i, and the corresponding right and left eigenvectors ξ_i and η_i, define the ith mode of the system. The ith modal reactive power variation is,

$$\Delta Q_{mi} = K_i \xi_i \tag{9}$$

where,

$$K_i^2 \sum_j \xi_{ji}^2 = 1 \tag{10}$$

with ξ_{ji} the jth element of ξ_i.

The corresponding i^{th} modal voltage variation is,

$$\Delta V_{mi} = \frac{1}{\lambda_i} \Delta Q_{mi} \tag{11}$$

The magnitude of each eigenvalue λ_i determines the weakness of the corresponding modal voltage. The smaller the magnitude of λ_i, the weaker the corresponding modal voltage. If $|\lambda_i| = 0$, the ith modal voltage will collapse. In (8), let $\Delta Q = e_k$, where e_k has all its elements zero except the kth one being 1. Then,

$$\Delta V = \sum_i \frac{\eta_{ik} \xi_i}{\lambda_i} \tag{12}$$

with η_{ik} the kth element of η_i.

V-Q sensitivity at bus k,

$$\frac{\partial V_k}{\partial Q_k} = \sum_i \frac{\xi_{ki} \eta_{ik}}{\lambda_i} \tag{13}$$

$$= \sum_i \frac{P_{ki}}{\lambda_i}$$

If all the eigenvalues are positive, V-Q sensitivities are also positive for all the buses, and the system is voltage stable. Negative eigenvalues of J_R will cause some buses to have negative V-Q sensitivities and therefore voltage instability. Zero eigenvalue of J_R is indicative of a system on the verge of voltage instability.

The participation factor of bus k to mode i is defined as,

$$P_{ki} = \xi_{ki} \eta_{ik} \tag{14}$$

From (13), P_{ki} indicates the contribution of the ith eigenvalue to the V-Q sensitivity at bus k.   The bigger the value of P_{ki}, the more λ_i contributes to V-Q sensitivity at bus k.   For all the small eigenvalues, bus participation factors determine the areas close to voltage instability.

It is unnecessary to calculate all the eigenvalues of a practical system with several thousand buses.   On the other hand, calculating only the minimum eigenvalue of J_R is not sufficient because there are usually more than one weak mode associated with different parts of the system, and the mode associated with the minimum eigenvalue may not be the most troublesome mode as the system is stressed.   If we can determine the m smallest eigenvalues of J_R, we have obtained the m least stable modes of the system.   If the biggest of the m eigenvalues, say mode m, is deemed a strong enough mode, the modes which are not calculated can be neglected because they are known to be stronger than mode m.   In practice, it is seldom necessary to compute more than 5 to 10 smallest eigenvalues.

Reference 7 describes an efficient numerical method for computation of a selected number of smallest eigenvalues of the Jacobian matrix J_R associated with systems with several thousand buses. It also describes how the elements of J_R may be modified to include detailed models for generators, loads, and SVCs appropriate for voltage stability studies.

The results of the application of modal analysis to the test system of Figure 4 are summarized in Table II.   The results provide eigenvalues associated with two weakest modes and their top five bus participations for the base case and the critical case.   The areas affected by these modes are identified in Figure 8.

Table II. Two Weakest Modes and Bus Participations

Base Case				Critical Case			
$\lambda 1 = 27.8832$		$\lambda 2 = 41.1596$		$\lambda 1 = 0.0748$		$\lambda 2 = 25.3407$	
Bus #	P_{ki}	Bus #	P_{ki}	Bus #	P_{ki}	Bus #	P_{ki}
27	0.1923	12	0.9667	21	0.2721	27	0.2030
17	0.1657	13	0.0082	24	0.1591	17	0.1642
18	0.1518	27	0.0059	16	0.0119	18	0.1488
15	0.1130	11	0.0043	15	0.0077	15	0.1126
16	0.0990	14	0.0040	17	0.0056	16	0.0910

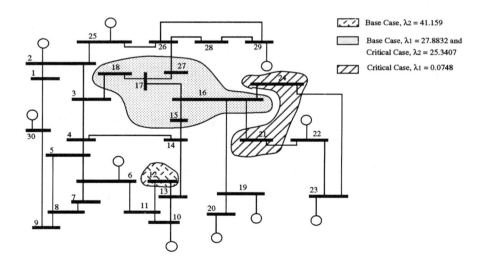

Figure 8: Areas Affected by Weakest Modes

As demonstrated in [7], the modal analysis approach, in addition to bus participation, can identify branch and generator participations. These provide useful information regarding factors influencing the voltage stability problem:

-       Branch participations show which branches are important in the stability of a given mode. This provides insight into possible remedial actions as well as contingencies which may result in loss of voltage stability.

-       Generator participations indicate which machines must retain reactive reserves to ensure stability of a given mode.

This approach has been applied successfully in [7] for voltage stability assessment of large complex systems.

Voltage stability indicator

In [10], a method for fast prediction of voltage instability and proximity to voltage collapse is proposed, based on the indicator L defined for load bus i as follows:

$$L_i = \left| 1 + \frac{V_{io}}{V_i} \right| \tag{15}$$

Where V_i is the actual complex bus voltage and V_{io} is the complex bus voltage when the generator bus voltages are kept the same as those under the present operating conditions and all the loads are removed.

L_i varies between 0 and 1. The bigger the value of L_i, the closer the bus is to voltage instability.

For the test system of Figure 4, Table III shows ranking of five load buses with highest L indices.

Table III. Five Buses with the Highest L Indices

Base Case			Critical Case		
No.	Bus #	L_i	No.	Bus #	L_i
1	27	0.0964	1	24	0.2361
2	15	0.0959	2	15	0.2079
3	24	0.0910	3	27	0.2014
4	17	0.0878	4	17	0.1862
5	18	0.0876	5	18	0.1826

Comparison with the results of sensitivity analysis (Table I) and modal analysis (Table II) shows that the L indicator approach does not identify all critical buses prone to voltage collapse, for example bus 21.

The advantage of the L indicator approach is its simplicity and computational speed. The modelling capability of the method is limited to that of conventional power flow analysis. This approach may be useful as a screening tool for rapidly identifying voltage weak buses.

D.   Dynamic Analysis of Voltage Stability

Full dynamic simulation, using time domain analysis provides a useful complement to static analysis, and is useful for the following applications:

- Ensuring that the system trajectory following a contingency will in fact reach a stable equilibrium point;

- Gaining an understanding of the dynamic interaction between the generating units, network controls, loads, and ULTCs;

- Coordinating protection and controls with power system requirements; and

- Developing and testing remedial measures, such as automatic load shedding and ULTC blocking.

Extended transient/midterm stability programs and longterm dynamics simulation programs are ideally suited for dynamic analysis of voltage stability. In order to be able to simulate slow phenomena extending up to several minutes, implicit integration methods need to be used. This would allow use of large time steps without causing numerical stability. Equipment models should include ULTC action, load dynamics, and over/underexcitation protections and controls.

IV.   CRITERIA

Most of the voltage stability criteria reported in the literature are based directly or indirectly on one or combination of voltage, reactive power, and active power margins. Others are based on indices that measure sensitivity of load bus voltage to change in active load power, load bus voltage to change in injected reactive power, or reactive power generated by all active sources to changes of load reactive power. Following are some of the reported criteria [4].

One of the criteria gaining wide acceptance in Europe is that developed in [8]. It is being implemented by Electricite de France (EDF) for on-line application [9]. The following is a brief description of the method:

-   The system algebraic equations are linearized around the initial operating point and then the sensitivities of bus angles, load bus voltages, and generator bus reactive power to incremental changes to system active and reactive power demands are calculated.

-   The demand is incremented until a generator reaches its control limit at which time the increment is recorded. This generator bus is converted to a load bus and the system equations are linearized around the new operating point.

-   The process is repeated until voltage instability is detected. The active and reactive power margins are equal to the sums of increments in the respective components of power demands calculated in all the transition steps.

In addition, EDF has one of the most comprehensive on-line voltage control schemes. The system operating states are divided into three categories:

i) Normal state where "n" and "n-1" security levels are assessed,

ii) Alert state where only "n" security level is assessed, and

iii) Emergency state where some "n" voltages are out of their bounds.

Under normal or alert state, the control is done on the secondary and the tertiary levels. The secondary control action (regional) is decided based on on-line study of "n" and "n-1" contingencies, the data of which is derived using state estimation. The discrete control actions, e.g. topology changes, unit start-up, generation rescheduling, etc. are coordinated manually. Voltage profile control, however, is done automatically: The EHV system is divided into about 30 zones. The secondary voltage regulator of each zone modifies the AVR set points of the voltage controllers in their respective zone based on feedback measurement of the so called "pilot-bus" which is representative of the voltage profile of its zone. Each regional control centre is usually able to control its voltage profile using 3 to 5 pilot buses. The tertiary level control is usually coordinated between the national and the regional operators to coordinate the set points of the pilot buses or change in the major system topology. This decision is usually forecasted ahead of time for load changes.

Under emergency state, the operator may operate the generators at their maximum reactive power limit, block the on-load tap changer operation, reduce the distribution voltage by up to 5%, or order load shedding to avoid voltage collapse.

An example of a criterion based on the V-P relation of Figure 1 is that of New York Power Pool (NYPP) [4,13]. The pre and post-contingency voltage performance as a function of power flow across a major transmission interface is calculated using off-line load flow studies. Pre-contingency and post-contingency high and low voltage limits are monitored on many of the 345 kV buses and selected 230 kV buses in off-line studies as basis for NYPP planning criteria. The low limit is typically set at 95% of nominal voltage. The high limit ranges from 105 to 110% of nominal voltage. In addition, the leading edge of the V-P curve is identified as in Figure 9. The power transfer corresponding to that edge is then reduced by 5%. This reduced transfer level is compared to the pre-contingency transfer level corresponding to the point at which the post contingency voltage equals 95%. To ensure that a voltage constrained transfer limit is determined with a safe margin, the lower of the two power transfer levels from the foregoing comparison is selected as the transfer limit. Figure 9 shows a condition in which the allowable transfer level is controlled by the edge of the V-P curve rather than the 95% lower voltage limit.

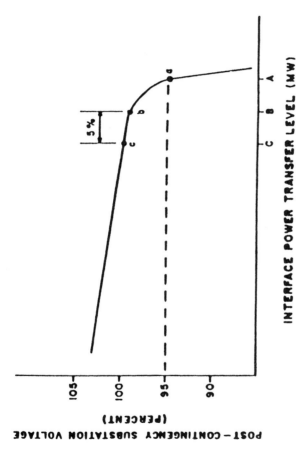

INTERFACE POWER TRANSFER LEVEL (MW)

(1) SMALL LETTERS a,b & c DENOTE POINTS ON THE CURVE, WHERE:
 . . . a IS THE POINT OBTAINED AT 95% VOLTAGE;
 b IS THE POINT AT THE "LEADING EDGE OF THE CURVE"; AND
 c IS THE POINT OBTAINED WHEN POWER TRANSFERS ARE REDUCED BY 5% FROM THE LEADING EDGE

(2) CAPITAL LETTERS A,B & C DENOTE POWER TRANSFER LEVELS CORRESPONDING TO POINTS a,b, & c RESPECTIVELY.

(3) FOR THIS EXAMPLE, C WOULD BE THE VOLTAGE - CONSTRAINED INTERFACE TRANSFER LIMIT.

Figure 9

An example of a criterion based on the Q-V relation is that of B.C. Hydro, Canada [14]. The same criterion is used for planning and operation. Referring to Figure 10, a hypothetical variable reactive power source is assumed at a central bus (INGLEDOW) representing the area of concern under a predefined contingency. The variation in the post-contingency bus voltage (V) as a function of the injected reactive power (Q) is recorded in the form of the Q-V relation shown in Figure 11. A criterion was then established as follows:

i)   A variable reactive power margin "A" must be provided between the voltage stability limit (the bottom of the curve) and the operating point. This margin is dependent on power transfer to the area under study. For B.C. Hydro's main load centre, the value of "A" must be at least 4.0 MVAr per 100 MW of total power transfer to the region.

ii)   The voltage margin "C" between the voltage stability limit and the operating point must be at least 5%.

iii)   The bus voltage corresponding to the stability limit must be less than 0.95 pu.

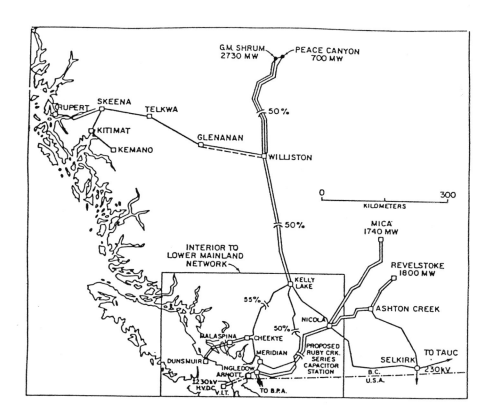

Figure 10: B.C. Hydro 500 kV Transmission System

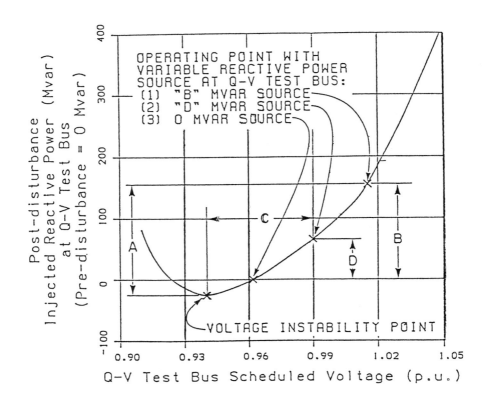

Figure 11: Q-V Curve Analysis Criteria

Recent planning studies concluded that item (iii) of the criterion results in inconsistent results, especially when the characteristics are such that the bottom of the curve is flat over a wide range of voltage. Another contributing factor to the inconsistency is the fact that voltage stability is limiting in more than one region of the system, which makes it difficult to base the criterion on the characteristic at only one representative bus. Accordingly, the criterion was recently modified to include items (i) and (ii) only while (iii) was eliminated.

One of the first utilities to investigate methods of analyzing voltage stability and implement related criteria is CPTE of Belgium. Reference [5] compares the various criteria considered by CPTE. The assessment is done using a load flow based program with the generators represented by a quasi steady state model linearized around the operating point. In this model, effects of incremental variations in active current, reactive current, and excitation voltage on terminal voltage are modelled taking excitation limits into account. The proximity to voltage instability could be assessed by any of the following criteria:

i) $Vi/Ei > 0.5/\cos((a-b)/2)$

where "Vi" is the set of bus voltages calculated from a power flow solution or by measurement, "Ei" is the set of open circuit voltages calculated by solving the network with all the loads removed and the generator active powers (except the slack bus) are reduced to zero, "a" is the system equivalent impedance angle calculated at bus "i" and "b" is the load power factor angle at the same bus.

ii) $(dQ_L/dQ_G) > 0$ at constant active power

This criterion indicates that for every operating point, voltage stability is ascertained if an incremental increase in reactive power demand is matched by a finite increase in generated reactive power. The criterion is tested by incrementing the reactive power demand at one bus or a group of buses while monitoring the change in the reactive power generated by all the sources (generators, SVC's, synchronous condensers, etc.).

iii) $(d|V|/|V|)/(dQ_L/Q_L) < 0$ at constant active power

This criterion is applied by incrementing the reactive power demand at one or more buses and the corresponding bus voltages are calculated. Both voltage and reactive power increments are normalized with respect to their corresponding initial values and then the above ratio is calculated.

iv) $P_L < P_{max}$ and $Q_L < Q_{max}$

The maximum active and reactive load limits of the network are similar to those explained earlier using the active dipole example. More details about the method are given in reference [5].

V. INCIDENTS OF VOLTAGE INSTABILITY/COLLAPSE

Invaluable lessons could be learned from the voltage instability and collapse incidents reported in the literature as to their causes and consequences. A number of these were selected for the purpose of this publication. The incidents are classified, below, according to the initiating cause which may or may not be voltage related but eventually would evolve to voltage collapse.

A. Load Increase

Load increase at a fast rate is considered to be the principal cause for the French system collapse of 19 December, 1978 [4], and the Tokyo system collapse of 23 July, 1987 [11].

On the morning of December 19, 1978, greater than anticipated temperature drop in France resulted in a rapid load increase of 4600 MW between 7 a.m. and 8 a.m. The resultant increase in power transfer led to increase in MW losses and to a remarkable increase in reactive power losses on the transmission lines connecting the Paris area to Eastern France. Some 400 kV lines were overloaded and severe voltage deterioration was observed for over 25 minutes. To ease the situation, some EHV/HV tap changers were blocked and 5% drop in distribution voltage was ordered in some areas. These actions were partly effective as the voltage on parts of the 400 kV system continued to decline and stabilized at about 350 kV. Shortly after, a 400 kV line was tripped by the action of its overload protection which, in turn, caused cascaded tripping of other 400 kV and 225 kV lines followed by wide spread strong oscillations over the entire French network.

A similar incident took place on a hot Summer day in Japan, where the temperature reached a record high of 39°C in the Tokyo area. The morning demand on the Tokyo Electric Co. (TEPCO) system of that day reached 39,100 MW and dropped to 36,500 MW shortly afternoon (lunch hour). By 1:00 p.m. the demand reached 38,200 MW. Additional generation and reactive power support were called upon to balance the increase in active and reactive power demand until the 500 kV voltage stabilized between 510 and 520 kV. Starting at 1:00 p.m. the load began to increase at a remarkable 400 MW per minute. As a result, the voltage started to decay gradually even though additional shunt capacitors (eventually all of them) were brought into service. This is primarily because the increase in reactive power losses was always higher than the discrete increase in shunt compensation especially that the reactive power contribution of the shunt capacitors is reduced by the square of the voltage ratio. By 1:10 p.m. the power demand reached 39,300 MW accompanied by continuing decay of voltage down to 74% in the western part and 78% in the central part of the system by 1:19 p.m. when three major substations were tripped by the operation of protective relays. As a result, 8168 MW of load was lost affecting 2.8 million customers. 60% of the lost load was restored in about 17 minutes and the rest were restored in about 3 hours and 20 minutes.

B.   Load Tripping

Load tripping, in comparison to load increase, may be surprising to many as a possible cause of voltage collapse. It should be noted, however, that the reason load increase initiated the voltage problems described above is the increase in power transfer to the load area without proper voltage support especially if the load characteristics at the receiving end is stiff (e.g. constant power). If load tripping happens towards the sending end and result in an increase in the transfer to remote areas, the result can be the same if proper voltage support is not provided. In this case, however, a soft load characteristics at the sending end (e.g. constant impedance) can aggravate the situation.

A classic example is the July, 1979 collapse of the North Coast region of B.C. Hydro, Canada (the western part of the radial section out of Williston substation, Figure 10). For better display, a single line diagram of the affected area is shown in Figure 12. Prior to the disturbance, a trouble shooting procedure made it necessary to switch the excitation of seven out of eight generators at Kemano station to manual. The main load of Kemano station is an aluminum smelter at Kitimat substation (600 MW). The power transfer out of Kitimat to the integrated system was 150 MW.

The disturbance was initiated by the tripping of 100 MW potline load at the smelter, causing an immediate increase of transfer out of Kitimat to 250 MW. With the generators at Kitimat on manual excitation voltage support for the increased transfer was inadequate, bringing the voltage in the area down. The smelter load is practically constant impedance which caused the power consumption at the smelter to drop by the square of the voltage. This, in turn, caused the power transfer to increase further bringing the voltage down even further. The process continued slowly in a monotone manner until the power transfer reached 390 MW and Skeena 500 kV voltage reached about 250 kV in one minute from the beginning of the disturbance and before line relaying isolated the region. Disturbance records captured at the Kemano generating station showed a fairly steady station

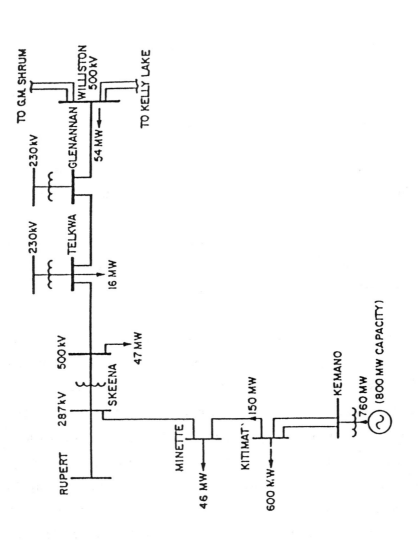

Figure 12: North Coast System - Single Line Diagram and Power

Flow Prior to July 1979 Voltage Collapse Incident

output within the observed duration of the disturbance which excludes the possibility of machine angle stability as one may suspect.

C.   Loss of Generation

Loss of generation in a certain zone of a network may deteriorate the voltage stability in two ways:

1.   The balance of the power may have to be called upon from remote locations which may increase the transmission reactive losses considerably depending on the prior loading of the interconnections.

2.   Loss of generators may result in severe shortage of reactive power sources which could leave the system with inadequate voltage support.

The January 12, 1987 incident on the French system, which affected the whole western part of France, is a good example for the latter: The pre-disturbance operating reserve was 5900 MW (approximately 7% of the system capacity). Between 10:55 a.m. and 11:42 a.m., three out of four thermal units at a generating station (Cordemais) located on the western region of the French system tripped by protective relaying action. Thirteen seconds later, and before emergency gas turbines were brought in service, the fourth unit tripped upon excitation overcurrent protection action as the fourth unit tried to control the voltage solely. Additional voltage support duties were picked automatically by other generators resulting in a sequence of generator trippings because of excitation overload. Nine units were tripped in 5 minutes following the outage of the first four generators resulting in a total of 9000 MW of lost generation. By 11:50 a.m., the voltage level of the 400 kV western part of the system stabilized at less than 300 kV and as low as 240 kV in some parts. It should be noted that in spite of the generation deficiency, the frequency did not change significantly which indicates a significant dependence of load characteristics on voltage. The voltage profile was restored by tripping 1500 MW of load at the western most part of the system. Figure 13 summarizes the results of the incident [4].

An example of loss of generation that result in large MW transfer and voltage instability is the December 28, 1982 in Florida (Refer to Figure 14): The East Coast 500 kV system was at an early stage of development in 1982. There were two 500 kV tie lines between Florida and Georgia but there were no 500 kV transmission between North and South Florida yet. At 11:33 a.m., Turkey Point #3 carrying 700 MW was tripped. At that time, Florida load was approximately at 60% peak and maximum import. The increase in power transfer over the tie lines to Florida caused voltage drop of 8.6% on the 230 kV system in North Florida. One minute later, Sanford #3 (in the North) tripped on maximum excitation limiter / over excitation protection resulting in an additional loss of 89 MW and 104 MVAr. As modest as it is, the latter caused an additional voltage drop of 9% on the 230 kV northern system. As the voltage deteriorated, out of step relays operated and separated Northeast Florida followed by a loss of another unit (Putnam #1). The separation and loss of generation resulted in the loss of about 11% of the system load for 36 minutes.

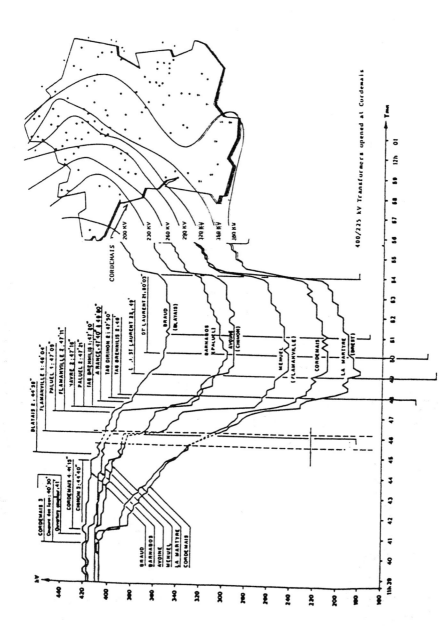

Figure 13: Details of the Power Failure of January 12, 1987

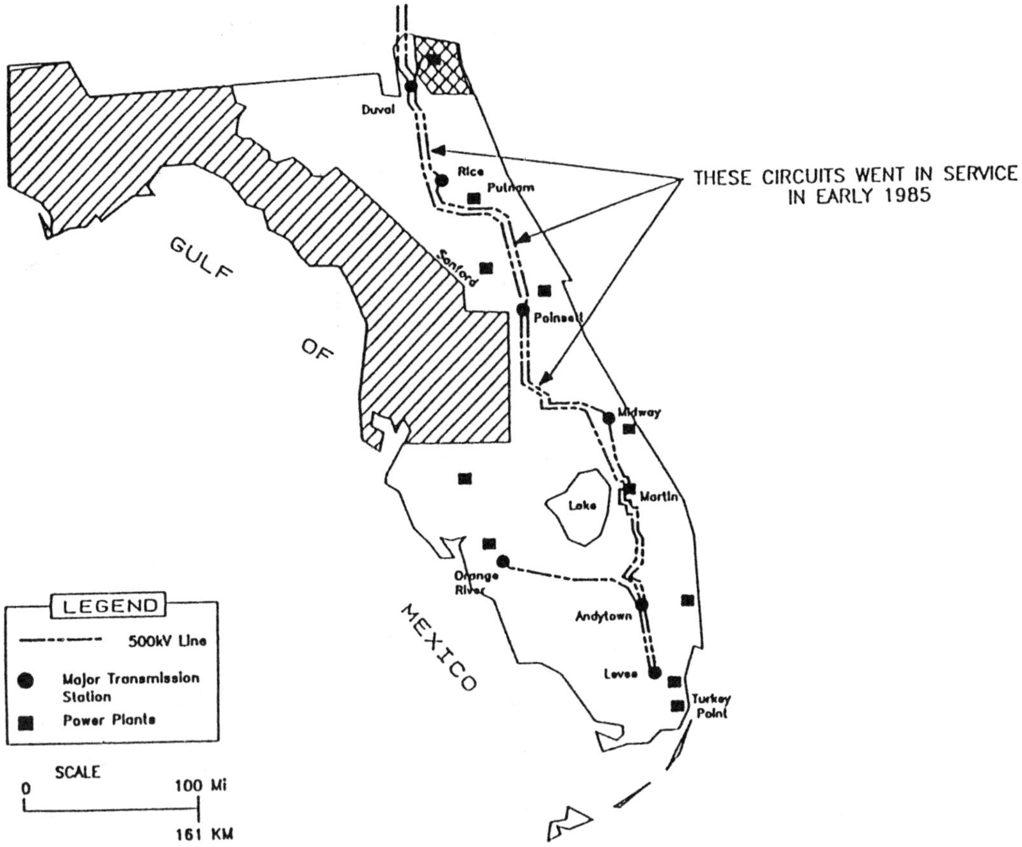

Figure 14: The FPL 500 kV Transmission System

D.   Loss of Transmission Line(s)

Loss of parts of the transmission system at a time of heavy transfer may result in overloading the remaining transmission far beyond its surge impedance loading and result in voltage collapse.

There are many documented examples, but we picked the May 17, 1985 one in Florida because of the familiarity with the system that the readers might already have built up in the preceding section: By the time of the incident, the 500 kV system was completed. The pre-disturbance loading of Southeastern Florida was 4294 MW with 53% of the power being imported.

At 11:36 a.m., an intense grass fire spread across the 500 kV transmission corridor causing a series of phase to ground faults. All three 500 kV circuits north of Andytown substation tripped within 10 minutes. This caused a shift in the power flow to three 230 kV and two 138 kV lines. Approximately 0.5 second later, one of the 230 kV lines tripped causing a large voltage drop. The remaining transmission circuits and two generators were tripped within the next 3.5 seconds, mostly by impedance protection. This resulted in isolating the Southeast Florida area with a generation deficiency of approximately 80% for about 3.5 hours. The voltage in the isolated area dropped to 40-50% of nominal. It should be noted that the underfrequency load shedding scheme did not operate properly because of the cut-off voltage characteristics of the underfrequency relays. This scheme was upgraded in 1986.

VI.   LESSONS LEARNED FROM THE DISTURBANCES

A.   The voltage instability phenomenon generally manifests as a slow dynamic decay in voltage. The rate of decay depends mainly on load characteristics, and the responses of voltage regulators and protective devices. The time frame to collapse could range from few to many minutes.

B.   In just about all the cases, the situations were aggravated by at least one protective device operation based on local equipment related signal. Tripping the equipment made the situation only worse from a system view point. Therefore, there is an urgent need for better system wide coordination philosophies between protective and control devices to better reflect overall system, rather than just equipment, remedial needs.

C.   No abnormal faults are necessary to initiate voltage instability/collapse. The Japanese and the French incidents are good examples.

D.   Proper load modelling is essential for accurate analysis of voltage collapse. Load characteristics could be such that voltage and load could decay until the system voltage stabilizes at low magnitude. This low magnitude depends to a large extent on the voltage regulator limits. The Japanese incident alerted the industry to the high reactive power demands of modern air conditioners at low voltage. On the other extreme, the B.C. Hydro incident showed the adverse effect of constant impedance loads when located near the sending end.

E.   Excessive dependence on shunt capacitors for reactive power support and voltage control could have detrimental effects. In the Japanese incident, switching a massive amount of shunt capacitors did not stop the voltage decay. Shunt compensation can be made most effective by the proper coordination of a mixture of shunt capacitors, static VAr systems, and synchronous condensers [3].

F.   Load shedding and blocking of automatic tap changers are the most effective ways to avoid voltage collapse. The B.C. Hydro case, however, showed that load tripping must be done at the proper location in order to be effective. Also, power system analysts have to ensure that extensive load tripping does not result in high voltage and subsequent tripping of facilities by over voltage protection. This happened in the July 13, 1977 incident in New York [4].

VII.   PREVENTION OF VOLTAGE COLLAPSE

This section identifies measures that can be undertaken in the planning and operation of power systems to mitigate voltage stability problems.   In addition, one application of a special protection scheme for prevention of voltage collapse is described.

Planning measures [12]

A.   Reactive power compensation: Voltage stability characteristics of power systems are significantly affected by reactive power compensation of the transmission network.   Adequate voltage stability margins should be ensured for the most onerous system condition (for which the system is to be designed to operate satisfactorily) by proper selection of the type, size, and location of reactive power compensation.

If generating stations are far removed, it may be necessary to use regulated compensators (synchronous condensers, SVCs) or series capacitors.

B.   Load compensation of generator AVRs: Depending on the relative location of generating units with respect to load centres, it may be advantageous to provide load (or line crop) compensation to automatic voltage regulators. This would in effect move the point of constant voltage electrically closer to the loads, thereby improving voltage stability.

C.   Coordination of protection and controls: Examination of actual system disturbances indicate that one of the causes of voltage collapse is lack of coordination between generating unit protections/controls and power system requirements. Adequate coordination should be ensured by performing dynamic simulation studies.

D.   Undervoltage protection schemes: To cater for unplanned or extreme situations, it may be desirable to provide special protection schemes such as under-voltage load shedding and ULTC blocking. Such schemes would prevent widespread system collapse.

Operating measures

A.   Stability margin: System should be operated with adequate voltage stability margin by appropriate scheduling of reactive power resources and voltage profile. There are at present no universal guidelines for selection of the degree of margin and the system parameters to be used as indices. These are likely to be system dependent and may have to be established based on the characteristics of the individual system.

If the required margin cannot be met by using available reactive power resources and voltage control facilities, it may be necessary to limit power transfers and startup generating units to provide voltage support at critical areas.

B.   Spinning reserve: Adequate spinning reactive power reserve must be ensured, if necessary, by operating generators at moderate or low excitation and switching-in shunt capacitors to maintain desired voltage profile.

C.   Operators' action: Operators must be able to recognize voltage stability related symptoms and take appropriate remedial actions such as voltage and power transfer controls and, possibly as a last resort, load curtailment. Operating strategies that prevent voltage collapse need to be established. On-line monitoring and analysis to identify potential voltage stability problems and possible remedial measures would be invaluable in this regard.

Voltage collapse protection scheme [4,12]

In special cases, it may be necessary to provide protection schemes for prevention of voltage collapse. The following is an example of such a scheme which has been implemented by Ontario Hydro.

Because of delays in obtaining approval to build 500 kV transmission lines, the Ottawa area in Eastern Ontario is at present supplied largely by 230 kV transmission. In order to prevent voltage collapse due to loss of a critical 230 kV circuit under heavy load periods, a coordinated scheme consisting of the following has been used as a stop-gap measure:

-      Fast auto reclosure of major 230 kV circuits supplying the area;
-      Automatic load rejection;
-      Automatic switching of shunt capacitors;
-      Automatic blocking of load tap changers.

The fast auto reclosure (0.9 s to 1.3 s) is used as a first measure to maintain voltage within acceptable limits. If reclosure is successful, voltage recovers and the system returns to normal. If reclosure is unsuccessful, depending on how low the voltage drops, load rejection may be triggered, followed by capacitor switching and possibly LTC blocking.

The automatic load shedding provides for up to 9 blocks for loads totalling 750 MW to be rejected, depending on load level. It consists of a two-tiered rejection scheme:

-       Fast rejection can be armed by operator action from an attended master station and will trip if local station voltage drops below a preset value for a minimum time period (1.5 s);

-       Normal rejection will always be armed and will trip load if voltage and time limits are violated (10 s);
        Used for protection in the event of unforeseen circumstances.

The automatic capacitor switching provides for switching (on/off) of a total of 36 capacitor banks in 17 transformer stations. the capacitors are switched in staggered blocks with time settings ranging from 1.8 s to 8.0 s, so that only the required amount of compensation is switched.

The automatic LTC blocking at 14 transformer stations provides for blocking of tap changers when the local HT voltage drops below a present value for a specified time, and unblocking when voltages recover for a specified time.

VIII. <u>REFERENCES</u>

1.   B.M. Weedy, "Electric Power Systems", book, John Wiley & Sons, New York.

2.   V. Venikov, "Transient Processes in Electrical Power Systems", book, Mir Publishers, Moscow.

3.   CIGRE Paper 38-01, TF 03, "Planning Against Voltage Collapse", 1986.

4.   IEEE Special Publication, 90TH0358-2-PWR, "Voltage Stability of Power Systems: Concepts, Analytical Tools, and Industry Experience", 1990.

5.   CIGRE Paper 38-11, "Voltage Stability - Fundamental Concepts and Comparison of Practical Criteria", 1984.

6.   H.K. Clark, "Voltage Control and Reactive Supply Problems", IEEE Tutorial Course on Reactive Power: Basics, Problems & Solutions, 87EH0262-6-PWR.

7.   B. Gao, G.K. Morison, and P. Kundur, "Voltage Stability Evaluation Using Modal Analysis", paper submitted for presentation at the IEEE PES Summer Meeting, August 1991.

8.   N. Flatabo, R. Ognedal, and T. Carlsen : "Voltage Stability Condition in a Power Transmission System Calculated by Sensitivity

Methods". IEEE Transactions on Power Systems, Vol. 5, No. 4, pp. 1286-1293, November 1990.

9.   C. Lemaite, et. al., "An Indicator of the Risk of Voltage Profile Stability for Real Time Control Applications", 1989 IEEE PES Summer Meeting, Long Beach, CA.

10.   P. Kessel, and H. Glavitch, "Estimating the Voltage Stability of a Power System", IEEE Transaction on Power Delivery, Vol PWRD-1, No. 3, pp. 346-354, July 1986.

11.   CIGRE Paper 37-87 : "The Power System Failure on July 23, 1987 in Tokyo", 1987.

12.   P. Kundur, "Voltage Stability - Practical Considerations", IEEE Panel Session on Voltage Instability, IEEE/PES Summer Meeting, San Francisco, CA, July 1987.

13.   L. Eng, "Voltage Instability - New York Power Pool's Criteria, Experience, Studies", IEEE Panel Session on Voltage Instability, IEEE/PES Summer Meeting, San Francisco, CA, July 1987.

14.   Y. Mansour and P.G. Harrington, "Voltage Instability - B.C. Hydro's Practice and Experience", Panel Session on Voltage Instability, IEEE/PES Summer Meeting, San Francisco, CA, July 1987.

ACKNOWLEDGEMENT

The authors wish to acknowledge the assistance provided by Mr. B. Gao of Ontario Hydro in running test cases presented in Section III.

RELIABILITY TECHNIQUES IN LARGE ELECTRIC POWER SYSTEMS

Lu Wang   J. Endrenyi

Ontario Hydro Research Division
Toronto, Ontario, Canada

1. POWER SYSTEM RELIABILITY CONCEPTS

1.1 Significance of Reliability

The reliability of an electric power system is, and always has been, a prime concern of utility management and engineers. In many parts of the world, the public has grown accustomed to a service of extremely high quality and is unwilling to lower the existing standards to any significant degree. While a growing number of constraints are placed on the planning and operation of electric power systems, utilities are obliged to maintain the customary reliability of their service.

For many years, decisions concerning the reliability of electric power supply were based on engineering judgment and experience. The results were satisfactory because power systems were comparatively small and simple. As the networks and their components became more complex, however, the empirical approaches had to be replaced by more rigorous methods, based on formal mathematical approaches and reliability theory.

This development is not unique: it is typical of the way scientific methods are applied in many areas. Devices and systems become more and more complicated, to the point that mathematical models and abstract theories must be employed to master them, instead of the instinctive solutions used in the early stages of development. The progress from the simple to the complicated seems to be a general attribute of the human condition.

Essentially, reliability studies provide predictions. They predict the system's future behaviour, based on past experience or a mathematical evaluation of an appropriate model (even in this case, model data are based on past performance). Since predictions cannot be made with certainty, they are inherently probabilistic. While this is well acknowledged, some reliability studies to this day are not based on probability concepts, just as most engineering design techniques (also involving predictions of future performance) are not. Yet the trend, however slow because of the many real difficulties, is towards probabilistic thinking and, therefore, in the following probabilistic methods will be highlighted.

1.2 Basic Concepts

The classical definition of reliability, equally applicable to components and systems, is the following:

> *Reliability* is the probability of a device or system performing its function adequately, for the period of time intended, under the operating conditions intended.

CONTROL AND DYNAMIC SYSTEMS, VOL. 42

The definition follows the everyday concept of reliability quite closely, except that reliability is measured, as it should, in terms of a probability. In studies of power generation and transmission reliability (the primary topic of this Chapter) this definition is seldom used; instead, reliability is considered an umbrella concept with no quantitative interpretation. It is broken down into two attributes, both still umbrella concepts: adequacy and security. These are defined as follows [5, 12].

> *Adequacy* is the ability of the system to supply the aggregate electric power and energy requirements of the customers, within component ratings and voltage limits, taking into account planned and unplanned component outages.

> *Security* is the ability of the system to withstand sudden disturbances such as the unanticipated losses of system components.

Note that the "ability to withstand sudden disturbances" is interpreted as the ability to avoid system instability caused by such disturbances.

Another aspect of reliability is system *integrity*, the ability to maintain interconnected operations. Integrity is violated if uncontrolled separation occurs in the presence of severe disturbances.

Figure 1.1 illustrates the domains covered by adequacy and security studies. Clearly, security assessments involve transient stability evaluations, while adequacy assessments concentrate on the steady state conditions after an incident, such as a component failure. In Figure 1.1, dotted lines indicate areas where probabilistic methods are still in an early stage of development. One of them, labelled "Only hot reserve available", identifies a range

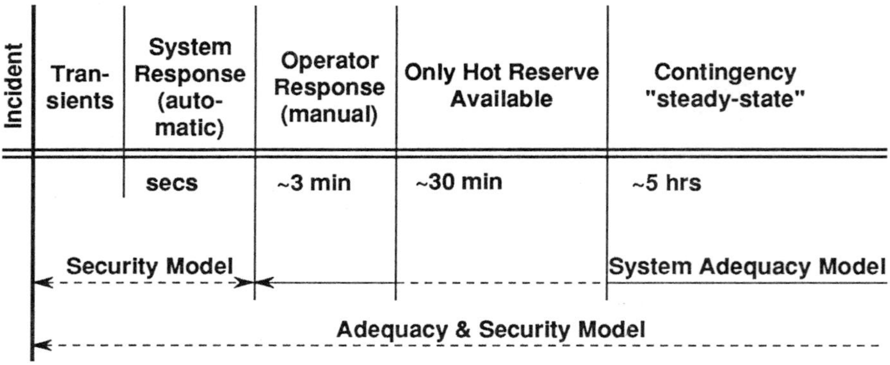

Incident	Tran-sients	System Response (auto-matic)	Operator Response (manual)	Only Hot Reserve Available	Contingency "steady-state"
		secs	~3 min	~30 min	~5 hrs
	← Security Model →				System Adequacy Model
	← Adequacy & Security Model				

Figure 1.1   The areas of adequacy and security studies

where the availability of thermal reserves is limited to the units kept spinning, an effect which is not usually considered in adequacy models. Since the magnitude and consequences of this effect depends on the operating policy, its consideration may not be part of the planners' mandate.

Note that system planners carry out mostly adequacy evaluations. True, this is often coupled with stability checks, but not in any integrated, probabilistic way: planners tend to combine adequacy and security studies on a deterministic basis. So do operators, who consider security aspects particularly in studies involving a time-frame of only a few hours. While an integrated adequacy-security evaluation method which is fully probabilistic has not yet been developed (see Figure 1.1), efforts are being made to extend the scope of reliability evaluations [13-19]. This Chapter is devoted, almost exclusively, to adequacy methods; for the above reasons, probabilistic security approaches will not be discussed.

1.3 The Aims of Power System Reliability Studies

The overriding concern is, both for planners and operators, to make the system as reliable as economically feasible. Common is a desire to maintain adequacy, security, and system integrity at satisfactory levels, and avoid widespread outages. However, many of the questions addressed by planners and operators are different.

Planners think about long-term goals and want to ensure, at the very least, that plans of future systems or extensions are sufficiently adequate. The simplest and most often performed reliability study is a comparison of alternatives. This is also the most practicable, since it requires only relative measures of reliability, easier to obtain than absolute ones whose accuracy may be heavily compromised by systematic errors from approximations in modeling and computations. With the increasing availability of absolute measures, it will be possible to make assessments against probabilistic reliability criteria. At present, this is practiced only in generating system reliability evaluations; for transmission systems, mainly deterministic criteria are used. More accurate measures will also facilitate cost-benefit assessments, and the design of economically optimal systems.

Operators' goals depend on the time horizon of the study. In the shortest time frame which may be up to only a few hours, typical questions are how to provide adequate spinning reserve to keep the risk of system failure below predetermined levels, how to maintain the ability to withstand sudden changes, and how to choose corrective action in a contingency. If the time horizon is up to one year, topics to deal with may include the expected unavailability of hydro-electric energy, the impact of the uncertainty in the forecast of river flows, economic operating schedule, scheduling energy sales and purchases, and scheduling component maintenance and overhaul. For a time horizon of 2 to 4 years, the topics considered may include mothballing and restoring units, and long-term fuel contracts, sales, and purchases.

Probabilistic models to deal with operating problems are not as well developed as those used in adequacy evaluations. Operational reliability will be discussed in Section 8; the rest of the sections will focus on probabilistic adequacy models.

1.4 Deterministic and Probabilistic Approaches

As already mentioned, reliability assessments are estimates of the future performance of a device or system. It was also noted that such assessments can be arrived at by either deterministic or probabilistic methods. This is not unique: the same choice of approaches is available in many other engineering design problems. While probabilistic methods provide more insight, they also require more complex modeling and are, in general, more difficult to apply. Only in recent years, and only in some applications, have these methods reached the level of maturity where they are viable alternatives. In this section, the probabilistic and deterministic approaches to estimating future system behavior will be compared.

In the traditional deterministic approach, design criteria are given in the form of tests which a device or system of acceptable design must withstand. It is required, for example, that certain combinations of system and load conditions should not result in system breakdown, or even in excessive component stress. Often "worst-case" conditions are analyzed where the stresses and strengths are calculated and then set apart by a "safety factor". If the criteria are satisfied, the system is considered "perfectly safe", that is, assured of no breakdowns during its "useful lifetime", under appropriate operating conditions. Typical reliability criteria for power systems include requirements that the system must retain its ability to supply all loads during the loss of the largest generating unit, or of selected transmission circuits (n-1 rule).

The traditional approach has some disadvantages, including the following.

- The variability in input data is ignored: actual component performance data are not fixed numbers, but constitute random variables with distributions.

- The selection of "worst-case" conditions is arbitrary; important conditions may be omitted, unlikely conditions included.

- The implied notion of no failure risk present in designs satisfying traditional criteria is misleading; it is commonly known that risk of failure always exists, and the deterministic approach provides no idea how safe the design actually is.

- The effort to stay on the safe side often results in overdesign.

In the probabilistic approach, the existence of failure risk in all design is acknowledged. Reliability criteria simply require that the risk of failure should not exceed a pre-selected limit. Reliability is measured through appropriate probabilistic indices, and the criteria are set as thresholds (limits) on the indices. In a mechanical design, for example, failure occurs if the load (stress) exceeds the strength (resistance) of the part in question. Let the reliability index be the probability of failure, P_F. If the distribution of stress is given by the probability density $s(x)$, and the distribution of resistance by $r(x)$, then P_F is computed by convolving the two densities:

$$P_F = \int_0^\infty [s(x) \, dx \int_0^x r(x) \, dx]$$

The reliability criterion is expressed through the requirement that P_F should not exceed a given limit ϵ. In contrast, deterministic design would be based on a selection of a "worst-case" load S and a "worst-case" strength R and it would be required that R > kS where k>1 is a safety factor. It is then tacitly implied that $P_F = 0$.

A further comparison between the deterministic and probabilistic approaches involving the adequacy evaluation of a generating system is provided in Table 1-1. It is assumed that failure occurs when the system load L exceeds the generating capacity G, resulting in a loss of load (LOL). The chosen index is the probability of loss of load, LOLP. Otherwise, the comparison in the table is self-explanatory.

1.5 Probabilistic Indices and Criteria

As noted above, probabilistic indices are measures of reliability. They are physical quantities and the values they take on for a given system are indications of its expected reliability performance. Deterministic reliability assessments are not based on the use of such indices (see Table 1-1).

Table 1-1

Comparison of Approaches
to System Adequacy Assessment

	Deterministic	Probabilistic
Failure event	Loss of load (LOL)	LOL
State	Adequate if LOL did not occur	Adequate if LOL did not occur
Index	-	P[LOL] = P [inadequate states] = LOLP
Reliability criterion	Selected states must be adequate	LOLP < ε

Probabilistic indices fall into the following categories.

- Probabilities (e.g., of a component failure)

- Frequencies of occurrences (e.g., of two generating units out of service)

- Mean durations

- Expectations (such as the expected number of days in a year, when the peak load demand is not met)

- Severity measures (a subset of expectations, these measures register higher or lower according to the severity of the incidents involved; a typical example is the expected yearly total energy not supplied by the system due to component failures).

- Cost indices (these, too, are expectations)

As mentioned before, reliability criteria are standards defined in terms of the indices. For example, a criterion in terms of the expectation index quoted may be that the expected number of days per year when the peak load is not met must not exceed 0.1 day/yr (or 1 day in 10 years). While the concept is simple, defining actual criteria is not. In principle, it can be done by using judgment based on past experience, by relating to already existing criteria, or by cost-benefit optimization.

The last approach to setting reliability criteria appears to be the most exact, but is also the most difficult. To increase reliability by whatever measure, investment costs need to be increased; in turn, the costs of outages, including damages to customers are expected to decrease. Where the incremental increase of investment and decrease of outage costs are equal, the total costs are at minimum and the reliability is optimal. This reliability value may then be considered the target, and used as a criterion for the given index. However, the cost functions involved are very difficult to establish; moreover, intangible factors, such as the customers' insistence on a level of reliability possibly higher than the optimal on the one hand, and political and social pressures to constrain new investments on the other, can seriously distort the optimization process.

1.6 System Failure Conditions

Before embarking on reliability analysis, the conditions, or criteria, for system failure must be established. In Table 1-1, for example, failure was defined as the event that the system load exceeds the available generating capacity. Reliability indicies are always defined in terms of system failure events. A flow chart indicating the steps of probabilistic reliability analysis and the roles of both system-failure and reliability criteria is shown in Figure 1.2. Much of the analysis consists of the so-called failure effects analysis (FEA), in which the system states are classified into failed or working states according to the failure criteria. This will be further discussed in Section 6.

Typical failure conditions for power generation and transmission systems include the following.

- Insufficient generation to meet the load demand
- Transmission system components overloaded
- Bus voltages outside tolerances
- Customer interruptions (loss of load)
- System break-up or complete collapse

Note that some of these conditions can be alleviated by corrective action, automatic or manual. Nevertheless, appropriate indices measure the extent of violations against any of the above conditions. After corrective action, only the last two remain: loss of load and system collapse - the most important indices are related to these conditions.

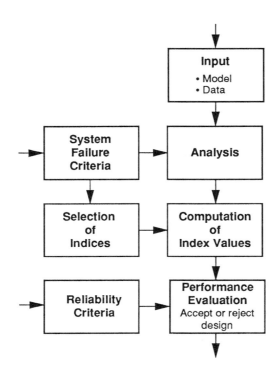

Figure 1.2 Steps of probabilistic reliability analysis

Occasionally, failure criteria other than those listed above are employed, especially if only a part of a transmission system such as a transmission corridor is being studied. In that case, system failure could be defined as, say, the loss of two or more circuits, with the loss of a single circuit considered tolerable. While this definition may superficially resemble the (n-1) reliability criterion used in deterministic studies, the two must not be confused: probabilistic system design is based on reliability criteria defined through probabilistic indices.

It should be observed that even if deterministic reliability criteria are employed, separate system failure criteria usually exist. For example, in Table 1-1 system failure is defined as the loss of load in both the deterministic and probabilistic approaches.

1.7 Studies of Large Composite Systems

At the present state-of-the-art, a comprehensive reliability analysis for the entire power system, from the generating units to the customer load points, cannot be performed. Instead, the system is broken into parts, representing generation, transmission, area supply networks including transformer stations, and distribution, and these parts are studied separately. Yet, there is some interaction between the parts, and therefore, it is generally not possible to simply add up the findings of the individual studies to produce overall reliability measures for the entire system [20]. This will be explained for generation and transmission, the parts of the system to which this entire Chapter is primarily devoted.

Failure events in the combined generation and transmission systems (called, as mentioned, the composite or bulk power system) can be classified according to the diagram in Figure 1.3. Area N in the diagram represents

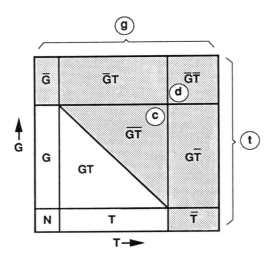

Figure 1.3 Classification of generation and transmission failure events

th state where all components are working; region T includes the collection of states where transmission elements are out (increasingly severe as one proceeds to the right), but the entire system has still not failed, while $\overline{T}$ includes the states where the system has failed due to transmission outages. Regions G and $\overline{G}$ can be similarly interpreted for generating unit outages. States in GT contain both generating unit and transmission line outages but not enough to cause system failure; the interpretation of regions $\overline{G}T$, $G\overline{T}$ and $\overline{G}\overline{T}$ is self-explanatory. Of particular interest is the domain $\overline{G}T$ (region c) where both generating unit and transmission line outages occur which by themselves would not fail the system, but in combination do.

The group of states $\overline{G}\overline{T}$ represents the interdependence between generation and transmission which does not allow to evaluate the two systems separately and then add up the results for the combined system. For example, if P_g is the probability of system failure due to generator outages alone (the probability of states in domains $\overline{G}$, $\overline{G}T$, and $\overline{G}\overline{T}$ - see Figure 1.3), P_t the same for transmission line failures, P_c and P_d are probabilities associated with domains c and d, respectively, then the system failure probability P_F is given by

$$P_F = P_g + P_t + P_c - P_d. \tag{1.1}$$

Only if P_c and P_d are negligible can the results of separate generation and transmission studies be added. In practice, P_d is often negligibly small, but not P_c. Table 1-2 indicates the distributions of the terms in (1.1) obtained in a few case studies. They include an analysis of the Brazilian South-Southeastern System (BSS), the Bonneville Power Administration (BPA) System, and the Georgia Power (GP) System [21, 22]. The contribution of region d is negligible in all cases. It is clear from Table 1-2, however, that the contribution of region c is significant.

Table 1-2

Relative Contributions of Regions g, t and c to the System Failure Probability P_F

	BSS	BPA	GP
P_g/P_F	0.45	0.07	0.24
P_t/P_F	0.18	0.31	0.62
P_c/P_F	0.37	0.62	0.14

Generating system reliability is often studied by itself. This is quite legitimate, because the predictions are used to determine the necessary reserve margin in capacity planning. For the above reasons, however, transmission systems are usually studied in conjunction with generating systems. So will be the case in the following discussions of large composite systems.

References

1. R. Billinton, *Power System Reliability Evaluation*, Gordon and Breach, New York, 1970.
2. J. Endrenyi, *Reliability Modeling in Electric Power Systems*, John Wiley & Sons, Chichester, England, 1978.
3. R. Billinton and R.N. Allan, *Reliability Evaluation of Power Systems*, Pitman, London, 1984.
4. R. Billinton and R.N. Allan, *Reliability Assessment of Large Electric Power Systems*, Kluwer Academic Publishers, Boston, 1988.
5. *Power System Reliability Analysis - Application Guide*, Prepared by CIGRE Working Group 38.03, Paris, 1987.
6. R. Billinton, "Bibliography on the Application of Probability Methods in Power System Reliability Evaluation," *IEEE Transactions* **PAS-91**, 646-660 (1972).
7. "Bibliography on the Application of Probability Methods in Power System Reliability Evaluation, 1971-1977," Report of the IEEE PES Application of Probability Methods Subcommittee, *IEEE Transactions* **PAS-97**, 2235-2242 (1978).
8. R.N. Allan, R. Billinton, and S.H. Lee, "Bibliography on the Application of Probability Methods in Power System Reliability Evaluation, 1977-1982," *IEEE Transactions* **PAS-103**, 275-282 (1984).
9. R.N. Allan, R. Billinton and S.M. Shahidehpour, and C Singh, "Bibliography on the Application of Probability Methods in Power System Reliability Evaluation, 1982-1987," *IEEE Transactions* **PWRS-3**, 1555-1564 (1988).
10. M.Th. Schilling, R. Billinton, A.M. Leite da Silva, and M.A. El-Kady, "Bibliography on Composite System Reliability," *IEEE Transactions* **PWRS-4**, 1122-1132 (1989).
11. M.Th. Schilling, A.M. Leite da Silva, R. Billinton, and M.A. El-Kady, "Bibliography on Power System Probabilistic Analysis (1962-1988), " *IEEE Transactions* **PWRS-5**, 1-11 (1990).
12. "Reliability Indices for Use in Bulk Power Supply Adequacy Evaluation," Report of the PROSD Working Group, IEEE PES Application of Probability Methods Subcommittee, *IEEE Transactions* **PAS-97**, 1097-1103 (1978).
13. R. Billinton and P.R.S. Kuruganty, "A Probabilistic Index for Transient Stability," *IEEE Transactions* **PAS-99**, 1953-206 (1980).
14. R. Billinton and P.R.S. Kuruganty, "Probabilistic Assessment of Transient Stability in a Practical Multi-Machine System," *IEEE Transactions* **PAS-100**, 2163-2170 (1981).
15. P.M. Anderson and A. Bose, "A Probabilistic Approach to Power System Stability Analysis," *IEEE Transactions* **PAS-102**, 2430-2439 (1983).
16. K.J. Timko, A. Bose, and P.M. Anderson, "Monte Carlo Simulation of Power System Stability," *IEEE Transactions* **PAS-102**, 3453-3460 (1983).

17. J.C. Dodu and A. Merlin, "New Probabilistic Approach Taking into Account Reliability and Operation Security in EHV Power System Planning at EDF," *IEEE Transactions* **PWRS-1**, 175-181 (1986).
18. C.H. Grigg, D.T. Tsai, and C.C. Fong, "Probabilistic Stability Model for the Evaluation of Operating Risk," *Proc. of the 17th Inter-RAM Conference for the Electric Power Industry*, Hershey, Penn. 369-376, 1990.
19. B. Porretta, D.L. Kiguel, G.A. Hamoud, and E.G. Neudorf, "A Comprehensive Approach for Adequacy and Security Evaluation of Bulk Power Systems", IEEE 1990 Summer Power Meeting, Paper No.90 SM 385-5-PWRS, Minneapolis, 1990.
20. *Composite System Reliability Evaluation Methods*, EPRI Report EL-5178, 1987.
21. S.H.F. Cunha, G.C. Oliveira, M.V.F. Pereira, V.L. Arienti and A.C.G. Melo, "Composite Generation/Transmission Reliability of the Brazilian Southern/Southeastern System," *Proc. of the 2nd International Symposium on Probability Methods Applied to Electric Power Systems*, EPRI Report EL-6555, 1988.
22. *Development of a Composite System Reliability Evaluation Program*, EPRI Report EL-6926, 1990.

2. POWER SYSTEM RELIABILITY INDICES AND CRITERIA

In Section 1, the concepts of probabilistic reliability indices and deterministic and probabilistic reliability criteria were introduced. In this section, a brief review of the actual indices and criteria used in power system reliability studies is offered. Since methods, indices, and criteria are different for generation and transmission, they will be reviewed separately for the two parts of the power system.

2.1 Generating System Indices and Criteria

2.1.1 Background

For many years, long-term expansion planning of generating systems was based on the size of the reserve margin considered necessary to maintain reliable operation. Reserve margin is defined as the difference between the installed capacity and the forecast annual peak load, usually given as a percentage of the latter. The primary shortcoming of this planning approach is that the reserve margin cannot be directly related to the reliability performance of a generating system: two systems with identical reserve margins may have vastly different probabilities of not being able to supply all loads, and vice versa. Consider, for example, two generating systems of the same size, one consisting of hydraulic units and the other of fossil-fired and nuclear units. To achieve the same level of customer supply reliability, the latter system would require a larger reserve margin than the former because hydraulic units, in general, are more reliable than thermal or nuclear units.

The application of probability theory to the determination of spare capacity requirements started in 1934 [1,2]. The loss of load concept was developed in 1947 [3,4] in an effort to put generation capacity expansion planning on a more rational basis. These developments marked the beginning of the application of probability methods to power system reliability assessment.

2.1.2 Adequacy Indices

As already discussed in Section 1.5, the basic indices of system adequacy are the *probability*, *frequency*, and *mean duration of system failure*. For a generating system, failure is defined simply as the condition when the available generating capacity is less than the load demand. Thus, the probability of system failure is called, in this case, the *loss-of-load probability*, LOLP. Its computation is described in Section 5.1.

Another index, particular to generating systems, is the *loss-of-load expectation*, LOLE. It is defined as the expected number of days in a year when the peak load demand is not met. Its computation, based on the use of a load-duration curve made up of daily peaks, is also described in Section 5.1.

Yet another, increasingly prominent index is the *expected energy not supplied* in a given period T, EENS. Clearly, this is a severity index (see Section 1.5). The period T can be the duration of a disturbance, in which case EENS is measured in units of MWh, or it can be interpreted cumulatively for, say, a year, in which case the unit is MWh/yr.

Often, EENS is normalized through dividing it by L_s, the system's yearly peak load; this way index values obtained for different systems become comparable. The index so derived is the *energy curtailment index*, ECI. If T = 1 yr, ECI is measured in units of MWhr/yrMW = hr/yr.

The severity of a disturbance j can also be expressed in *system-minutes* (SM_j), where

$$SM_j = 60 \ ECI_j \quad (sy\text{-}min)$$

The amount of energy not supplied during the disturbance is the same as the amount that would arise if the entire system would be interrupted during peak load conditions for SM_j minutes. It follows, that the severity of interrupting the entire system for 1 min at the time of peak load is 1 sy-min. The total system minutes accumulated in a year, SM, is given by

$$SM = \sum_j SM_j \quad (sy\text{-}min/yr)$$

where the summation is over all disturbances j which occurred in the year.

2.1.3 Security Indices

As already mentioned, probabilistic security methods are in an early stage of development; they are, therefore, not discussed in this chapter. Nevertheless, the following list contains a few of the indices that have been proposed for security evaluations.

- Probability of lower than critical system frequency during a transient

- Probability and frequency of system instability

- Average value of load suddenly disconnected

- Response risk: average energy unsupplied after spinning reserve is exhausted and before further operating reserve can be put into service

2.1.4 Reliability Criteria

Of the deterministic criteria prominent for many years, only one has survived. This is the percentage reserve criterion - a poor one to go by as explained at the outset of this Section, but retained for non-technical reasons.

For many years now, most utilities around the world have used the LOLE index to establish a generating system reliability criterion [5]. The most frequently mentioned LOLE criterion has been 0.1 day per year (or one day in ten years). A recent publication [6] indicated that electrical utilities in various countries have adopted different risk criteria, ranging from 0.1 to 5 days per year. However, it should be pointed out that direct comparison of numerical figures could be misleading since different computational methods may have been used, resulting in different interpretations.

Most recently, criteria have been developed in terms of the SM index. In Ontario Hydro, such a criterion was determined by balancing the reduction of customer costs if fewer interruptions occur against the costs of installing and operating additional capacity to achieve improved reliability. The outcome was a criterion of 25 sy-min/yr. However, in deriving criteria, and also, when comparing the reliability performance of different systems, it must be remembered that important factors, such as the modeling of generation and load, load forecast uncertainty, the assistance from interconnections, the effects of voltage reduction and public appeals may be considered differently by different utilities.

2.2 Transmission System Indices and Criteria

2.2.1 Indices

Several sets of adequacy indices have been proposed for measuring the predicted and the past operating performances of bulk power systems. The predicted performance indices are used mainly for planning the bulk transmission system. The past performance indices, derived from system operating and outage records, can be used as feedback information in comparison with specified performance targets.

The types of indices employed in the two applications are essentially the same. It is customary to subdivide them into load-point and overall system indices. A selection of load-point indices are listed in Table 2-1. The system indices are either the sums (in case of the basic values) or the averages (for the average values) of all load-point indices in the system.

Of the overall system indices, the most frequently used is the *Bulk Power Energy Curtailment Index,* BPECI, which is defined as

$$\text{BPECI} = \frac{\sum_i \sum_j L_{ij} T_j f_j}{L_s} = \frac{\text{EENS}}{L_s} \quad (\text{hr/yr})$$

where L_{ij} = average load curtailment at bus i due to disturbance j, MW

        T_j = mean duration of load curtailment due to disturbance j, hr

f_j = frequency of disturbance j, 1/hr

L_s = maximum system load, MW

EENS = expected energy not supplied in the system per year, MWhr/yr

The BPECI index can be easily converted into system minutes through multiplying it by 60.

It can be assumed that security indices would be seldom computed for transmission systems alone. They would certainly be determined for bulk power systems which combine generation and transmission. At this time, it is foreseen that these indices would take the same form as the ones described in Section 2.1.2.

Table 2-1
Load-Point Indices

Basic Values

Probability of failure
Frequency of failure
Expected number of voltage violations/year
Expected number of load curtailments/year
Expected amount of load curtailed/year
Expected energy not supplied/year
Expected total duration of load curtailment/year

Average Values per Curtailment

Average load curtailed/disturbance
Average energy not supplied/disturbance
Average duration of load curtailment

2.2.2 Reliability Criteria

At the present stage of development, probabilistic criteria are seldom if ever employed in transmission reliability studies [7]. This is so, partly, because the index values calculated from models containing a number of approximations are not considered sufficiently accurate; partly, because historical development and local conditions in various countries are different; and partly, because of the difficulties of verifying reliability predictions. A world-wide survey [5] indicates that most utilities still use deterministic criteria in their transmission and bulk system planning process.

The deterministic criteria are often in the form of the so-called (n-1) rule. In this procedure, one or several base cases (differing in the load conditions and the generating scheme) are subjected to a number of single outages involving a predetermined set of lines, cables and, in some countries, generating units; the criterion specifies that line flows and bus voltages must remain within given limits in all of these contingencies. Occasionally, requirements go beyond the (n-1) rule and a more severe (n-2) criterion is also imposed. This is sometimes achieved by simply considering the contingency states identified by the (n-1) rule as base cases and applying the (n-1) rule once again.   In addition, special combinations may be included such as the losses of two principal lines, the loss of a double circuit linking a generating station to the transmission network, or the losses of one circuit and a generating unit.

2.3 Classification of Disturbances by Severity

Customer interruptions may occur during power system disturbances. The amount of energy not supplied during a disturbance appears to be a natural yardstick for measuring the severity of system disturbances. However, since the impact of a given amount of unsupplied energy on the power system depends on the size of the system, its normalized form, the system minutes, is a more suitable index for the purpose.

With the help of this index, the severity of system disturbances can be classified as follows [8].

Degree 0 - less than 1 sy-min, an incident resulting in a condition normally considered acceptable

Degree 1 - between 1 and 10 sy-min, an incident with significant impact on customers but not considered serious

Degree 2 - between 10 and 100 sy-min, an incident with serious impact on customers

Degree 3 - between 100 and 1000 sy-min, an incident with very serious impact on customers

Degree 4 - over 1000 sy-min, an incident with catastrophic impact on the entire system

Based on the above classification, a frequency index can be computed for each of the severity categories.

References

1.   S.A. Smith Jr., "Spare Capacity Fixed by Probabilities of Outages," *Electrical World* **103**, 222-225 (1934).

2.   G. Calabrese, "Generating Reserve Capability Determined by the Probability Method," *AIEE Transactions* **66**, 1439-1450 (1947).

3.   H.P. Seelye, "Outage Expectancy as a Basis for Generator Reserve," *AIEE Transactions* **66**, 1483-1488 (1947).

4.   E.S. Loane and C.W. Watchorn, 'Probability Methods Applied to Generating Capacity Problems of a Combined Hydro and Steam System," *AIEE Transactions* **66**, 1645-1657 (1947).

5. R. Juseret, "Reliability Criteria Used in Various Countries," (Summary of papers presented to CIGRE Study Committee 37, Oslo, June 1983), *Electra*, No. 110, 67–101 (1987).

6. *Power System Reliability Analysis - Application Guide*, prepared by CIGRE Working Group 38.03, Paris, 1987.

7. "Bulk Power System Reliability Concepts and Applications," Report of the Task Force on Bulk Power System Reliability, IEEE PES Application of the Probability Methods Subcommittee, *IEEE Transactions* **PWRS-3**, 109–117 (1988).

8. W.H. Winter, "Measuring and Reporting Overall Reliability of Bulk Electricity Systems," *CIGRE Proceedings*, Paper No. 32–15 (1980).

3. STATE-SPACE REPRESENTATION OF SYSTEMS

3.1 Introduction

Probabilistic system reliability methods endeavour to compute system reliability measures from component reliability information. Many methods to accomplish this are described in the literature [1-5]. They include approaches based on logic diagrams, state space analysis, fault trees, and simulation. However, not all of these methods are used in the reliability analysis of large power systems. Common to those that are is that the system in question is represented by its state space. Therefore, a brief description of this approach is given in this Section.

In the state-space representation, the system is described by its states (every component failure, maintenance, or repair event, or relevant environmental changes such as those in load or weather will change the state of the system), by the possible transition between states, and the rates of these transitions. These concepts are illustrated in Figure 3.1 and will be further explained in Section 3.2. If it can be assumed that the transition rates are constant (independent of previous history of transitions and of time), the process of the system changing its states constitutes a homogeneous Markov process with continuous time parameter and discrete state space. The solution of the Markov model provides the state probabilities both as a time-function and in the steady-state, and also the frequencies of occurrence for each state. This will be discussed in Section 3.3.

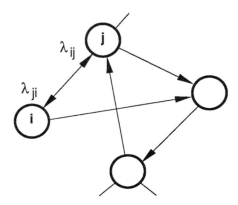

Figure 3.1 System states, possible transitions and their rates

The rates of transitions between system states (Figure 3.1) are equal to the component failure, repair, maintenance, etc. rates causing those transitions. Thus, the state-space model for the entire system is indeed based on component data.

3.2. The Markov Process

In a stochastic (that is, probabilistic) process, the system's whereabouts in the state space is determined by a set of random variables, one for each moment in time (in some applications, the independent variable can be other than time). In the general case, the probabilities of the system being in one state or another at a given moment depend on the entire previous history of the system's travels between the states. In a Markov process, the probability of finding system X in state j at time $t+\Delta t$ depends only on the state i the system is occupying at time t, but not on the history of the system previous to t.

Because of the above property of the Markov process one can define transition probabilities $p_{ij}(\Delta t)$ such that

$$p_{ij}(\Delta t) = P[X(t + \Delta t) = j \mid X(t) = i]$$

which in a homogeneous process do not depend on t either. The transition intensities λ_{ij} are related to the p_{ij} as follows:

$$p_{ij}(\Delta t) \simeq \lambda_{ij} \Delta t$$

where λ_{ij} is the expected number of transitions (hence: rate of transitions) from i to j in unit time, given that the system is immediately returned to i after each departure. It follows that the reciprocal of λ_{ij}, M_{ij} is the mean time of those stays in i, T_{ij}, which are followed by a transition to j.

With the rates λ_{ij} being constants, it can be shown that the durations T_{ij} are exponentially distributed, that is,

$$P[T_{ij} \leq t] = F_{Tij}(t) = 1 - e^{-\lambda_{ij}t}$$

This close relationship between the Markov process and the exponential distribution allows for a more convenient mathematical solution of the model, but such a model may be inaccurate in some practical cases where the transitions clearly do not follow the exponential law. However, the model is often considered an acceptable approximation even in such cases.

The transition rate can be conveniently condensed into a matrix, called the transition intensity matrix **A**. This is defined as follows:

$$
\mathbf{A} =
\begin{bmatrix}
-\sum_{j \neq 1} \lambda_{1j} & \lambda_{12} & \lambda_{13} & \cdots \\
\lambda_{21} & -\sum_{j \neq 2} \lambda_{2j} & \lambda_{23} & \cdots \\
\vdots & \vdots & \vdots &
\end{bmatrix}
$$

3.3. Computation of State Reliability Indices

To calculate system reliability indices, the probabilities of the individual system states are needed. It can be shown that the time-dependent state probabilities $p_i(t)$ are the solutions of the matrix differential equation

$$\dot{p}(t) = p(t)\ A$$

where $p(t)$ is a row vector containing the elements $p_i(t)$, and $\dot{p}(t)$ a row vector containing $\dot{p}_i(t)$, the time-derivatives of $p_i(t)$. To find the long-run state probabilities p_i, their row-vector, p, is obtained from the equation

$$0 = pA \tag{3.1}$$

The equations in this system of linear equations are not independent, therefore, one of them must be replaced by

$$\sum_i p_i = 1 \tag{3.2}$$

which is obviously true.

Additional indices required in reliability evaluations are the various frequencies associated with the system states. The frequency of an event is the expected number of occurrences of the event per unit time (relating, if not otherwise noted, to steady-state conditions). The frequency of transitions from state i to state j, f_{ij}, is given by the formula

$$f_{ij} = p_i\ \lambda_{ij}$$

The frequency of encountering state i, f_i, is equal to the frequency of leaving state i which, in turn, is the total frequency of transitions out of state i to any other state. Therefore,

$$f_i = p_i \sum_{j \neq i} \lambda_{ij} \tag{3.3}$$

Another useful index is the mean time spent in state i, T_i. This can be calculated from the equation

$$f_i\ T_i = p_i \tag{3.4}$$

which is self explanatory.

If the failure, repair, etc. processes of all components are independent, the probabilities p_i can be directly obtained without solving equations 3.1 and 3.2. Assume that the condition of each component alternates between two states, working and failed, and let the mean up-time of component k be T_{Uk} and its mean down-time (repair time) T_{Dk}. The failure rate of the component, λ_k, is the reciprocal of T_{Uk}, and its repair rate, μ_k, is that of

T_{Dk}. The availability A_k of the component is defined as $\mu_k/(\lambda_k+\mu_k)$, and its unavailability, $\overline{A}_k$, as $\lambda_k/(\lambda_k+\mu_k)$. In a system state i, some components are working and some are in repair; let the set of working components be U_i and the set of failed components be D_i. If all components are independent, the probability of state i is given by the simple expression

$$p_i = \prod_{k \in U_i} A_k \prod_{k \in D_i} \overline{A}_k$$

3.4 Closing Remarks

The state-space representation of systems is underlying all the reliability assessment methods described in this Chapter. However, the methods themselves differ according to the application. The main differences are in the following:

- how to classify the system states into acceptable and unacceptable categories according to the system failure criteria;

- how to partition the state space into working and failed domains;

- how to compute the appropriate indices.

In generating system studies, the first two tasks are comparatively simple and are carried out by constructing and evaluating capacity outage tables or equivalents. These techniques, along with the computations of indices, are described in Sections 4 and 5.

In transmission and bulk power system studies, state classification usually requires detailed load flows. The state space is partitioned through a state selection process; this involves either enumeration or simulation. Sections 6 and 7 are devoted to these topics.

It should be noted that simulation methods may be useful in generating system reliability evaluations, too, particularly if energy constraints must also be considered.

References

1. M.L. Shooman, *Probabilistic Reliability: An Engineering Approach*, McGraw-Hill, New York, 1968.
2. C. Singh, R. Billinton, *System Reliability Modeling and Evaluation*, Hutchinson, London, 1977.
3. J. Endrenyi, *Reliability Modeling in Electric Power Systems*, John Wiley & Sons, Chichester, England, 1978.
4. E.J. Henley and H. Kumamoto, *Reliability Engineering and Risk Assessment*, Prentice Hall, Englewood Cliffs, N.J., 1981.
5. G.J. Anders, *Probability Concepts in Electric Power Systems*, John Wiley & Sons, New York, 1990.

4. GENERATING SYSTEM RELIABILITY MODELING

In generating system reliability studies, the only components considered are the generating units dispersed in the power system. It is assumed that the transmission and distribution network between the generating units and the loads is of infinite capacity and zero failure rate. This assumption is not as unrealistic as it first seems; in these studies the adequacy of the generating system is assessed without any transmission constraints; those constraints are acknowledged in bulk power system reliability studies, to be discussed in Sections 6 and 7.

In generation system reliability studies, mathematical models are needed to represent the generating units, the generating system as a whole, and the load demand. Some of the most frequently used models are described in this Section.

4.1 Generating Unit Models

An electric power generating system consists of many generating units of different ages, sizes, and primary sources of energy. Each of these units is a complex system in itself, with many possible modes of operation and failure. However, the ability of a generating unit to supply system load can be represented by simplified models. These models are described in the following.

4.1.1 The Two-State Model

In this model it is assumed that a generating unit is either operating in full capacity, C, or failed. The operating and outage cycles can be represented by a two-state model, as shown in Figure 4.1.

The long-term state probabilities of the unit are given by

$$P[up] = \frac{\mu}{(\lambda+\mu)} = p_1$$

$$P[down] = 1 - P[up] = p_2$$

where λ is the failure rate (which is equal to the reciprocal of the mean operating time), and μ is the repair rate (which is equal to the reciprocal of the mean repair time). Scheduled maintenance outages are not included in this model. The probability P[down] is also known as the unavailability of a unit. In the utility industry it is known as the *forced outage rate* (FOR), which is calculated from recorded operating and forced outage hours in a given time interval by the following equation:

$$FOR = \frac{FOH}{SH + FOH}$$

where FOH and SH are the total forced outage hours and service (or operating) hours, respectively, in the period of interest. Note that the term "forced outage rate" is a misnomer, because the concept does not represent a rate (such as the failure or repair rate), but describes a probability value or its estimate.

In the model of Figure 4.1 the probability distributions of the uptimes and downtimes need not be specified. If the state residence times are exponentially distributed, the model becomes Markovian. The two-state Markov model has been extensively used to represent other components in power system reliability studies.

4.1.2 The Multi-State Model

Generating units are often operated at less than their full capacities. The effect of partial outages on generation system reliability can be significant and, therefore, must be properly accounted for. The two-state model does not recognize this mode of derated operations. A method of including the effect of partial capacity operations is to add one or more derated states to the two-state model. The three-state model of a generating unit is shown in Figure 4.2. The capacity level of the derated state, denoted by C_2 in Figure 4.2 (b), is usually fixed at a suitably selected value.

Although the probability distributions of the state residence times in the multi-state model are usually different from the exponential, they are assumed to be that to avoid unwieldy mathematics. The resulting three-state Markov model can be solved easily for the state probabilities. They are:

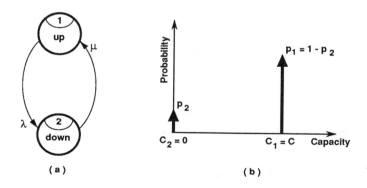

Figure 4.1 A two-state model of a generating unit:
(a) state-space diagram, (b) probability mass function

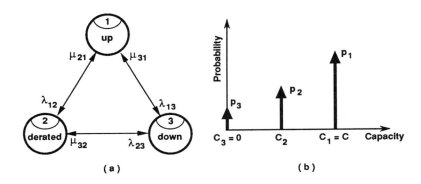

Figure 4.2 A three-state model of a generating unit:
(a) state-space diagram, (b) probability mass function

$$P[\text{up}] = \frac{n_1}{d} = p_1$$

$$P[\text{derated}] = \frac{n_2}{d} = p_2$$

$$P[\text{down}] = \frac{n_3}{d} = p_3$$

where $n_1 = \lambda_{23}\mu_{31} + \mu_{21}\mu_{31} + \mu_{21}\mu_{32}$

$n_2 = \lambda_{12}\mu_{31} + \lambda_{12}\mu_{32} + \lambda_{13}\mu_{32}$

$n_3 = \lambda_{12}\lambda_{23} + \lambda_{13}\lambda_{23} + \lambda_{13}\mu_{21}$

$d = n_1 + n_2 + n_3$

The three-state model does provide satisfactory results in most generating
system reliability studies.

4.1.3 Model of Units for Peaking and Cycling Operations

It has been recognized that in generating capacity planning applications
the forced outage rate is not an adequate risk index for peaking or cycling
units. This is because, for a unit not constantly in demand, the forced outage
hours may contain periods when the unit is not needed. A more appropriate

measure is obtained by separating the forced outage hours into demand and non-demand portions and compute the ratio

$$FOR(demand) = \frac{FOH(demand)}{SH + FOH(demand)}$$

where SH is the actual in-service hours and FOH(demand) is the portion of the forced outage hours when the unit is needed for service.

In 1972, an IEEE Task Group proposed a four-state Markov model [1] to represent gas turbine units and combustion turbine units for peaking operations. This model was modified later [2], as shown in Figure 4.3, so that it can be applied to conventional thermal units in peaking and cycling operations. The parameters of this model are:

T = average time between periods of required operations (excluding scheduled outages,

D = average duration of required operating periods,

p_s = probability of starting failure,

r = average repair time,

m = average time between occurrences of forced outages while in service,

n = average time between occurrences of forced outages while in reserve shutdown.

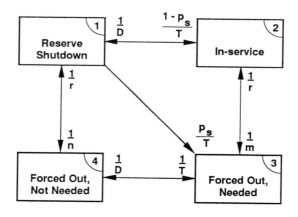

Figure 4.3 A four-state model for peaking and cycling units

In the original IEEE model it was assumed that the generating unit cannot fail while in the reserve shutdown state. This assumption is acceptable for gas or combustion turbine units because any failure that may have developed during reserve shutdown will not manifest itself until a startup attempt is made, and the failure event will be recorded as a starting failure. The modified model of Figure 4.3 can be used for such units by setting $n=\infty$. The same assumption cannot be made for conventional thermal units which are usually monitored while in hot reserve. The occurrences of failures will be recognized and repairs made as soon as the maintenance crew is available for the work.

4.2 Generating System Model

The probability model of a generating system is the amalgamation of all the unit models in the system. It can be represented by a capacity outage probability table, which consists of entries of unavailable generating capacities due to forced outages and their corresponding probabilities. The capacity outage probability table of a generating system can be obtained by a recursive formula, adding one generating unit at a time. This formula is given by the expression

$$P(X) = \sum_i p_i\, P'(X - C_i) \tag{4.1}$$

where $P(X)$ is the probability of capacity outage of X MW after a unit is added,

$P'(X)$ is the probability of capacity outage of X MW before a unit is added,

C is the capacity of the unit being added,

n is the number of states in the generating unit model, including the full capacity state and the outage state,

i is an index of the state of the unit being added, from 1 (full operation) to n (outage),

C_i is the unavailable capacity from the unit when in the ith state, with $C_1 = C$ and $C_n = 0$,

p_i is the probability of the unit being in the ith state.

In the above equation $P'(X-C_i)$ is zero if X is less than C_i since it is impossible to have negative capacity outage states.

In practice, the increments in capacity outages between adjacent entries of the capacity outage probability table are set to be equal. Therefore, the available capacity in each of the states of a generating unit model should be rounded to an integer multiple of the step size. To initialize the computation of the capacity outage probability table, set $P(0)=1$ and the probabilities of all other capacity outage states equal to 0.

In many applications it is more convenient to use the cumulative capacity outage probability table, which can be easily obtained from the capacity outage probability table described above. An entry at the X_i MW level of the cumulative capacity outage probability table gives the probability that the unavailable generating capacity due to forced outages is greater than or equal to X_i.

Transitions from a given capacity state to higher or lower capacity states occur when generating units fail or return to service. The rates of these transitions can be obtained by combining the appropriate failure and repair rates of the generating units involved in the transitions.

4.3 System Load Models

The load demand in a power system is usually approximated by hourly average values. It is a common practice in planning studies to assume that all the weekly load profiles are identical, and that the daily load profiles for the days of the week are all different, with the exception of holidays. The daily load profile is described by expressing the hourly loads as a fraction of the daily peak. Therefore in making load forecasts, it is sufficient to forecast the daily peak loads. Several load models have been used in generating system reliability studies. These are described as follows.

4.3.1 The Load Duration Curve

The load duration curve is obtained by plotting hourly loads in a descending order, in percentage of the time for which load exceeds a given level, as shown in Figure 4.4.

In some applications the hourly loads can be replaced by daily peak loads of the study period, and the resulting load model is known as the daily peak load variation curve. The implication in this representation is that the daily peak load is assumed to last all day.

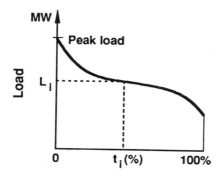

% of time for which load exceeds L_i

Figure 4.4 The load duration curve

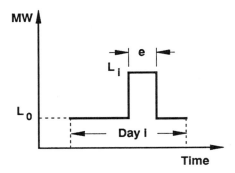

Figure 4.5 Two-level daily load model

4.3.2 The Chronological Load Model

A more detailed load representation, needed for the computation of several of the reliability indices, is provided by chronological load models. Such a model was proposed by Ringlee and Wood [3]. In this representation, the daily load cycle is represented by a high load interval and a low load interval, as shown in Figure 4.5. The high load level, L_i, of the i-th day may be different from those of the other days in a given study period. The low load level, L_o, is the same for every day. The mean duration of the daily high load interval is e, which is known as the exposure factor, and that of the low load interval is 1-e. The transition rates into the high and low load intervals are $1/(1-e)$ and $1/e$, respectively.

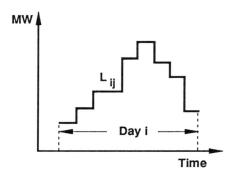

Figure 4.6 Multi-level daily load model

As a generalization to the above load model, the continuous load variations of the day can be approximated by several discrete load levels [4], as shown in Figure 4.6. In this model the day is subdivided into several load intervals which may or may not be equal. The modeled system load assumes a new level at the beginning of each load interval. The low load level of a given day may be different from that of the previous or the following day, and the load level in the last interval of a given day may or may not be equal to the load level of the first interval in the following day. In applications, larger number of load levels in the model requires longer computing time. The choice of the number of levels used is usually a compromise between the accuracy of load representation and the computational burden.

4.4 Summary

The most frequently used models of generating units and system load have been briefly described in this Section. Some of these models represent considerable simplifications of the real system. The simplifications are necessary in most cases in order to keep the computational burden in the applications manageable. Further refinements to the models would give results with improved accuracy; however, the improvements in accuracy are often system dependent, and may not be sufficiently significant to justify the increased complexity of the analysis.

Special models are required at times to represent unique situations, such as generating units that have limited supply of primary energy sources or units that are operated on a standby basis. These models will be described in the next Section when their applications are discussed.

References

1. "A Four-state Model for Estimation of Outage Risk for Units in Peaking Service," Report prepared by IEEE Task Group on Models for Peaking Service Units, *IEEE Transactions* **PAS-91**, 618-627 (1972).
2. L. Wang, N. Ramani, and T.C. Davies, "Reliability Modeling of Thermal Units for Peaking and Cycling Operations," *IEEE Transactions* **PAS-102**, 2004-2011 (1983).
3. R.J. Ringlee and A.J. Wood, "Frequency and Duration Methods for Power System Reliability Calculations - Part II - Demand Model and Capacity Reserve Model," *IEEE Transactions* **PAS-88**, 375-388 (1969).
4. R. Billinton and C. Singh, "System Load Representation in Generating Capacity Reliability Studies, Part I - Model Formulation and Analysis," *IEEE Transactions* **PAS-91**, 2125-2132 (1972).

5. GENERATING SYSTEM RELIABILITY ASSESSMENT METHODS

There have been many publications in the literature on probabilistic assessments of generating system reliability. These methods are based on the simple concept of loss of load, a state in which the forecast system load exceeds the available generating capacity. As already listed in Section 2, the most frequently used indices are the loss of load probability (LOLP), the loss of load expectation (LOLE), and the frequency and mean duration (F&D) of loss of load. The methods of computing these indices are described in this Section.

Although the methods described here were originally developed for static generating capacity reliability evaluations, some of the basic ideas may also be applicable to operational reliability studies, which will be discussed in Section 8.

5.1 Loss of Load Methods

The concept of loss of load probability can be expressed as

$$LOLP = P[\text{load} > \text{available generating capacity}]$$
$$= \int_X P[X] \, P[L > (I - S - X)] \, dX \qquad (5.1)$$

where X is a random variable representing the unavailable generating capacity due to forced outages, L a random variable representing system load demand, I a constant representing the installed generating capacity, and S a constant representing the unavailable generating capacity due to scheduled outages (for maintenance, lack of primary energy sources, or any other cause).

In practical applications, both X and L are usually assumed to be discrete. The computations required in Equation (5.1) can be carried out by using the capacity outage probability table and a system load model. Both concepts were described in Section 4. The commonly used methods for computing the loss of load indices are given below.

5.1.1 Loss of Load Probability

This index is the probability of occurrence of the loss of load condition in a generating system during a specified study period. Several approaches to computing the LOLP index have been developed. They are described in the following.

A. USING CAPACITY OUTAGE PROBABILITY TABLE AND LOAD DURATION CURVE

This method of computing the LOLP makes use of the exact capacity outage probability table and the load duration curve. In this case P[X] in Equation (5.1) becomes discrete and the LOLP is given by

$$LOLP = \sum_j P[X_j] \, P[L > (I - S - X_j)]$$

$$= \sum_j P[X_j] \frac{t_j}{100} \tag{5.2}$$

where $P[X_j]$ is an entry of the capacity outage probability table at the X_j MW level, and $P[L>(I-S-X_j)]$ is the probability that the load level L is larger than the available generating capacity. The latter is equal to the fraction of time $t_j/100$ when load is larger than L_j and is obtained from the load duration curve (see Figure 4.4). Note that t_j is given in percentage of the total time in the study period.

Conversely, LOLP can be computed by using the cumulative capacity outage probability table and the probabilities of the exact load levels L_k. The probability of loss of load is now given by

$$LOLP = \sum_k P[L_k] \, P[X > (I - S - L_k)] \tag{5.3}$$

where the second set of probabilities on the right-hand side are the appropriate entries of the cumulative capacity outage probability table.

Equations (5.2) and (5.3) can be generalized. Observing, for example, that $P[L>(I-S-X_j)]$ in Equation (5.2) is a shorthand notation for $P[L>(I-S-X_j) \mid X=X_j]$, and using the definition of conditional probabilities, Equation (5.2) can be rewritten as

$$LOLP = \sum_j P[L > (I - S - X) \cap X = X_j]$$

$$= P[L > (I - S - X)]$$

where the last expression is quite obvious in its generality (see also the first statement in Equation (5.1)). Equations (5.2) and (5.3) actually provide methods for computing it. For a period of N hours, then,

$$LOLP = \frac{1}{N} \sum_{i=1}^{N} P[L_i > (I - S_i - X_i)] \tag{5.4}$$

where L_i is the load, S_i the scheduled outage, and X_i the capacity on forced outage at hour i. Note that S_i is not a random variable.

In the above calculations the load duration curve may be replaced by the daily peak load variation curve. The index so computed gives the probability of loss of load with the assumption that the daily peak load lasts for the entire day. Equation (5.3) still applies but now the subscript i refers to the ith day, not the ith hour. Because of the underlying assumption, the result will be on the conservative side.

In the foregoing approaches the LOLP is calculated in two steps. The first step involves the computation of the capacity outage probability table, using the recursive formula of Equation (4.1). This method of computing the probabilities of capacity outages is exact, with the exception that in applications the capacity outages in the table are rounded to a suitable selected step size. The second step involves the convolution of the capacity outage probability table and the load duration curve. The computations in these two steps may require considerable computing time, especially for large systems. In order to reduce computing requirements in LOLP calculations, methods have been developed in which the capacity outage probability table is approximated by a series expansion of the normal probability distribution. These methods are known collectively as the cumulant method which will be described next.

B. THE CUMULANT METHOD

In this approach, it is assumed that the available capacities of individual generating units in a power system are independent random variables. The available generating capacity of the system, itself a random variable, tends to have a normal distribution if the system is large. This, when approximated by a series expansion, can be expressed in terms of its cumulants, which, in turn, are given by the sums of the corresponding cumulants of the individual generating units. The cumulants are a set of descriptive parameters for a distribution, just as the moments are; in fact, they can be expressed in terms of the moments [1]. For the series expansion, previous published methods have made use of the Gram-Charlier expansion [2], the Edgeworth expansion [3], and the Hermite polynomials [4]. The method presented in the following is based on the Gram-Charlier expansion for an N-unit system, using the multi-state generating unit model of Figure 4.2(b). Let

p_{ni} = probability of the nth unit operating in state i,

C_{ni} = available capacity from the nth unit when operating in state i,

M_{jn} = the jth moment (about 0) of the nth unit,

$$= \sum_i p_{ni} \, C_{ni}^j$$

The first 4 cumulants of the nth unit are given by [1]

$$k_{1n} = M_{1n}$$

$$k_{2n} = M_{2n} - M_{1n}^2$$

$$k_{3n} = M_{3n} - 3M_{2n}M_{1n} + 2M_{1n}^3$$

$$k_{4n} = M_{4n} - 4M_{3n}M_{1n} - 3M_{2n}^2 + 12M_{2n}M_{1n}^2 - 6M_{1n}^4$$

The uth cumulant of the available capacity of an N-unit system is given by

$$K_u = \sum_{n=1}^{N} k_{un}$$

The Gram-Charlier expansion of the probability density function of the available system capacity X is

$$f(X) = N(X) - \frac{K_3}{6\sigma^3}N^{(3)}(X) + \frac{K_4}{24\sigma^4}N^{(4)}(X) - \frac{K_5}{120\sigma^5}N^{(5)}(X)$$

$$+ [\frac{1}{72}(\frac{K_3}{\sigma^3})^2 + \frac{K_6}{720\sigma^6}]N^{(6)}(X) + \cdots$$

where $\sigma^2 = K_2$

$N(X)$ = normal distribution of standardized capacity X,

$N^{(i)}(X)$ = the ith derivative of $N(X)$.

To compute the LOLP for a given system load level L, one must obtain first the standardized variable $Z = (L-K_1)/\sigma$. The probability of loss-of-load for load level L is given by

$$P[(I - S - X) < L] = \int_{-\infty}^{Z} f(X)dX \qquad (5.5)$$

Equation (5.5) can be evaluated very easily from the the standardized normal distribution and the cumulants K_u.

The cumulant method is computationally efficient because the LOLP is calculated directly at each load level. It is not necessary to compute the capacity outage probability table which requires substantial computational effort. The major drawback of the cumulant method is that the error of the approximation is system dependent, and it may not be acceptable in some applications. Examples of efforts to improve the accuracy of the cumulant method can be found in References 3 and 4.

5.1.2 Loss of Load Expectation

The loss of load method is probably the most widely used technique in generating system reliability studies. It seems that when the loss of load concept was first developed, it was sufficient for generating system planners to obtain an estimate of the expected number of days in which loss of load conditions could occur in a given study period. A conservative estimate of this measure is the LOLE, which is obtained by assuming that the daily peak load lasts for the whole day.

The loss of load expectation can be defined, therefore, as the expected number of days in which the available generating capacity is insufficient to supply the peak load in a given study period, usually a year. It can be calculated using the following formula:

$$LOLE = \sum_{i=1}^{N} P[L_i > (I - S_i - X_i)] \qquad (5.6)$$

where the probabilities on the right-hand side can be obtained from the cumulative capacity outage probability table, L_i is the peak load and S_i is the scheduled outage capacity on day i, and N is the number of days in the study period. Equation (5.6) is based on the version of Equation (5.4) using the daily load curve through the connection

$$LOLE = N \cdot LOLP$$

which is an application of the general rule in probability theory stating that for an event A, the expected value of its occurrence in N trials is

$$E[A] = N \, P[A]$$

As noted, the values of L_i are taken from a load duration curve made up of daily peak loads. If a load duration curve of hourly entries is used, the interpretation of LOLE changes to mean the expected number of hours in the study period when the load demand is not satisfied. Now N is, of course, the total number of hours in the study period. If it can be assumed that there is at most one hour in a day when supply cannot meet demand, the LOLE based on the hourly load duration curve can be re-interpreted as the expected number of days in the study period when loss of load occurs (not necessarily at the time of the daily peak) due to insufficient generating capacity.

5.2 The Frequency and Duration Method

In contrast with the loss of load methods which can produce only probability indices, this method also computes the frequency and mean duration of encountering situations when generating capacity cannot meet load demand [5-8]. Its application has been readily accepted by capacity planners not only because of the broader information it provides, but also because the computed indices can be easily compared with actual experience. The method will be described in the following, for a generating system with an installed capacity I, and using the two-level load model of Figure 4.5. Note that in the frequency and duration method, a chronological load model must be used because system failure can occur in two ways: either when the generating capacity on outage increases or, given the same outage condition, when the load increases, and both possibilities must be accounted for.

The general state of the system at any moment is determined by the capacity outage X_j and the load level L_i prevailing at that time. A margin M_k can be defined for every state k such that $M_k = I - X_j - L_i$; loss of load occurs if $M_k < 0$. If the states are divided into a working set W where $M \geq 0$ and a failed set F where $M_k < 0$, then the system failure probability P_F will be

$$P_F = \sum_{k \in F} p_k$$

where p_k, the probability of the state k, is given by

$$p_k = P[X_j] \, P[L_i].$$

The frequency of encountering capacity deficiency is

$$f_F = \sum_{k \in F} (p_k \sum_{h \in W} \lambda_{kh})$$

where λ_{kh} consists of transitions to working states where capacity outage and/or system load are lower than before.

The mean duration of capacity deficiency is

$$T_F = \frac{P_F}{f_F}$$

With the frequency and duration method, the expected energy not supplied, EENS, can also be calculated. For a single disturbance, it is given by the sum

$$EENS = - \sum_{k \in D} a_k T_k M_k$$

where D is the set of states visited during the disturbance, a_k is the number of times state k is travelled through in the same period, and T_k is the mean duration of state k in hours. As a result, EENS is obtained in MWh if M_k is expressed in MW. If the yearly EENS is required, it can be computed from

$$EENS = - \sum_{k \in F} f_k T_k M_k$$

where f_k is substitutied in occurrences per year. Now EENS is obtained in MWh/year. These indices can be converted into system-minutes as shown in Section 2.

5.3 Other Considerations

5.3.1 Interconnected Systems

The methods described in the previous Section are applicable to single-area systems with no interconnections. However, most modern power systems are interconnected with other systems for the purposes of reserve-sharing, economic interchange, and emergency assistance. The application of single-area methods to interconnected systems would yield results which may prove to be too conservative.

The simplest approach to incorporating the effect of interconnection assistance is to model the available capacity from neighboring systems by an equivalent generating unit. The capacity and availability of the equivalent unit are those of the tie line.

The concept of the LOLP method for two-area interconnected systems with a tie line of finite capacity T is illustrated in Figure 5.1, where it is assumed that one system will assist the other to the extend possible without curtailing its own load. In reality, interconnection agreements may have other possible arrangements. In Figure 5.1, the area of the rectangle represents probability arrays of margin states in the two systems. The area to the left of line A-A represents the LOLP in system A without assistance from system B. Similarly, the area below line B-B represents the LOLP in B without assistance from A. With the tie line in place, the LOLP in A will be reduced by $p_T P'_A$, and that in B by $p_T P'_B$, where p_T is the probability of the tie line in the operating state. The LOLP's in systems A, B, and the interconnected system are, respectively,

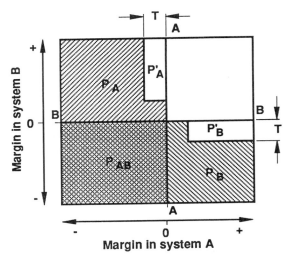

Figure 5.1 Probability arrays of margin states for
a two-area interconnection

$$LOLP_A = P_A + P_{AB}$$

$$LOLP_B = P_B + P_{AB}$$

$$LOLP_S = P_A + P_B + P_{AB}$$

The frequency and duration method has also been extended to two-area interconnected systems [9]. Details of the extension are not given here due to space limitations.

Multi-area generating system reliability assessment can also be carried out using the decomposition-simulation approach [10]. In this method, the system state space is decomposed into disjoint sets of acceptable (no loss of load), unacceptable (loss of load), and as yet unclassified states. The decomposition process is an iterative one, starting by taking the entire state space as unclassified, and continuing until a specified stopping point is reached. The indices, such as LOLP and EENS, can be readily computed for the acceptable and the unacceptable sets of states. The indices for the remaining unclassified states are estimated by using the Monte Carlo sampling process. The original decomposition-simulation approach described in [10] has been extended recently by Lago-Gonzalez and Singh [11] by introducing methods for multi-area load representations, a more accurate algorithm for computing the indices in the simulation phase, and a method for computing the frequency of loss of load indices.

5.3.2 Energy Limited Units

The output capacity of some generating units are limited by the availability of the primary energy source. A typical example is a run-of-the-river hydraulic unit, the output of which depends on the available amount of water flow. Such a unit can be modeled by a multi-state unit where the availabilities of the capacity states are conditioned by the probabilities of having specified amounts of flow. The model is then treated like that of an ordinary unit in the analysis.

For another type of energy limited units the storage capacities of the primary energy sources can sustain full capacity operation for only very short periods of time. Such units are usually scheduled to operate during peak hours. One approach to considering a unit of this type is to use its capacity and energy distributions to modify the system load model. The resulting load model is then used in the analysis with the rest of the units that are not energy limited.

A more rigorous approach to incorporating energy limited generating units in reliability studies is the application of Monte Carlo simulation. This method can be used to produce simulated operating records or a suitable number of simulated samples of system states. The desired risk indices can be derived from the simulated operating records or samples. Examples of applications of the Monte Carlo method can be found in [12-14]. A more detailed description of the Monte Carlo method is given in Section 7.

5.3.3 Operating Considerations of Generating Units

In general, operating considerations and constraints of generating units are not modeled explicitly in generating system reliability studies for long range capacity planning. This does not mean that operating considerations do not have significant impact on the results of these studies. Some of the important factors are:

- Generating unit duty cycles and their effects on operating economy and security.

- Operating reserve policies and requirements and their effects on duty cycles of peaking units.

- Postponability and management of unplanned outages.

- Economic interchange and emergency assistance from interconnected systems.

The operating considerations can be incorporated either analytically or by Monte Carlo simulation [14-17]. Studies on sample systems show that modeling of these factors can make the computed system reliability indices more accurate.

5.4 Summary

The fundamental formulations of the commonly used probability methods for generation system reliability assessment have been described in this Section. These methods are based on the generation and load models described earlier. Some of the detailed considerations in practical applications, such as uncertainties in forced outage data and load forecast, are not included because of space limitations. Readers are referred to References 18 and 19 for more exhaustive discussions on this and other subject matters.

References

1. M.G. Kendall and A. Stuart, *The Advanced Theory of Statistics, Volume 1 - Distribution Theory*, Charles Griffin & Company Limited, London, 1963.
2. N.S. Rau and J.P. Stremel, "The Cumulant Method of Calculating LOLP'" IEEE Paper A79 506-7 (1979).
3. D.J. Levy and E.P. Kahn, "Accuracy of the Edgeworth Approximation for LOLP Calculations in Small Power Systems," *IEEE Transactions* PAS-101, 986-996 (1982).
4. H. Duran, "A Recursive Approach to the Cumulant Method of Calculating Reliability and Production Cost," *IEEE Transactions* PAS-104, 82-90 (1985).

5. J.D. Hall, R.J. Ringlee, and A.J. Wood, "Frequency and Duration Methods for Power System Reliability Calculations - Part I - Generation System Model," *IEEE Transactions* **PAS-87**, 1787-1796 (1968).
6. R.J. Ringlee and A.J. Wood, "Frequency and Duration Methods for Power System Reliability Calculations - Part II - Demand Model and Capacity Reserve Model," *IEEE Transactions* **PAS-88**, 375-388 (1969).
7. C.D. Galloway, L.L. Garver, R.J. Ringlee, and A.J. Wood, "Frequency and Duration Methods for Power System Reliability Calculations - Part III - Generation System Planning," *IEEE Transactions* **PAS-88**, 1216-1223 (1969).
8. A.K. Ayoub and A.D. Patton, "Frequency and Duration Method for Generating System Reliability Evaluation," *IEEE Transactions* **PAS-95**, 1929-1933 (1976).
9. R. Billinton and C. Singh, "Generating Capacity Reliability Evaluation in Interconnected Systems Using a Frequency and Duration Approach, Part I - Mathematical Analysis," *IEEE Transactions* **PAS-90**, 1646-1654 (1971).
10. D.P. Clancy, G. Gross, and F.F. Wu, "Probabilistic Flows for Reliability Evaluation of Multi-Area Power System Interconnections," *Electrical Power and Energy Systems*, **5**, No. 2, 101-114 (1983).
11. A. Lago-Gonzalez and C. Singh, 'The Extended Decomposition-Simulation Approach for Multi-Area Reliability Calculations,' *IEEE Transactions on Power Systems* **5**, No. 3, 1024-1031 (1990).
12. L. Paris and L. Salvaderi, "Pumped-Storage Plant Basic Characteristics: Their Effect on Generating System Reliability," *American Power Conference Proceedings*, **35**, 403-418, 1974.
13. B. Manhire and R.T. Jenkins, "BENCHMARK: A Monte Carlo Hourly Chronological Simulation Model Which Includes Effects of Ramp-Rates and Reservoir Constraints," *Conference Proceedings on Generation Planning: Modeling and Decisionmaking*, 231-265, 1982.
14. A.D. Patton, J.H. Blackstone, and N.J. Balu, "A Monte Carlo Simulation Approach to the Reliability Modeling of Generating Systems Recognizing Operating Considerations," *IEEE Transactions* **PWRS-3**, 1174-1180 (1988).
15. A.D. Patton, C. Singh, and M. Sahinoglu, "Operating Considerations in Generation Reliability Modeling - An Analytical Approach," *IEEE Transactions* **PAS-100**, 2656-2671 (1981).
16. *Modeling of Unit Operating Considerations in Generating Capacity Reliability Evaluation*, EPRI Report EL-2519, 1982.
17. *Reliability Models of Interconnected Systems That Incorporate Operating Considerations*, EPRI Report EL-4603, 1986.
18. J. Endrenyi, *Reliability Modeling in Electric Power Systems*, John Wiley & Sons, Chichester, England, 1978.
19. R. Billinton and R.N. Allan, *Reliability Evaluation of Power Systems*, Pitman, London, 1984.

6. BULK POWER SYSTEM RELIABILITY MODELING

6.1 Introduction

Bulk power systems (or composite systems) consist of generating units and the high-voltage network (usually 200 kV and above) interconnecting the generating stations and delivering energy to the load buses throughout the system. Thus, in most bulk system reliability evaluations only two types of components are considered, generating units and transmission circuits. Nevertheless, probabilistic reliability methods applied in these evaluations are quite complex and require sizeable computing effort. Reasons for this are the following.

- Large system size (often several hundreds of buses)

- Modeling difficulties (dependent relationships, load representation, energy constraints, load flows, remedial action)

- Computational difficulties (the reconciliation of detail, accuracy and computing effort)

Modeling will be discussed in this Section, and the computational aspects in Section 7.

The existing programs for bulk power system reliability evaluations consider system adequacy only and are mostly planning-oriented [1-3]. This calls for two extensions in the discussions, the integration of security aspects into the evaluations, and a study of operational reliability. As mentioned earlier, however, probabilistic security assessment is in the very early stages of development and, for this reason, it is not discussed here. The concepts and tools developed for operational reliability are reviewed in Section 8.

As already noted, probabilistic approaches are not yet commonly used in bulk system reliability evaluations. Many of the models are considered overly approximate, and feasible solution methods not sufficiently accurate [4-6]. The models described in this Section will give the reader an appreciation of the complexities involved. The application of more accurate models is contingent on continuing refinements in the solution methods, as discussed in Section 7.

6.2 Special Modeling Techniques

6.2.1 Failure Modes of Overhead Transmission Lines

An overhead transmission line may fail due to hardware breakdowns, or it may develop ground faults due to arcing caused by flash-overs of insulators or other abnormalities. In either case the line will be removed from service very quickly by the protection system. In cases of hardware failures, the failed component must be repaired or replaced before the line can be returned to service. Such outages are commonly called *permanent outages*.

The protection systems of most high-voltage transmission lines are designed with a feature called "auto-reclosure", which allows the line to be re-energized automatically after a pre-selected dead time has elapsed. The dead time is designed to allow de-ionization of the flash-over path when a ground fault develops due to flash-over. In most cases the de-ionization process can be completed within the dead time, and the line can be successfully returned to service. In such instances the line is out of service for very short durations only, usually from a fraction of a second to several seconds, and these interruptions are termed *transient outages*.

If the de-ionization process is not complete by the end of the dead time, the automatic re-energization of the line will re-ignite the arcing and the line will be tripped out again. The line protection is usually designed to reclose once or twice only, after which the line will be locked out. However, an attempt can be made by a system operator to return the line to service by closing the circuit breakers manually after a few minutes of the lock-out. If the manual reclosure is successful, the transmission line outage is called a *temporary outage*.

If the flash-over causes permanent damage to the insulator or other components, the manual reclosure will also fail, and the interruption becomes a permanent outage.

To simplify analysis, the three types of outages described above are assumed to be independent, and as such they can be modeled separately. The types of outages that should be incorporated in a bulk power system reliability assessment depends on the specific applications.

6.2.2 Common-Mode Failures

In addition to independent failures of power system components, events may occur which affect several components simultaneously. A common-mode failure is one where a single event causes multi-component failures which are not consequences of one another. An example of the common-mode failure is the collapse of a transmission tower which supports two transmission circuits. Some common-mode failures in power systems are caused by external events, such as fire, flood, and motor vehicle and aircraft accidents.

The most frequently used model for incorporating common-mode failures is shown in Figure 6.1, where the model represents a two-component combination [7]. In this model, a transition rate λ_{AB} is introduced to represent common-mode failures of both components in addition to state transitions

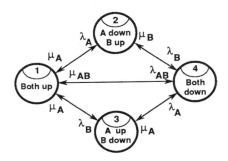

Figure 6.1 Model for common-mode failure of two components

due to the independent failures and repairs of the components. Following a
common-mode failure, the transition to the state where both components are
working can be made either by returning one component to service at a
time, or by returning both components to service at the same time.

The model of Figure 6.1 can be easily extended to systems involving
more than two components. The practical difficulty of applying such models
lies in the scarcity of data on common-mode failure rates.

6.2.3 System Load Models

The load models described in Section 5 can also be applied in bulk power
system reliability studies to represent the loads of individual buses, but not
the load of the entire system. The commonly used models for representing
system load can be characterized by the following two attributes.

- The customer loads supplied from a given substation are aggregated
  into a substation bus load. Each bus load is considered a separate entity. The aggregation of all bus loads in the system constitutes the system load.

- Substation bus loads vary with time in cycles. The system load model
  must specify whether the bus loads are independent, completely correlated, or partially correlated [8].

The true nature of the relationships among bus loads is not known, and
has not been studied. It seems reasonable to assume that bus loads are
neither fully independent nor completely correlated. The bus load relationships may be system specific.

The variations of bus loads in daily cycles cannot be fully recognized. Reliability assessments have been made in the past under the assumption that all bus loads are fixed at their respective annual peak levels. The system reliability (or risk) indices obtained under this assumption would be pessimistic, because (1) a given system state classified as failure under peak load condition may represent success at a lower load level; and (2) it is very unlikely that the peak loads of all the buses will occur coincidentally. To adjust for (1), reliability assessments can also be carried out at one or more lower load levels, and proper weights are applied to the results for each load level to obtain the final system indices. To compensate for (2), a snapshot of all loads at the time of system peak may be taken, and the lower load levels determined by assuming that all bus loads are completely correlated.

6.2.4 Effects of Adverse Weather Conditions

Overhead transmission lines are constantly exposed to the weather environment. It is well established that the hazard rates of overhead lines under "severe" weather conditions, such as strong wind, lightning, snow, and freezing rain, are much higher than during periods of clear skies with very little or no wind ("normal" weather).

Severe weather conditions occur randomly within the context of seasonal and climatic factors. They usually last for short durations (a few hours) and their intensities vary over wide ranges. The prevailing types of severe weather and the sizes of the affected areas may depend on the climatic and geographical characteristics of the region. Past experience shows that the effects of severe weather conditions on the hazard rates of overhead lines change gradually with their intensities. It would be very difficult, if not impossible, to develop a model which could incorporate the effects of all the meteorological parameters accurately.

A more practical modeling approach would be to represent the various weather conditions by discrete states. The most frequently used method is a two-state Markov model, as shown in Figure 6.2, where all the adverse weather conditions are amalgamated into one state. Successful application of

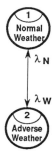

Figure 6.2 Model for adverse weather conditions

this model depends on careful analyses of meteorological and line outage data to arrive at a set of criteria for determining severe weather conditions and the necessary model parameters.

In some places it is also required to considered extreme weather conditions, such as tornadoes and hurricanes, which can cause line failures with probabilities close to one. The effects of extreme weather conditions can be incorporated separately with very little difficulty [9]. For more information about weather modeling and the derivation of parameters, the reader is referred to Appendix I in [3].

6.2.5 Effects of Protection System Malfunctions

The protection system in a power system is designed to serve two major functions: quick isolation of faults after their occurrences and protection of major power system components from possible damages by excessive voltages or currents. Their reliable functioning is of utmost importance to the secure operation of the power system. Failures of protection system components may result in two different modes of malfunction of the protection system: (1) failure to operate when it is required, and (2) operation when not required to do so. Some of the unnecessary operations may be harmless; others may have serious consequences. The models and methods for incorporating the effects of protection system failures on bulk power system reliability will be summarized in this section.

A. ADJUSTING THE FAILURE RATES OF THE PROTECTED COMPONENTS

This is a very simplistic approach in which the malfunctions of a protection system are accounted for by increasing the outage rates of the protected components. For example, an additive constant failure rate is often assigned to each transmission line terminal to account for false trips, or to represent failures of station equipment which are not modeled, such as transducers, relays, and communications channels. Most early studies of stations and transmission system reliability used this approach.

Unfortunately, the additional rates are very arbitrary estimates of the effects they purport to represent. Also, the possibility of multiple effects due to a single cause is overlooked.

B. MARKOV MODELS

Markov models have been applied to incorporate the effects of protection system failures on bulk power system reliability. Two different approaches have been employed.

- Modeling the circuit breaker only. This approach is based on the fact that all failures and malfunctions of the protection system are manifested by the operation (or non-operation) of circuit breakers. The modes of operation and failure of a circuit breaker are represented by a Markov model. The effects of the malfunctions in the protection system are incorporated by assigning appropriate transition rates to this model. An example of this approach can be found in [8, 10].

The disadvantage of this approach is that the breakers are assumed to be independent of each other, whereas in reality their malfunctions, if originated in the protection system, may be interrelated.

-   Modeling the protected component and the circuit breaker. In this approach, undetected failures in the protection system are accounted for by a multi-state Markov model of the protected component. Undesired trips and correct trips in normal fault-clearing are incorporated in another Markov model, that of the circuit breaker at the boundary of the protection zone. The system model is obtained by combining the component models with the circuit breaker models, based on the assumption of component independence. Bulk power system reliability indices can be computed using the system model and the concept of minimal failure events. The general concept of this approach is described in [11], and an example of its application can be found in [12].

This modeling approach is capable of correctly representing the effects of the major modes of failure in protection systems on power system reliability. However, it is not clear if false trips related to the occurrences of power system faults can be properly represented in this method.

C. EVENT TREE METHOD

A protection system consists of four major groups of components: the transducers, the relays, the communications channels, and the circuit breakers which constitute the boundary of the protection zone. After the occurrence of a fault in the protected power carrying component, the normal operations or failures to operate of the protection system components can be described in a detailed tree structure. The end events of the tree give indications on the resulting status of the circuit breakers, and hence the effects of failures in the protection system. The probabilities of the end events can be easily calculated. A description of applying the event tree method to protection system modeling is given in [13].

The event tree method can recognize the sequential logic of protection system operations and the analysis can be extended to analyze the system at increasing depth. Conceivably, the construction of event trees and the computation of event probabilities can be performed through an appropriate algorithm.

Records show that false trips do occur with no clear indication as to which component is to blame. The problem of how to incorporate false trips using the event tree method needs to be resolved. Also, unlike the previous methods, this approach requires a detailed analysis of the protection system and its components. One may consider this an undesirable complication in the overall modeling task.

In either of the above methods, some effects of protection system operations or malfunctions have not been modeled, including the following.

-   Protection zones are designed to overlap, and some component failures (other than that of circuit breakers) or protection system malfunctions may affect two neighboring protection zones simultaneously.

- The protection system may also have an effect on power system reliability when it operates as it should. For example, cascading outages may be precipitated through protective actions which would not be identified by conventional reliability programs. In fact, some of the power blackouts in the past were the results of such operations before they could be blocked manually.

These effects deserve consideration in future investigations on bulk power system reliability.

6.2.6 Effects of Failures Originating in Stations

Failures of station equipment other than protection system components may also cause transmission circuit outages. In fact, a single equipment failure in a station, such as a ground fault in a circuit breaker, may result in the outages of several circuits. Many of these outages are of relatively short duration; they are terminated by appropriate switching in the station, which can often restore the transmission system well before the failed component is repaired.

Modeling station-originated transmission circuit outages can be quite complex because of the multiple effects just described, and because these effects vary with the station bus and breaker arrangements and with the way the individual circuits are tied to the station buses. The traditional method of accounting for such failures by adding a constant term to the failure rates of the transmission circuits is unsatisfactory because of the crude approximation it entails. To do full justice to the problem, the transmission system has to be augmented by the station networks [14], making the system under study, in effect, a composite of the transmission and station systems. It is clear that while such an extension is necessary if station effects are to be accounted for, it multiplies the computational burden.

The necessity of considering station-originated failures (and also protection system failures) in transmission system reliability studies may be questioned. It could be argued that, since such studies are performed to assess the reliability of the transmission system, only outages should be included that originate in the transmission system and whose effects would be remedied, if necessary, within that system. Station failure effects on transmission are relieved by changes made in the stations and not by, say, increasing the redundancy of the transmission system. Therefore, such effects should be of no concern in a transmission study proper, where station reliability evaluations are handled separately.

To include or not to include station-originated failure effects into bulk power system reliability assessment ultimately depends on how the system load points are defined. In a pure transmission system study, they will be considered at the termination of the transmission lines on the high-voltage buses of the transformer stations; therefore, no station components need to be included. On the other hand, if the load points are assumed to be at the secondary buses from where the load feeders are supplied, the failures of station components must also be taken into account and, in fact, the composite of transmission and station systems is then to be evaluated.

6.2.7 Effects of Scheduled Outages of Transmission Components

Power system components are routinely taken out of service for inspection, maintenance, and overhaul. These activities are planned in advance and are known as scheduled outages. They are carried out to correct any deterioration that may have occurred or accumulated while in service, and to restore the components to improved conditions.

Transmission systems are designed with a high degree of redundancy to ensure supply continuity. Most system failures are caused by the overlapping outage of two or more components. The possible combinations of overlapping outages of two components are: (1) two components fail independently; and (2) one component fails while another is on scheduled outage. Historical records show that component unavailabilities due to scheduled outages are much higher than those due to failures. Therefore, the accurate modeling of a component failure during the period when another component is on scheduled outage is an important factor in transmission system reliability assessment.

Scheduled outages cannot be assumed to occur randomly. An obvious constraint is that a scheduled outage will not be allowed to take place if it causes service interruptions. The durations of scheduled outages are also subject to operational and other practical constraints. Therefore, the Markov model, which is used often in power system reliability studies, may not be applicable. An intuitive modeling approach is described as follows.

Let $i_M j_R$ = a system failure state in which component j fails while component i is on scheduled outage,

λ_{Rj} = failure rate of j,

λ_{Mi} = scheduled outage rate of i,

t_{Mi} = scheduled downtime of i, a random variable,

T_{Mi} = the mean of t_{Mi},

t_{rj} = the repair time of j, an exponentially distributed random variable with mean r_j,

t_F = residence time in state $i_M j_R$, a random variable,

α = a random variable uniformly distributed in the interval $[0,1]$.

The frequency of occurrence of the state $i_M j_R$ is given by [8]

f_F = (failure rate of j) x (probability that i is on scheduled outage)

$$\simeq \lambda_{Rj} \lambda_{Mi} T_{Mi} \qquad (6.1)$$

It should be noted that Equation (6.1) is based on the assumption that the mean in-service times are much longer than the mean repair or scheduled outages times. The computation of f_F is straightforward.

The duration of the state $i_M j_R$ is equal to either the remaining portion of the scheduled downtime of i or the repair time of j, whichever is shorter. This concept can be expressed analytically by [15]

$$t_F = \min\{\alpha t_{Mi}, t_{Rj}\}$$

The mean duration of the state $i_M j_R$ is given by

$$T_F = E\{t_F\}$$

The duration T_F depends on the probability distribution of t_{Mi}. If the scheduled outage times can be approximated by an exponential distribution, T_F is given by

$$T_F = \frac{T_{Mi} T_{Rj}}{T_{Mi} + T_{Rj}}$$

A unique probability model was proposed for the scheduled outage durations of transmission system components to recognize the fact that maintenance procedures on these components are conducted mostly during the daylight hours. Details of this model and the method of computing T_F are given in [15].

6.3 Load Flow Methods in State Evaluation

When carrying out adequacy assessment of bulk power systems, contingencies must be classified into acceptable and system-failure categories. System failures are identified by applying a set of pre-selected criteria which may include the lack of service continuity, and measures of service quality, such as harmonic content and bounds on frequency and bus voltage variations, and overload of transmission lines.

The bus voltages (magnitudes and phase angles) in a transmission network are functions of power injections (both real and reactive) at generating station buses and power consumptions at the load buses. The functional relationships are described by a set of non-linear equations and the solutions of these equations are called load flows. A large number of papers and books have been published on load flow methods [16]. These methods can be classified into several categories, and their applications to state evaluations in bulk power system reliability studies are summarized below.

6.3.1 Full AC Load Flow

The power flow equations of a transmission system can be solved iteratively on computers using the Gauss-Seidel [17] or the Newton-Ralphson [18] method. The solution procedure consists of assuming an initial solution for bus voltage magnitudes and phase angles, and computing corrections to the previous solution in subsequent iterations until convergence is reached. Special programming techniques, such as sparsity oriented triangular factorization [19], have been developed to reduce computing time.

A full AC load flow gives accurate solutions consisting of the real (or active) and reactive power generations, the magnitues and phase angles of bus voltages, real and reactive power flows in transmission components, and line losses. The number of load flows required in an adequacy assessment can be very large and consequently, the computational burden can become excessive.

6.3.2 Decoupled AC Load Flow

Practical power systems are characterized by strong interdependences between real power injections P and bus voltage phase angles θ, and between reactive power injections Q and bus voltage magnitudes V. On the other hand, the interdependences between P and V, and between Q and θ are relatively weak. These physical characteristics can be utilized to transform the original power flow equations into two separate sets, with the first set relating P to θ and the second set, Q to V. The "decoupling" of the system parameters makes it possible to solve the two sets of equations separately using matrix inversion methods. In a further approximation, linearization of the decoupled system equations gives what is known as the *fast decoupled load flow* [20].

In the solution process of the decoupled load flow, the results of the "P-θ" equations are used in the solution of the "Q-V" equations, and vice versa. Several iterations are required to reach an acceptable load flow solution. Nevertheless, the fast decoupled load flow method offers considerable savings in computer time in comparison with the full AC load flow. The solution accuracy is very close to that of the full AC load flow.

6.3.3 DC Load Flow

A further simplification to the load flow problem can be made by assuming that all bus voltage magnitudes stay constant at 1 per unit and that the resistive components of the transmission line impedances are negligible in comparison with their reactive counterparts. Under these assumptions the original set of non-linear equations are reduced to a single set of linear equations relating P to θ. This approximation is commonly referred to as the DC load flow.

The DC load flow gives approximate solutions of the bus voltage angles from which the flows of real power in the transmission network can be calculated. However, it does not provide information on bus voltages and reactive power. The solution process involves a simple matrix inversion with no iterations required. Therefore, the computing time of the DC load flow is much less than that required by the decoupled load flow.

6.3.4 Transportation Model

This model (also called the network flow model) consists of a source node representing system generation, a sink node representing system load, and a network of branches of finite capacities representing the transmission system. Flows in the network are constrained by Kirchhoff's current law, which states that the net flows into or out of a node is zero, except at the source and sink nodes. However, Kirchhoff's voltage law is disregarded. Solutions of this power flow model can be computed by the so-called *max-flow/min-cut* algorithm [21].

The accuracy of the power flows obtained by the transportation model is low, but it may be acceptable in some applications. The model has been used as a tool for a fast first orientation, and in many multi-area studies.

6.3.5 Miscellaneous Approximate AC Load Flow Methods

As mentioned earlier, adequacy assessment of bulk power systems may include evaluations of thousands of contingency states. If information on bus voltages is required, the DC load flow method cannot be used; on the other hand, the computational burden may be considered excessive even for fast decoupled load flow calculations. Several approximate load flow methods have been developed specially for applications to contingency analysis. One method uses injected powers at appropriate buses to simulate load and generation changes and outages of transmission elements [22]. The injected powers are calculated using the sensitivity matrix of the base case load flow, and then proceed to calculate the changes in voltage magnitudes and phase angles for all the buses in the system. Another method uses a hybrid model [23] which is a mixture of the fast decoupled load flow and the nodal iterative model of [17]. Both methods claim to be faster than the fast decoupled load flow method.

6.4 Remedial Measures in State Evaluation

The state of a bulk power system after the loss of one or more components may be acceptable or unacceptable. It is unacceptable if load buses are left without supply, or the results of a load flow study show that lines are overloaded or bus voltages are out of bounds. With the possible exception of loss of load, these conditions are not allowed to persist: system operators will carry out remedial action to alleviate overloads and restore bus voltages; at the same time, they will attempt to minimize the loss of load.

There are several measures system operators will consider when remedial action is called for. They include:

- generation adjustment and redispatch

- transformer tap adjustment

- phase-shifter adjustment

- capacitor switching

- load transfer
- curtailment of interrutible load
- curtailment of firm load

Most state-of-the-art reliability programs undertake to model at least some of these measures. By simulating the actions of system operators, the index values derived become more realistic, since it is recognized that many of the unacceptable states are rendered acceptable again as a result of such actions.

Modeling remedial measures is not simple, and is never complete or without approximations. Most programs will select just a few of the above options, and model the operator action in terms of a constrained optimization problem. For example, the objective function may be to minimize generation redispatch and load shedding under the constraint that no overloads should appear in the load flow solution. Linear programming approaches are employed to obtain feasible solutions. The methodology is capable of accommodating pre-set priorities in the remedial action.

For more details, the reader should turn to Reference 24. In Reference 25, the features of several bulk power system reliability programs are compared, including their capabilities for remedial actions such as overload alleviation and generation rescheduling.

References

1. R. Billinton and S. Kumar, "Adequacy Evaluation of a Composite Power System - A Comparative Study of Existing Computer Programs," *CEA Transactions*, **24** (1985).
2. *Composite-System Reliability Evaluation: Phase 1 - Scoping Study*, EPRI Report El-5290, 1987.
3. *Power System Reliability Analysis - Application Guide*. Prepared by CIGRE Working Group 38.03, Paris, 1987.
4. J. Endrenyi, P.F. Albrecht, R. Billinton, G.E. Marks, N.D. Reppen, and L. Salvaderi, "Bulk Power System Reliability Assessment - Why and How? Part I: Why?," *IEEE Transactions* **PAS-101**, 3439-3445 (1982).
5. J. Endrenyi, P.F. Albrecht, R. Billinton, G.E. Marks, N.D. Reppen, and L. Salvaderi, "Bulk Power System Reliability Assessment - Why and How? Part II: How?," *IEEE Transactions* **PAS-101**, 3446-3456 (1982).
6. "Bulk Power System Reliability Concepts and Applications," Report of the Task Force on Bulk Power System Reliability, *IEEE Transactions* **PWRS-3**, 109-117 (1988).
7. IEEE Task Force Report, "Common Mode Forced Outages of Overhead Transmission Lines," *IEEE Transactions* **PAS-95**, 859-863, (1976).
8. J. Endrenyi, *Reliability Modeling in Electric Power Systems*, John Wiley and Sons, Chichester, England, 1978.
9. G.L. Anders, P.L. Dandeno, and E.G. Neudorf, "Computation of Frequency of Right-of-Way Losses Due to Tornadoes," *IEEE Transactions* **PAS-103**, 2375-2381 (1984).

10. J. Endrenyi, "Reliability Models for Circuit Breakers with Multiple Failure Modes," *IEEE Paper C 74-138-4, PES Winter Meeting, New York, 1974.

11. C. Singh and A.D. Patton, "Models and Concepts for Power System Reliability Evaluation Including Protection-system Failures," *Electrical Power and Energy Systems*, **2**, No. 4, 161-168 (1980).

12. R. Billinton and J. Tatla, "Composite Generation and Transmission System Adequacy Evaluation Including Protection System Failure Modes," *IEEE Transactions* PAS-102, 1823-1830 (1983).

13. R.N. Allan and A.N. Adraktas, "Terminal Effects and Protection System Failures in Composite System Reliability Evaluation," *IEEE Transactions* PAS-101, 4557-4562 (1982).

14. R. Billinton, P.K. Vohra, and S. Kumar, "Effect of Station Originated Outages in Composite System Adequacy Evaluation of the IEEE Reliability Test System," *IEEE Transactions* PAS-104, 2649-2656 (1985).

15. L. Wang, "The Effects of Scheduled Outages in Transmission System Reliability Evaluation," *IEEE Transactions* PAS-97, 2346-2353 (1978).

16. B. Stott, "Review of Load Flow Calculation Methods," *IEEE Proceedings* 62, 916-929 (1974).

17. A.F. Glimn and G.W. Stagg, "Automatic calculation of load flows," *AIEE Transactions* 76, 817-828 (1957).

18. W.F. Tinney and C.E. Hart, "Power Flow Solution by Newton's Method," *IEEE Transactions* PAS-86, 1499-1456 (1967).

19. W.F. Tinney and J.W. Walker, "Direct Solutions of Sparse Network Equations by Optimally Ordered Triangular Factorization," *IEEE Proceedings* 55, 1801-1809 (1967).

20. B. Stott and O. Alsac, 'Fast Decoupled Load Flow," *IEEE Transactions* PAS-93, 859-869 (1974).

21. L.R. Ford and D.R. Fulkerson, *Flows in Networks*, Princeton University Press, Princeton, N.J., 1962.

22. M.S. Sachdev and S.A. Ibrahim, "A Fast Approximate Technique for Outages Studies in Power System Planning and Operations," *IEEE Transactions* PAS-93, 1133-1142 (1974).

23. K. Behnam-Guilani, "Fast Decoupled Load Flow: The Hybrid Model," *IEEE Transactions on Power Systems* 3, 734-742 (1986).

24. *Reliability Evaluation for Large-Scale Bulk Transmission Systems*, EPRI Report EL-5291, 1988.

25. L. Salvaderi, R. Allan, R. Billinton, J. Endrenyi, D. McGillis, M. Lauby, P. Manning, and R. Ringlee, "State of the Art of Composite-System Reliability Evaluation," *CIGRE Proceedings*, Paper No. 38-104 (1990).

7. BULK POWER SYSTEM RELIABILITY ASSESSMENT METHODS

7.1 Introduction

In transmission system and bulk power system studies, two reliability assessment methods are predominantly used, one based on state enumeration and the other on Monte Carlo simulation. The general process of evaluation, applicable for either method, is illustrated in the flow chart of Figure 7.1. Note that this logic provides the details of the step "Analysis" in the diagram of Figure 2.1.

The steps of the process need very little explanation. Of particular interest are the first and the last steps, because that is where the two methods differ. State selection may occur by systematically considering one state after another (state enumeration), or by Monte Carlo simulation. Likewise, the computation of indices may be performed by using an analytical approach based on the solution of the Markov process involved, or by some counting algorithm directly associated with the Monte Carlo simulation used for state selection. Accordingly, there are three possibilities:

- state selection by state enumeration, indices computed by analytical methods;

- state selection by simulation, indices estimated through an associated counting process;

- a hybrid approach where state selection is by simulation, but indices are computed analytically.

In the following sections all of these approaches will be discussed. Methods used in state evaluation were already described in Sections 6.3 and 6.4.

7.2 State Enumeration

7.2.1 State Selection and the Computation of System Reliability Indices

In this approach, states are selected for evaluation (failure effects analysis) one by one, in some logical sequence, until (at least theoretically) the state space is exhausted. The most frequently used sequence is to examine first all the single contingencies (states in which only one component is failed), then all the double contingencies, and so on. In bulk power system studies, this may mean the removal of transmission lines one-by-one and the evaluation of generation system adequacy under each of these conditions, then similar evaluations while pairs of lines are removed, and so on.

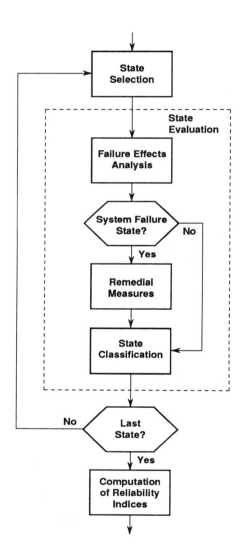

Figure 7.1 Flow chart for composite system reliability evaluation

Since, for any large system it is impossible to consider all the states in the state space, the procedure must be terminated before the computing effort becomes unreasonably large, even though accuracy is thereby reduced. Methods of doing this are discussed in Section 7.2.4.

The state evaluation process (see Figure 7.1) common to all approaches of state selection, is aimed at measuring each system state against the system failure criteria applicable in the given study, and labeling the states working or failed according to the outcomes of these comparisons. As mentioned in Section 2.3, system failure criteria are particular to the part of the system under study. State evaluation in bulk power system adequacy assessment involves load flow calculations for each state examined to determine whether all loads are supplied without exceeding permissible line loadings and within bus voltage tolerances. Methods of load flow analysis were discussed in Section 6.

Indices are computed from the solution of the Markov model based on the state space and transition rates representing the system. In Section 3, the long-term state indices, including the probability p_i of state i and its frequency, f_i, were derived. With the state-related indices computed, the steps of obtaining system reliability indices can be charted. Referring to Figure 7.1, first each state is classified as working or failed in the state evaluation process. Working states belong to the domain W, and failed states to the domain F. The probability of system failure, P_F, then simply becomes

$$P_F = \sum_{i \in F} p_i \tag{7.1}$$

The frequency of system failure, f_F, is given by the formula

$$f_F = \sum_{i \in F} (p_i \sum_{j \in W} \lambda_{ij}) \tag{7.2}$$

and the mean duration of system failure, T_F, by

$$T_F = P_F / f_F \tag{7.3}$$

Another important system index, defined in Section 3, is the expected value of the energy not supplied in a year (EENS). If in a failure state i the load curtailment is L_i (this information is a by-product of the failure effects analysis in the state evaluation process), then

$$EENS = \sum_{i \in F} L_i T_i f_i$$

with the result usually obtained in MWhr/yr. To facilitate comparisons between different power systems, this index is often normalized through being divided by L_s, the maximum system load in the year.

7.2.2 Techniques to Reduce Computing Effort

If all the states of a large system had to be considered in the state enumeration process, the computing effort even on a large mainframe computer could take years if not decades.   Several approaches have been developed, therefore, for terminating the enumeration before the computing burden becomes unreasonable while retaining acceptable accuracy in the results. In the following, three of these approaches will be discussed.

A. TRUNCATION OF STATE SPACE

The most obvious method of reducing the number of states to be examined is to omit those not likely to occur. In order to avoid the need for calculating all state probabilities for the purpose, the state space is truncated by disregarding states beyond a given number of multiple outages. The implicit assumption is that states representing r failures have much smaller probabilities than states with r-1 failures. While this is generally true, the comparison may not hold for the total number of r-fold versus the total number of (r-1)-fold contingencies [1].   The diagrams in Figure 7.2 serve as illustrations.

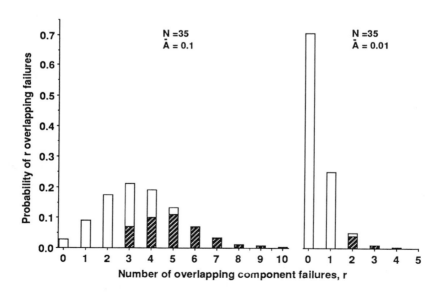

Figure 7.2 Failure probability distributions in a 35-component
system for different component unavailabilities

The first diagram shows the probability distribution of exactly r (r=0, 1,2,...) out of a total of 35 identical independent components being on outage if component unavailability is 0.1 (a rather poor value typical of thermal generating units). The probabilities follow the Binomial distribution and illustrate, for example, that the probability of all units being in service is only 2%. However, only part of the states for each r value represents system failure and this part is indicated by the shaded ordinates (in the diagram, no single and no double failure state is assumed to be system failure). Clearly, if truncation is carried out after the double failures, no system failure state is captured, and the false illusion is created that the system failure probability is about zero. To identify the majority of system failure states, states up to r=6 had to be considered and only the rest could be truncated. In generating unit studies, therefore, truncation of the state space is, while necessary, somewhat controversial.

The second diagram in Figure 7.2 illustrates the same distributions if component unavailability is 0.01 (typical of transmission lines). The proportion of failure states for each value of r is now different, indicating conditions more representative of transmission lines. In this case, truncation beyond r=2 will likely be satisfactory because evaluation of the first two levels of contingencies will in most cases provide a good estimate for the system failure probability. It is, in fact, a widely accepted practice.

B. RANKING FAILURE STATES BY SEVERITY

A very effective way of reducing the computational burden consists of ordering the system failure states by the severity of their impact on the system, defining a cut-off point in the sequence and neglecting all states beyond that point. This method has been implemented in several bulk power system reliability programs [2-4].

Central to this approach is the application of a measure of system stress. A performance index, PI, is used for the purpose; for a given system state it is defined as

$$PI = \sum_i w_i \left(\frac{L_i}{L_{oi}}\right)^n$$

where w_i is a weighting factor for line i, L_i is the real power flow in the line, L_{oi} its rated power, and n is an even integer. Similar performance indices have been defined for generating unit outages and bus voltage violations.

Ranking is based on finding the gradient (maximal change) of PI. If line k is removed, the change in PI is

$$\Delta PI = \frac{\partial PI}{\partial B_k} \Delta B_k + \cdots$$

where B_k is the susceptance of line k. If higher-order terms are neglected, the computation of ΔPI is very fast. On the other hand, the omission of higher-order terms may lead to misranking, that is, to errors in the ranking order.

Another undesirable effect is masking which occurs when several lines are underloaded counterbalancing the effect of overloaded lines. As a result, the PI value may be lower than in other states where no overload occurs but most lines are loaded to near capacity. The danger of masking is reduced if a higher value is chosen for the exponent n; however, this leads to mathematical difficulties in determining ΔPI. To avoid complications, the value of n=2 is used in practical cases.

C. UPPER AND LOWER BOUNDS

If for a reliability index upper and lower bounds can be defined such that the interval in-between reduces as more and more states are evaluated, the state enumeration process can be substantially accelerated. Instead of having to evaluate all states, the process is terminated when the bounds are close enough to provide an acceptably accurate estimate of the index.

Consider, for example, the probability of system failure, P_F, as the index in question. Initially, the upper and lower bounds are 1 and 0, respectively. If in the evaluation process state i is identified as a success state, the upper limit is reduced by p_i (it adds to the system success probability and thus lowers the limit of system failure probability) while there is no change in

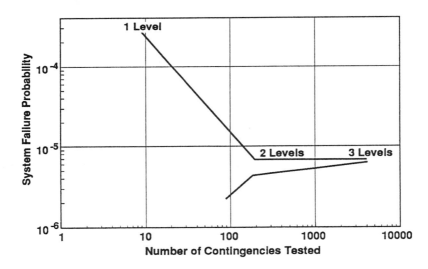

Figure 7.3 Probability bounds for the IEEE 24-bus test system
Copyright © 1982. Electric Power Research Institute, EPRI EL-2526, *Transmission System Reliability Methods*. Reproduced with permission.

the lower limit; conversely, if state j is identified as system failure state, the lower limit is raised by p_j while the upper limit remains unaffected. The two bounds meet when the state space is exhausted, identifying the exact value of P_F. However, it is much sooner that they get close enough for obtaining a reasonable estimate of P_F.

An example is provided in Figure 7.3 [3] where the system failure probability of the 24-bus IEEE test system [5] is the index to be estimated. It appears that after testing two levels of contingencies (127 tests) the bounds are so close that a good estimate of P_F can be obtained. Considering that most three-level contingencies are expected to represent system failure, this estimate would be nearer to the upper bound, so it could be 6×10^{-6}. One has also to remember that most reliability indices need not be worked out to three or four-digit accuracy; in many applications 10 or 20 percent difference in an index value will make very little difference.

The method of upper and lower bounds can also be used for estimating the accuracy of results when other approaches are used for terminating the state enumeration process.

7.3 Monte Carlo Simulation

In a simulation approach, systematic state enumeration is replaced by computer-generated sequences of system states. In the state selection step of Figure 3.2, only those states which are generated in the simulation process are considered. A simulated sequence represents a sample of realizations. The reliability of the system is assessed after accumulating a sufficient number of realizations. The events to be simulated in power system reliability studies are usually considered random, and hence the simulation process is called *Monte Carlo simulation*. The essential elements of Monte Carlo simulation for power system reliability assessment applications are described in this section.

7.3.1 Generation of Random Numbers

The Monte Carlo simulation process requires the random selection of numbers from specified distributions. The key to this selection is the generation of uniformly distributed random numbers. This task was a difficult one in the early days. One of the most well-known methods for generating random numbers was the use of an electronic roulette wheel constructed solely for this purpose [6]. Nowadays the task is performed by algorithms implemented on digital computers. These algorithms employ deterministic rules, and therefore the sequences obtained, if the process is initialized the same way, are identical. Numbers so generated are called *pseudorandom* numbers. One of the most frequently used algorithms is the linear congruential method, which is described by the following two steps:

1. Generate a series of numbers S_i, i=1,2,3, ... , using the equation

$$S_{i+1} = (a\, S_i + b)(\mathrm{mod}\ p) \qquad (7.4)$$

where a,b, and p are suitably chosen positive integers, with a and b in the interval $[0,p)$. Equation (7.4) states that S_{i+1} is the remainder of $(a \ S_i + b)/p$. The number S_1 is call the seed, which must be provided by the user at the beginning of the process.

2. Compute $R_i = S_i/p$. The series of numbers R_i, i=1,2,3, ... , are uniformly distributed in $[0,1)$.

Once the uniformly distributed random numbers are generated, the corresponding random numbers from another given distribution can be obtained by a suitable transformation [1].

7.3.2 Basic Simulation Processes

A power supply system consists of a large number of components, each of which may go through various changes of states, such as failures, repairs, start-ups, reserve shutdowns, and scheduled maintenance outages. The time distributions of these changes may or may not be governed by the exponential law. The operation of the power system can be replicated on a digital computer by simulating the changes of component states through random sampling. The basic simulation processes can be explained by way of the following simple example.

Consider an electric power generation system consisting of 4 identical 25 MW units, each of which can be represented by a two-state (operating and failed) Markov model with known unavailabilities and exponentially distributed up times and downtimes. The generation system is operated to supply a constant load of 75 MW. It is desired to estimate the probability of failing to supply the load by Monte Carlo simulation. This can be accomplished by using one of two different approaches, the snapshot (sometimes called random or non-sequential) or the sequential sampling.

Snapshot sampling - Each realization of this simulation process consists of making independent random draws of four numbers, one for each generator, from a uniform distribution between 0 and 1. A given generating unit is operating if its corresponding randomly drawn number is larger than the unavailability. Otherwise the unit is failed. The realization, a snapshot of the system condition in a randomly chosen moment, represents a failure to supply the load if two or more generating units are failed. Repeating the process N times, let F failures to supply the load be obtained. An estimate of the probability of failing to supply the load is given by F/N. The accuracy of the estimate improves as the total number of realizations N is increased (see Section 7.3.3).

Sequential sampling - In this approach the computer is used to generate simulated operating histories of the four generating units. At the beginning of the simulation process, the initial conditions (or states) of the generating units are established by a snapshot sample. The simulated operating history of each of the 4 units is then created by randomly drawing the uptimes and downtimes from their respective distributions alternatively, in a chronological order. The simulated operating histories of the 4 generating units are then examined to obtain time intervals in which 2 or more generating units have failed. Let T be the length of the simulated operating history (the same for all 4 units), and T_F the sum of all the time intervals of failing to supply the load. An estimate of the probability of failing to supply the load is given by T_F/T. The accuracy of the estimate improves as the length of the simulated operation period T is increased.

The choice of the simulation approach to use depends on the system to be simulated and the desired results. The snapshot simulation approach usually gives estimates of probabilities only, unless it is combined with analytical techniques as described in Section 7.4. In general, there is no restriction on the type of time distributions governing the state transitions. On the other hand, it is often implicitly assumed that system components are independent of each other. There are two exceptions: if the procedure for interpreting the random draws from the uniform distribution between 0 and 1 is based on historical data which incorporate the dependent condition, and if it is based on the prior solution of a Markov model which, again, incorporates the dependent relationship. In the latter case, however, exponential distributions must often be assumed for all the state transition times. Note that in the above example of 4 units, the failure and repair histories of the units are assumed independent.

Since the sequential simulation approach creates artificial operating histories of the simulated system, all aspects and parameters of the operation of the system can be recorded during the simulation process. All of the commonly used reliability indices can be derived from this information, there are no restrictions on the time distributions, neither must component independence be assumed. Therefore, the approach is particularly useful in simulating situations where the parameters of interest are correlated. A typical example is the simulation of power and energy generation from several hydraulic generating stations along a river system, with a large reservoir located upstream and other reservoirs of limited storage capacities located downstream. Water released for power generation from the upstream reservoir will reach the downstream generating stations with appropriate time delays; some of it can be stored in the downstream reservoirs, the rest is used for power generation at the downstream stations. The amount of power generated in each station and the storage level in the reservoirs are all correlated. While under certain assumptions snapshot sampling can still be used to determine the generating capacity and available energy at any point in time, in most cases this problem is better handled by sequential simulation.

In general, sequential simulation is more flexible than snapshot sampling, but it is also more complex and usually requires longer computing time. Each technique has its areas of application; both have been used widely in power system reliability studies.

The simulation process can be terminated after a specified number of samples has been obtained, or after the specified level of accuracy has been achieved. In the case of sequential simulation, the process can also be terminated after some observed parameters become stable, or their incremental changes per unit simulated operation time are less than the specified values. The relationship between sample size and accuracy will be discussed in the next Section.

7.3.3. Sample Size and Accuracy

In applying the Monte Carlo simulation technique, it is often required to determine the accuracy of the simulation results, or alternatively, the number of realizations needed to achieve a specified level of accuracy. Since Monte Carlo simulation is a sampling process, the accuracy of the simulation results depends on the number of simulation realizations, or sample size. This can be demonstrated by way of the following example.

Assume that the loss of load probability, P_L, is to be estimated by simulation. The simulation process is terminated after N realizations. The result of the i-th realization, P_{Li}, is a function of the system state X, a random variable (in the following, random variables are denoted by bold-face symbols). P_{Li} is 0 if there is no loss of load, and it is 1 if loss of load does occur. The P_{Li} values constitute another random variable, denoted by Y.

An estimate of the loss of load probability is given by the sample mean which is

$$\hat{\mathbf{P}}_L = \frac{1}{N} \sum_i P_{Li}$$

In another simulation run, one would likely obtain a different estimate for P_L. Therefore, the estimate $\hat{\mathbf{P}}_L$ is also a random variable. Its mean is given by

$$E[\hat{\mathbf{P}}_L] = E[Y] = \lim_{N \to \infty} \frac{1}{N} \sum_i P_{Li}$$

which is the exact value of P_L. This value can be estimated but not calculated because its calculation would require the generation of an infinite number of realizations. A measure of the inaccuracy of an estimate of P_L from a finite sample size is given by the variance of $\hat{\mathbf{P}}_L$, $\text{Var}[\hat{\mathbf{P}}_L]$, which is computed from

$$\text{Var}[\hat{\mathbf{P}}_L] = \frac{1}{N} \text{Var}[Y]$$

where $\text{Var}[Y]$ is the sample variance. Clearly, as $N \to \infty$, the inaccuracy tends to zero.

In applications, considerable savings in computing time and cost can be achieved by reducing the sample size required for a specified level of accuracy. This can be accomplished by reducing the sample variance. Special techniques have been developed for the purpose, some of which are described in the following Section.

7.3.4. Variance Reduction Techniques

A. IMPORTANCE SAMPLING

The basic idea of this technique is to distort the probability distribution of the sampling space so as to reduce the sample variance without altering the the desired sample mean [7]. This can be illustrated by the following derivation: Let $Y=f(X)$, where X is a discrete random variable describing the system states and Y is, as in the example above, a random variable indicating the results of a reliability "test" for each state. The expected value of Y is given by

$$E[Y] = \sum_i f(x_i)\, P(x_i)$$

where x_i is the i-th outcome of a Monte Carlo simulation process for estimating the expected value of Y, $P(x_i)$ is the probability of x_i, and $f(x_i)$ the result of an evaluation of the i-th outcome. The sample variance is given by

$$\text{Var}[Y] = \sum_i \{f(x_i) - E[Y]\}^2\, P(x_i)$$

$$= \sum_i f^2(x_i)\, P(x_i) - (E[Y])^2 \tag{7.5}$$

Define $Z = g(X)$ such that $g(x_i) = f(x_i)\, P(x_i)/P^*(x_i)$, where $P^*(x_i)$ is a suitably chosen probability distribution. The ratio $P(x_i)/P^*(x_i)$ actually distorts the original probability distribution by emphasizing   important events and de-emphasizing unimportant ones. It is easy to see that

$$E[Z] = \sum_i g(x_i)\, P^*(x_i) = \sum_i f(x_i)\, P(x_i)$$

$$= E[Y]$$

$$\text{Var}[Z] = \sum_i \{g(x_i) - E\{Y\}\}^2\, P^*(x_i)$$

$$= \sum_i \{f^2(x_i)\, P(x_i)/P^*(x_i)\}\, P(x_i) - (E[Y])^2 \tag{7.6}$$

By comparing Equations (7.5) and (7.6), it is easily seen that the estimated sample variance can be reduced by making proper choice of the ratio of $P(x_i)/P^*(x_i)$.

B. CONTROL VARIABLES

This method makes use of a known random variable which is correlated with the one to be estimated [8]. The concept of the technique is illustrated in the following derivation. Let $Y=f(X)$ as before. The expected value of Y is to be estimated by Monte Carlo simulation. Instead of doing so directly, define

$$Z = Y - W + E[W]$$

where $W = g(X)$ is called a control variable. It is easy to see that $E[Z] = E[Y]$. The variance of Z is given by

$$\text{Var}(Z) = \text{Var}(Y) + \text{Var}(W) - 2\text{Cov}(Y,W)$$

Variance reduction can be achieved by choosing a control variable W which is strongly correlated with Y such that $2\text{Cov}(Y,W) > \text{Var}(W)$. Such a random variable can often be identified from fast approximate studies using state enumeration [9].

C. ANTITHETIC VARIABLES

Let $Y = f(X)$ as above. The expected value of Y is denoted by $E[Y]$. Let Y' and Y'' be two unbiased estimators of $E[Y]$. Obviously $Y_A = (Y' + Y'')/2$ is also an unbiased estimator of $E[Y]$. The variance of Y_A is given by

$$\text{Var}(Y_A) = [\text{Var}(Y') + \text{Var}(Y'') + 2\text{Cov}(Y',Y'')]/4$$

If Y' and Y'' are uncorrelated, $\text{Cov}(Y',Y'') = 0$ and

$$\text{Var}(Y_A) = [\text{Var}(Y') + \text{Var}(Y'')]/4$$

However, if Y' and Y'' are negatively correlated, $\text{Cov}(Y',Y'') < 0$ and

$$\text{Var}(Y_A) < [\text{Var}(Y') + \text{Var}(Y'')]/4$$

A straightforward method of obtaining negatively correlated estimators Y' and Y'' is as follows [10]. Let $(u_1, u_2, u_3, ..., u_n)$ be the series of random numbers generated from a uniform distribution in the simulation process for obtaining Y'. If $(1-u_1, 1-u_2, 1-u_3, ..., 1-u_n)$ are used to obtain Y'', then Y' and Y'' will be negatively correlated.

The choice of the variance reduction technique to use and its effectiveness may depend on the application. In certain combinations, these techniques can reduce the sample size needed to achieve the same accuracy by one or even two orders of magnitude [11, 12].

7.4 Hybrid Approaches

Monte Carlo simulation with random sampling has the advantage of simplicity and speed. Among its disadvantages appears the limitation that if indices are computed through counting and averaging, only the state probabilities and the expected energy not served can be estimated, but no frequencies or mean durations of any kind. Admittedly, in many applications, the information so available is all that is needed.

It is not difficult, however, to compute frequency and duration indices if simulation and Markov techniques are combined. In one approach [13], state selection is through simulation with random sampling of component states and load as before, but while state probabilities are still estimated by counting, state frequencies and durations are determined analytically, using Equations (3.3) and (3.4). The system failure probability, frequency and duration are then computed from Equations (7.1), (7.2), and (7.3), respectively. It should be noted that frequencies are properly computed only if a chronological load model is used [13, 14].

Elegant as it is, there is a price to pay for using this approach. Since the equations employed are part of the Markov solution, the implicit assumption that all transition durations are governed by exponential distributions must be accepted. This imposes a constraint which, in general, is not postulated when using "pure" simulation. As a result, one of the attractive features of simulation, namely, that times to transitions between states can have any distribution is no longer valid.

Other hybrid approaches are based on combinations of simulation and enumeration. In one, it is proposed to apply enumeration to the transmission line outage states, and to associate these states with generating system outage events obtained through simulation. In this manner, both processes are used to their best advantage, as explained below in Section 7.5. In another approach, the system failure states are generated by simulation, but the load states are enumerated; the system indices obtained for each load level are then appropriately combined. This method is used in the program MECORE recently developed at the University of Saskatchewan.

To reduce computing effort in the above approaches, the techniques described in Sections 7.2.2 and 7.3.4 can still be applied.

7.5 Comparisons

None of the two approaches discussed, state enumeration and simulation, is superior to the other; both have applications where they are preferable. State enumeration is best applied when comparatively rare but high-impact component failures are of prime importance; at the same time, the number of components is not excessive. Such is the case with transmission line failures. Monte Carlo simulation is more suited to evaluating systems with a large number of less reliable components where the effects of higher-level multiple failures may not be negligible, as in the case of thermal generating units. With variance reduction techniques applied, it can no longer be said that Monte Carlo simulation is not a viable alternative in bulk power system reliability studies.

As a consequence, some programs which have been developed for bulk power system adequacy assessment are based on state enumeration and others on Monte Carlo simulation. Table 7-1 provides a listing of the better-known programs and indicates some of their features. All use load flow computations for state evaluation, and many, but not all, employ AC load flow. Most incorporate remedial action, although the models for it differ in sophistication. All Monte Carlo simulation methods use snapshot sampling. As far as computing effort is concerned, the programs exhibit significant differences even if the systems evaluated are similar in size and characteristics. Further details are available in references [15-17].

Table 7-1
Bulk Power System Reliability Programs

E - Enumeration, MC - Monte Carlo Simulation

Program Name	Origin	State Selection	Remarks	References
PCAP	PTI, USA	E		15,19
GATOR	Florida Power, USA	E	Large systems, low coincident outage levels	15,17
SYREL	EPRI, USA	E	Transmission primarily	15,17
RECS	Georgia Power, USA	E		15
TPLAN	PTI, USA	E	Transmission primarily	17
COMREL	Univ. of Saskatchewan, Canada	E		15,17
PROCOSE	Ontario Hydro, Canada	E	Transmission-constrained generation program	17,20
FIAPT	Hydro Quebec, Canada	E		15
RELACS	UMIST, UK	E		17
MEFISTO/ TRANQUEL	Belgium	E		21
ZUBER	TH Darmstadt, Germany	E	Transmission only	15
SICRET	ENEL, Italy	MC		15, 17, 22
MEXICO	EDF, France	MC		15,23
CONFTRA	CEPEL, Brazil	MC		15, 17
CREAM	EPRI, USA	MC		12
ZANZIBAR	EDP, Portugal	MC		
ESCORT	CEGB, UK	MC		

The features of an "all-inclusive" program are described in [18]. For a given application, however, models and program features should be selected according to the needs of that application. It is unlikely that any one program could serve all needs.

Reference 18 also identifies three points to be observed when developing a practical bulk power system reliability program. First, a consistent level of accuracy must be employed in all modules of the computations; next, modeling and computational accuracy must be matched; and finally, the level of accuracy must be balanced against the computational effort required. With continuing refinements in the solution methods, the computational burden for the same program features will undoubtedly decrease. This will allow for the introduction of more and more sophisticated models with greatly improved accuracy. As a result, users will apply probabilistic methods with increasing confidence.

References

1. J. Endrenyi, *Reliability Modeling in Electric Power Systems*, John Wiley & Sons, Chichester, England, 1978.

2. T.A. Mikolinnas and B.F. Wollenberg, "An Advanced Selection Algorithm," *IEEE Transactions* **PAS-100**, 608-617 (1981).

3. *Transmission System Reliability Methods*, EPRI Report EL-2526, 1982.

4. *Reliability Evaluation for Large-Scale Bulk Transmission Systems*, EPRI Report EL-5291, 1988.

5. "IEEE Reliability Test System," IEEE Committee Report, *IEEE Transactions* **PAS-98**, 2047-2054 (1979).

6. W.G. Brown, "History of RAND's Random Digits - Summary," in *Monte Carlo Method*, (A.S. Housholder ed.), National Bureau of Standards, Washington D.C., 1951.

7. M. Mazumdar, "Importance Sampling in Reliability Estimation," in *Reliability and Fault Tree Analysis - Theoretical and Applied Aspects of System Reliability and Safety Assessment*, (Edited by R.E. Barlow, J.B. Fussell and N.D. Singpurwalla), pp.153-163, SIAM, Philadelphia, 1975.

8. A.H. Ang and W.H. Tang, *Probability Concepts in Engineering Planning and Design, Vol. II - Decision, Risk and Reliability*, John Wiley & Sons, New York, 1984.

9. G.C. Oliveira, M.V.F. Pereira, and S.H.F. Cunha, "A Technique for Reducing Computational Effort in Monte Carlo Based Composite Reliability Evaluation," *IEEE Transactions* **PWRS-4**, No. 4, 1309-1315 (1989).

10. J.M. Hammersley and D.C. Handscomb, *Monte Carlo Method*, John Wiley and Sons, New York, 1964.

11. G.J. Anders, J. Endrenyi, M.V.F. Pereira, L.M.V. Pinto, G.C. Oliveira, and S.H.F. Cunha, "Fast Monte Carlo Simulation Techniques for Power System Reliability Studies," *Proceedings of CIGRE*, Paper No. 38-205 (1990).

12. *Development of a Composite System Reliability Evaluation Program*,
    EPRI Report EL-6926, 1990.
13. A.C.G. Melo, M.V.F. Pereira, and A.M. Leite da Silva, "Frequency and
    Duration Calculations in Composite Generation and Transmission
    Reliability Evaluation," submitted to the IEEE for presentation at the
    1991 Summer Power Meeting.
14. *Modeling of Unit Operating Considerations in Generating Capacity
    Reliability Evaluation*, EPRI Report EL-2519, 1982.
15. R. Billinton and S. Kumar, "Adequacy Evaluation of a Composite Power
    System - A Comparative Study of Existing Computer Programs," *CEA
    Transactions*, **24** (1985).
16. *Power System Reliability Analysis - Application Guide*. Prepared by
    CIGRE Working Group 38.03, Paris, 1987.
17. L. Salvaderi, R. Allan, R. Billinton, J. Endrenyi, D. McGillis, M.
    Lauby, P. Manning, and R. Ringlee, "State of the Art of Composite-
    System Reliability Evaluation," *CIGRE Proceedings*, Paper No. 38-104
    (1990).
18. "Bulk Power System Reliability Concepts and Applications," Report of
    the Task Force on Bulk Power System Reliability, *IEEE Transactions*
    PWRS-3, 109-117 (1988).
19. P.L. Dandeno, G.E. Jorgensen, W.R. Puntel, and R.J. Ringlee, "A
    Program for Composite Bulk Power Adequacy Assessment," *IEE Con-
    ference Publications No. 148:Reliability of Power Supply Systems*, 97-
    100, 1977.
20. B. Porretta and D.L. Kiguel, "Bulk Power System Reliability Evaluation,
    Part I: PROCOSE - A Computer Program for Probabilistic Composite
    System Evaluation," *Proceedings of the 14th Inter-RAM Conference for
    the Electric Power Industry*, Toronto, 1987.
21. H. Baleriaux, D. Brancart, and J. VanKelecom, "An Original Method
    for Computing Shortfall in Power System (Mefisto Method)," *CIGRE
    Proceedings* Paper No. 32.09 (1974).
22. P.L. Noferi, L. Paris, and L. Salvaderi, "Montecarlo Methods for Power
    System Reliability Evaluations in Transmission and Generation
    Planning," *Proceedings of Reliability and Maintainability Symposium*,
    Washington, 1975.
23. J. Auge, J. Barbey, J. Bergougnoux, J. de Calbiac, J. Delpech, and J.
    Pouget, "Risk of Failure in Power Supply and Planning of EHV
    Systems," *CIGRE Proceedings* Paper No. 32.21 (1972).

8. OPERATIONAL RELIABILITY

8.1 Time Horizons in Operational Reliability Studies

Operational reliability assessments deal with the prediction of system performance up to a specified lead time T. This time also indicates the study horizon. As already noted, the goals and methods of operational reliability investigations depend on the lead time. The time horizon of these studies can be divided into three categories:

- *Near-term time frame.* It extends from the present (t=0) to a few hours into the future. The risks at t=T are evaluated given the state of the system at t=0, and the results assist the system operator to decide what actions to take.

- *Short-term time frame.* It extends up to a year into the future. Studies in this category are usually aimed at setting operating policies.

- *Mid-term time frame.* It may cover a period of several years and, as such, its methods overlap with those in the planning function. These studies evaluate the effects on operations of long-term sales, purchases, fuel contracts, mothballing, recommissioning, and so on.

Note that the designations *near-term*, *short-term*, and *mid-term* are not universally adopted.

Although probabilistic reliability techniques have been routinely applied in the long term planning of some parts of the power system, their applications to power system operations have yet to gain wide acceptance. The majority of power systems are still operated by deterministic rules, which have proven to give satisfactory results. The reluctance in adopting the probabilistic approach can be attributed to the lack of operating experience and the fact that any change in operating policy may have enormous impact on customer supply reliability. However, more and more utilities are considering the feasibility of adopting probabilistic operating criteria and are looking into practical ways of testing the new approach without risking operational reliability.

8.2 Near-Term Power System Reliability

The purposes of assessing system reliability in this time frame are to maintain operating risks within specified bounds, to minimize the likelihood of high risk operating states, and to avoid or forestall potential emergency situations. The modeling considerations and solution techniques used here are different from those in the previous Sections where the systems are assumed to be in the steady state. The length of the lead time is determined by the time required to place standby components into service following component failures, and by the time required to make orderly transitions of system operating configurations.

A power system operating in a satisfactory state at present may encounter difficulties in the future if one of the following conditions occurs: (1) loss of generating capacity which cannot be replaced with the available spinning reserve; (2) losses of generation, transmission, or load causing system instability; and (3) loss of generation or transmission components, leaving the system in a highly vulnerable state. The probabilities of occurrence of these events change with the lead time. This can be illustrated by way of an example where the probability of failure, $P_F(t)$, of a component which is operating at t=0, increases with t from $P_F(0)=0$ to $P_F(\infty)=1$. Repair is neglected, which is a fair assumption if t is not too large. The probability of system risk is also a function of time, and in general it increases with the lead time.

The time function of the near-term reliability of power systems has been called the security function in the literature. It should be noted that the term security is used here somewhat differently from the way it is defined in Section 1 and used throughout the other Sections. This is unfortunate, but security seems to be a useful term and has yet other definitions in other applications. The conceptual formulation of the security function is as follows [1]:

$$S(t) = \sum_i [p_i(t)q_i(t) + \sum_{j \neq i} p_j(t-\Delta t)p_{ji}(t-\Delta t,t)q_{ji}(t)] \qquad (8.1)$$

where i and j are indicators of system states,

   $p_i(t)$ is the probability of the system in state i at time t,

   $q_i(t)$ is the probability that state i constitutes a steady-state breach of security at time t,

   $p_{ji}(t-\Delta t,t)$ is the probability of transition from state j to state i in the time interval {t-Δt,t}, given the system is in state j at t-Δt,

   $q_{ji}(t)$ is the probability that transition from state j to state i during {t-Δt,t} results in transient breach of security.

A transient breach of security occurs when a transition of system state, usually caused by the failure or sudden outage of one or more system components, induces a serious disturbance, such as instability or excessive frequency excursion. A steady-state breach of security is an unsatisfactory system operating condition, such as generating capacity deficiency or component overload, regardless of the cause or duration of this condition. Note that by the terminology of the previous sections, a steady-state breach of security would be called a breach of adequacy.

Equation (8.1) has two sets of terms. The first set gives the contributions to system risk from steady-state breaches of security, and the second set represents the contributions from transient breaches of security.

It is clear that Equation (8.1) requires the use of the state enumeration approach. In practical applications it is not necessary nor is it possible to carry out an exhaustive enumeration of all breaches of system security. Methods and criteria for truncating the enumeration process must be developed. It seems logical that the criteria should be in terms of the severity and the probability of occurrence of breaching system security. Also, it appears possible to rank the potential breaches by either probability or severity in ways rather similar to those discussed in Section 7.2.2. System operators could make use of this information to alleviate the potential risk or to plan for control actions.

The computed risk index is a probability measure. The severity of breaches of security is not considered in the formulation. However, a weighting scheme can be easily incorporated in the formulation to take severity into account. When weighting is applied, the computed risk index is no longer a pure probability.

The formulation of the security function in Equation (8.1) is general and incorporates breaches of system security by events in the generation and transmission systems. The application of the concept may be limited to the generation system only, in which case the problem concerns the adequacy of operating reserve. This is the subject of discussion in Section 8.4.

8.3 Component Modeling for Near-Term Reliability

Near-term reliability is concerned with system risks in the next few hours into the future. Two types of components need to be considered in this time frame, those that are in operation at the time of reliability evaluation and those that are on standby. For an operating component the probability that it will become unavailable during the interval of interest is required. For a standby component one needs to know the probability of its being in service at time t=T, given that procedures to place the component in service started at t=0.

8.3.1 Modeling Operating Components

Consider a component whose operating-repair cycles can be represented by a two-state (up or operating, and down or failed) Markov model with failure rate λ and repair rate μ. The probability of finding this component in the failed state at t=T, given that it was operating at t=0, is

$$P_{down}(T) = \frac{\lambda}{\lambda + \mu} \left(1 - e^{-(\lambda+\mu)T}\right) \qquad (8.2)$$

If the lead time T is very short such that $(\lambda+\mu)T \ll 1$, and the probability of repairs is neglected in this short time interval, Equation (8.2) becomes

$$P_{down}(T) \approx \lambda T \qquad (8.3)$$

The product λT is called the outage replacement rate of the unit.

Certain power system components, such as generating units and HVDC transmission lines, can be operated in derated capacity states. The three-state model of Figure 4.2(a) can be used to represent such components. The approximate probabilities of finding a component in the derated or in the down state at t=T, given that it was in the up state at t=0, are, respectively,

$$P_{\text{derated}}(T) \approx \lambda_{12}T$$

$$P_{\text{down}}(T) \approx \lambda_{13}T$$

The above equations require the assumptions that $\lambda_{12}T \ll 1$ and $\lambda_{13}T \ll 1$.

8.3.2 Modeling Standby Components

In modeling a standby component for near-term reliability assessment, it is assumed that (1) the component parameters are exponentially distributed, (2) the component is ready for startup (i.e., it is not in the failed state), (3) the startup procedure begins at t=0. As before, repairs are not considered in this time frame. The model in Figure 8.1 can be used to represent the standby component in the near-term time frame. In this Figure, p_{SF} is the starting failure probability (a constant), $1/\lambda_s$ is the mean startup time, and $1/\lambda$ is the mean up time.

The system of differential equations of this model can be solved to obtain the state probabilities for t>0 . Namely,

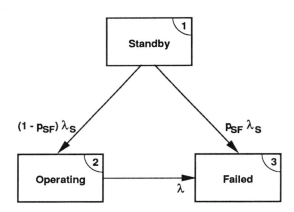

Figure 8.1 Near-term model of a standby component

$$p_1(t) = e^{-\lambda_s t}$$

$$p_2(t) = \frac{\lambda_s(1-p_{SF})}{\lambda_s-\lambda} (e^{-\lambda t} - e^{-\lambda_s t})$$

$$p_3(t) = 1 - p_1(t) - p_2(t)$$

Other special models, such as the one for thermal generating units in hot or cold reserve [2], may also be needed in near-term reliability assessment.

8.4 Operating Reserve of Generating Systems

Excess generating capacity is allocated to ensure operating reliability and to account for unforeseen events, such as the sudden outage of a generating unit or the unexpected increase in system load. It is impossible physically to startup and load a generating unit momentarily. Time is required to bring a generating unit from standstill to a ready state for taking up load. This time varies from a few minutes for hydraulic and combustion turbine units to several hours for thermal units in cold reserve. Therefore, part of the reserve is kept in a state where it is almost instantaneously available, in anticipation of a sudden need for more capacity. The extra generating capacity so allocated is called operating reserve. The most reliable way of operating the generating system is, of course, to keep all available units ready to be loaded. But this mode of operation is not feasible because of high operating costs.

Operating reserve may consist of two parts. The first part consists of units kept spinning and ready to take up load momentarily. It is called spinning reserve. The second part is made up of quick-start units which can be synchronized to the system in a few minutes. This part of the operating reserve is often called the 10-minute or 15-minute reserve. In some utilities, emergency assistance from interconnected systems, interruptible load, and other emergency control actions are also considered as part of the operating reserve.

Deterministic methods have been used in the utility industry in setting the size of operating reserves. The usual practice is to set the operating reserve not smaller than the capacity of the largest unit in operation. Using deterministic methods for setting operating reserve may result in operating risks which are either too high for customer supply reliability, or too low for operating economy.

Several probability methods for determining operating reserve have been suggested. These methods are briefly described in the following.

8.4.1 The PJM Method

The application of probability methods to the determination of spinning reserve was first proposed by a group associated with the Pennsylvania-New Jersey-Maryland (PJM) Interconnection in 1963 [3]. The proposed method was used to determine the amount of generating capacity that should be scheduled ready for operation so that the risk of capacity deficiency at a lead time T would not exceed the specified level. The basic concept of this method is similar to that of the LOLP method. The risk index is the probability of just carrying or failing to carry a given load at the lead time T, and is defined by

$$P_{CL}(T) = \sum_Y P[C(T) \leq Y] \, P[L(T) = Y] \tag{8.4}$$

where C(T) and L(T) are, respectively, the available capacity (in-service plus operating reserve) and system load at the lead time T. The method for computing the capacity outage probability table, as described in Section 4.2, can be applied here, too, to obtain P[C(T)≤Y], except that in this application the forced outages rates of generating units must be replaced by the outage replacement rates (see Equation 8.3). The term P[L(T)=Y] accounts for short-term load forecast uncertainty, which can be approximated by a normal distribution in discrete steps.

At the time when this method was conceived the generating capacity in the PJM system consisted mainly of fossil-fired units. The logical choice of T was the average startup time of such units. The computation of $P_{CL}(T)$ in Equation (8.4) is straightforward.

Since then, many power companies have installed combustion turbine units with much shorter startup times than fossil-fired units. Startup times of fossil units can be reduced by keeping them in various stages of hot-reserve status. Under these circumstances, the use of a single startup time for all reserve units is unsatisfactory. Other factors, such as the probabilities of failure to start of combustion turbine units, partial forced outages of large steam units, and non-exponential distributions of state residence times in the near-term time frame, can have significant effects on operating reserve requirements. Several modifications to the PJM method were introduced to incorporate these considerations [1,4,5]. The new operating reserve program developed for the PJM system was described in [4]. The main disadvantage of the PJM method is that the computed index gives an indication of system risk at the end of the lead time only; it does not account for the risks attributable to the states that the system may travel through during the interval.

8.4.2 The Security Function Method

As mentioned earlier, the security function method for near-term power system reliability described in Section 8.2 can be applied to determine the operating reserves of generating systems. This is accomplished by considering the breaches of system security in the steady-state only and limiting the contributions to those originating from the generating system. In this case the security function of Equation (8.1) becomes

$$S(t) = \sum_i p_i(t) q_i(t) \tag{8.5}$$

Here, $p_i(t)$ is the probability of the generating system in capacity state i at lead time t, and $q_i(t)$ is the probability that capacity in state i constitutes a breach of security at t. Note that in this application the breach of security must be defined in terms of the operating policy and the size and range of the operating reserve margin. In the simplest case where the breach is defined as having zero or negative operating reserve margin, the method becomes identical to the PJM method.

The generating system capacity state probabilities can be obtained in two steps. First, the appropriate models of generating units described in Section 8.3 are solved for the state probabilities of individual generating units at a lead time t. In the second step the system capacity state probabilities are obtained by using the algorithm for computing the capacity outage probability table. The probabilities of breaches of system security must be evaluated according to the chosen operating policy.

An application of the security function method can be found in [6].

8.5 Short- and Mid-Term Reliability of Generating Systems

The generating capacity of a large power system may consist of hydraulic, fossil-fired, and nuclear units. These units have diversified operating constraints, operating costs, and maintenance requirements. The effective and economic utilization of the available generating resources require careful planning well in advance of actual operations. Operational planning in the short- and mid-term involves the scheduling of generating capacity to meet forecast load demand and, at the same time, observing the requirements for operating reserve, load forecast uncertainty, availability of hydraulic generation, planned outages, and contracted sales and purchases. A reliability assessment method for operational planning of hydro-thermal generating systems is described in the following.

8.5.1 Reliability Criteria

Operational planning has been carried out using deterministic criteria, such as defining operating reserve requirements in terms of the sizes of the first few largest units. This approach has resulted in high operating reliability, but experience seems to suggest that the criteria may be more conservative than necessary. Criteria in terms of the LOLP index has also been used in capacity scheduling. A major limitation of this approach is that it cannot adequately consider the energy constraints of peaking hydraulic generation.

It has been suggested that the reliability criteria for operational planning be defined in terms of the expected unsupplied energy [7, 8]. The advantages of using this risk index are:

- it can be normalized to system minutes, as discussed earlier, for easy comparisons among systems of different sizes,

- it can be used in both capacity and energy reliability studies,

- it is also used in long-term planning, which provides consistency and continuity in the planning philosophy.

8.5.2 A Method for Short- and Mid-Term Reliability Assessments

A probability method was proposed for computing the expected unsupplied energy of a hydro-thermal generation system over a specified time interval [7]. Hydraulic energy limitations are included in the computations. The following assumptions are made:

- Generating units are in the steady state in the short- and mid-term time frame, and the unit models described in Section 4 can also be used in this application. Note that the peaking unit model of Figure 4.3 to be used here is an extension of the near-term model in Figure 8.1.

- The uncertainties of hydraulic generation due to forced outages and the availability of water are combined into a single probability distribution.

- The load forecast uncertainty is normally distributed.

- Planned outages are known and are not included in the reliability computations.

The formulation of the method is as follows.

Let G_i = Total available capacity from thermal and nuclear units for hour i,

X_i = thermal and nuclear capacity on forced outage in hour i,

$P(x)$ = probability that $X_i = x$,

Y_i = forecast load in hour i,

$P(y)$ = probability that $Y_i = y$,

Z_i = available hydraulic capacity in hour i,

$P(z)$ = probability that $Z_i = z$,

S_i = contract sales in hour i,

C_i = firm capacity purchase in hour i,

R = interruptible load.

Define $Q_i(y,z)$ as the firm load to be carried by thermal and nuclear units, also called the equivalent thermal system load. Then

$$Q_i(y,z) = y - z + S_i - C_i - R$$

The expected unsupplied energy in hour i is given by

$$EENS_i = \sum_z \sum_y \sum_x P(x) \{x - [G_i - Q_i(y,z)]\} \, P(y) \, P(z)$$

where the summation over x is limited to the region of $X_i \geq [G_i - Q_i(y_i, z_i)]$. Therefore, the expected unsupplied energy in a given study interval is

$$EENS = \sum_i EENS_i$$

Note that the probabilities $P(x)$ are entries in the capacity outage probability table discussed in Section 5. In the case where the generating capacity in a plant is subjected to possible bottling by transmission outages, the probabilities of transmission outages can be combined with those of the generating units in the plant.

The above method of computing the expected unsupplied energy may require long computing times for extended study intervals. A fast approximate computing method was described in [7].

8.5.3 Applications

The proposed method can be applied to assess the reliability of the operating schedules of hydro-thermal generation systems. In addition, it can also be used to assess opportunities for economic sales, purchase requirements, operating reserve planning studies, and maintenance scheduling. Examples of these applications can be found in [9].

Note that operating considerations in long-term planning were discussed in Section 5.3.3.

8.6 Reliability-Constrained Unit Commitment

The consumption of electric energy varies with time: it is higher during the day and early evening, lower at night; higher on weekdays, lower on weekends and holidays. As system load varies, generating units are put into or taken out of service. The shutdown of generating units during low load periods is carried out to reduce operating costs. Unit commitment is a process which determines the schedule for starting up and shutting down of thermal generating units to satisfy the forecast load demand plus spinning reserve requirements in a specified period. In addition, the commitment schedule must recognize such constraints as must-run units, minimum running times, minimum shutdown times, fuel restrictions, and the possible unavailability of crew to start up two or more units simultaneously in the same plant. The most commonly used techniques for unit commitment are the priority list method, the dynamic programming method, and the mixed integer programming method. Spinning reserve requirements are usually given in deterministic terms. An overview of the subject can be found in [10].

Probabilistic methods for assessing operating reliability of power systems have been developed in recent years. The resulting operating reliability requirements can be applied to unit commitment in place of the deterministic spinning reserve requirements [11-13]. The method proposed in [12] is briefly described here as an example. In this method, the unit commitment schedule is obtained in three steps. First, a generation schedule is prepared based on unit running costs using dynamic programming. Next, security calculations are carried out using Equation (8.5) to check the level of security provided by this schedule for the forecast load in the scheduling period. The schedule is then modified to achieve the desired level of operating security. Since the resulting schedule may contain startups of generating units which were shutdown earlier, it may be more economical to keep these units operating rather than shutting them down for relatively short durations, avoiding thereby added startup costs. This is examined in the last step of the process, and the schedule is modified accordingly. It is obvious that any change in the commitment schedule introduced in this final step will result in more running capacity, and, therefore, higher operating reliability. Note that this example does not include the scheduling of hydraulic generating units.

8.7 Summary

Although methods have been developed for operating reliability assessments, there appears to be very little information in the literature on risk criteria for operating reliability. This situation can improve only with increased application.

More and more power systems are operated with the assistance of modern computer systems to carry out data acquisition, processing and information display functions. Various software packages have been developed for applications in energy management to achieve operating economy. With the computing facilities needed for assessing operating reliability (or risk) becoming already available at system control centers, one of the important tasks for the operators is to become familiar with the new concept and to gain some operating experience in trial applications.

References

1. A.D. Patton, "A Probability Method for Bulk Power System Security Assessment, I - Basic Concepts," *IEEE Transactions* **PAS-91**, 54-61 (1972).
2. R. Billinton and A.V. Jain, "The Effect of Rapid-start and Hot Reserve Units in Spinning Reserve Studies," *IEEE Transactions* **PAS-91**, 511-516 (1972).
3. L.T. Anstine, R.E. Burke, J.E. Casey, R. Holgate, R.S. John, and H.G. Stewart, "Application of Probability Methods to the Determination of Spinning Reserve Requirements for the Pennsylvania-New Jersey-Maryland Interconnection," *IEEE Transactions* **PAS-82**, 726-735 (1963).
4. L.G. Leffler, R.J. Chambliss, G.A. Cucchi, N.D. Reppen, and R.J. Ringlee, "Operating Reserve and Generation Risk Analysis for the PJM Interconnection," *IEEE Transactions* **PAS-94**, 396-403 (1975).
5. R. Billinton and M. Alam, "Outage Postponability Effects in Operating Capacity Reliability Studies," IEEE PES 1978 Winter Meeting, New York, Paper A78 064-8, 1978.
6. J. Bubenko, E. Paulsson, D. Sjelvgren, and M. Anderson, "Security Assessment for Power System Operation," in *Computerized Operation of Power Systems* (S.C. Savulescu ed.), Elsevier, Amsterdam, 1976.

7. L. Wang, "A Probability Method for Operation Planning," *Proceedings of the 9th Annual Engineering Conference on Reliability for the Electric Power Industry*, Hershey, Pennsylvania, 1982.

8. W.H. Winter, "Measuring and Reporting Overall Reliability of Bulk Electricity Systems," *CIGRE Proceedings*, Paper No. 32-15 (1980).

9. L. Wang, K. Gallyas, and D.T. Tsai, "Reliability Assessment in Operational Planning for Large Hydro-Thermal Generation Systems," *IEEE Transactions* **PAS-104**, 3382-3387 (1985).

10. A.J.Wood and B.F. Wollenberg, *Power Generation, Operation, and Control*, John Wiley & Sons, New York, 1984.

11. J.D. Guy, "Security Constrained Unit Commitment," *IEEE Transactions* **PAS-90**, 1385-1390 (1971).

12. A.K. Ayoub and A.D. Patton, "Optimal Thermal Generating Unit Commitment," *IEEE Transactions* **PAS-90**, 1752-1756 (1971).

13. A.V. Jain and R. Billinton, "Unit Commitment Reliability in a Hydro-Thermal System," IEEE PES 1973 Winter Meeting, New York, Paper C73 096-5, 1973.

9. SUMMARY AND OUTLOOK

In this Chapter, an overview was given of the methods used in bulk power system reliability assessment.   Particular attention was paid to the development of probabilistic approaches which appear to be gradually gaining a foothold in reliability studies. At the present state of the art, the level of maturity reached by the various methods is quite uneven, and no attempt was made to conceal this fact or create the impression that all problems are solved and all questions answered. However, improvements in modeling and computing methods continue to occur, making them more credible and, thus, acceptable as tools in planning and operation.

A number of areas within the broad field of power system reliability were not discussed, mostly because they were outside the scope of this Chapter. They are

- Station system reliability

- Distribution system reliability

- Cost and worth of reliability and decision methods based on cost optimization

- Data bases and data collecting schemes

Other methods were not discussed because they are still in an embryonic stage of development. They include

- Combined probabilistic adequacy and security studies

- Preventive maintenance optimization for system components

- Verification methods

Clearly, much research remains to be done in the field. In 1989, an EPRI workshop on the research needs and priorities in power system planning and engineering identified the following research topics as having the highest priority in the area of power system reliability[1].

- Bulk power system adequacy assessment

- Impact of protection systems on reliability

- Bulk system cost/benefit evaluation

- Prediction of multiple facility outages

- Cost of interruptions

- Cascading outages

- Outage data collection and analysis

- Multi-area reliability assessment

- Reliability planning criteria

- System security

- Preventive maintenance

Several of these topics are presently under intensive study. Some, such as the effects of protection system failures, could already be given a brief discussion in this Chapter. The methods and programs now available have many uses and several are widely applied; but further concerted effort by the many researchers in the field is required to ensure that, ultimately, power system reliability methodology based on probabilistic models will become universally adopted.

Reference

1. *Proceedings: Power System Planning and Engineering - Research Needs and Priorities*, EPRI Report EL-6503, 1989.

Coordination of Distribution System Capacitors and Regulators: An Application of Integer Quadratic Optimization

Ross Baldick
University of California
Berkeley, CA 94720

I Introduction

External forces such as higher fuel costs, deregulation, and increasing consumer awareness are changing the role of electric utilities and putting pressure on them to become more 'efficient.' Until recently, increases in efficiency were mostly due to improving *generation* technology; however, the potential for such improvements has been almost completely exploited [1, chapter 1]. Efficiency improvements are increasingly due to *non-generation* technologies such as *distribution automation systems,* which increase the options for real-time computation, communication, and control. This technology will prompt enormous changes in many aspects of electric power system operation.

In this chapter, we investigate the potential of such technology to improve efficiency in a radial *electric distribution system,* through the coordination of switched capacitors and regulators. We consider

- an objective based on the minimization of losses;

- the inclusion of voltage constraints;

- careful treatment of the combinatorial aspects of discrete control; and,

- a formulation applicable to distribution systems with laterals.

This chapter is based on "Efficient Integer Optimization Algorithms for Optical Coordination of Capacitors and Regulators," by Ross Baldick and Felix F. Wu, which appeared in IEEE *Transactions on Power Systems*, 5(3):805-812, August 1990, © 1990 IEEE, and on "Approximation Formulas for the Distribution System: the Loss Function and Voltage Dependence," by Ross Baldick and Felix F. Wu, which appeared in IEEE *Transactions on Power Delivery*, 6(1):252-259, January 1991, © 1990 IEEE.

We also describe the relationship between:

1. the capacitor and regulator *expansion design* problem, and,

2. the coordination problem.

Electrical losses, operating constraints, and the role of distribution automation systems in the control of capacitors and regulators will be described in the next subsection. In subsection B we will explicitly define the capacitor and regulator coordination problem and review the literature. Then in subsection C the organization of the rest of the chapter will be outlined.

A   Losses, constraints, and distribution automation systems

About 5% of all electricity generated in the United States is lost due to resistive heating in the distribution system. Capacitors can be used to reduce losses by reducing reactive power flows. Ideally, the capacitors should be adjusted to exactly *compensate* [2] for the loads in the system; however, in most practical applications the capacitors cannot be controlled continuously: they can only be switched off or on, or to one of a few *tap settings*. Therefore, the best *switched* configuration should be chosen, which minimizes losses subject to constraints.

A variety of constraints affect the distribution system; however, we will concentrate on voltage constraints in this chapter. Transformers with variable taps can be used to adjust voltage: we will refer to these as 'regulators' [3, §9-4]. The control of the regulators can help in the satisfaction of voltage constraints, but as with capacitors, regulators are usually controlled in discrete steps.

Currently, capacitors and regulators are typically controlled on the basis of local information, to achieve local objectives such as voltage or reactive power support. The introduction of telemetering and remote control through distribution automation allows for the possibility of coordination for *global* objectives. In the next subsection, the capacitor and regulator problem will be defined precisely and the associated literature will be reviewed.

B   Capacitor and regulator coordination

There has been considerable work on the placement and sizing of capacitors and regulators [4, 5, 6, 7, 8]; however, the actual coordination of switched capacitors and regulators has received less attention. Following Wirgau [9], the optimal coordination problem, taking into account the discrete nature of capacitor and regulator control, can be stated as follows:

Problem 1 (Capacitor and regulator coordination) *Select the switching schedule for the capacitors and regulators so as to* minimize *losses in the system,* subject to *voltage constraints.* □

Remark 1 By switching schedule we mean a choice of discrete capacitor and regulator tap settings consistent with their installed ratings. In problem 1 we additionally require that the switching schedule satisfies the voltage constraints and refer to such a schedule as voltage feasible.

The *general* problem of minimizing a function over combinatorial possibilities is very difficult and has received considerable attention in the literature (see, for example, [10, 11, 12, 13]) and some progress has been made on this *particular* problem [14, 15, 16, 17]. However, previous formulations have, in general:

1. simplified the problem by assuming that all capacitors lie on the main feeder or that the distribution system is single-phase [4, 5, 16, 17, 18];

2. ignored the combinatorial problems associated with optimal coordination of switched elements by tacitly assuming that some enumerative technique such as branch and bound or even explicit search will be effective [4, 6, 16, 17];

3. ignored the switched nature of the capacitors and regulators by formulating a continuous relaxation of the coordination problem, where the capacitor and regulator settings are continuously controllable [5, 18, 19]; or,

4. neglected the voltage constraints in the coordination problem [4, 5, 18].

We formulate the general single-phase case where switched capacitors and regulators are placed arbitrarily on a radial distribution system. El-Kib et al. present progress on the three-phase case in [20]. For general single-phase systems with

- a small number of on-off switched capacitors, and,

- a small number of regulators having only a few tap positions,

it is reasonable to use a branch and bound or explicit search technique for problem 1. However, some researchers envisage the incorporation of many tapped capacitor banks into the distribution system [21]. Furthermore, regulators can have a large number of tap settings. In this case the combinatorial problem becomes computationally intensive to solve by enumerative techniques, particularly if it falls into a class of problems known as NP-hard [22, chapter 5].

The continuous relaxation of the coordination problem is much easier to solve than the switched problem because continuous optimization algorithms can be used. Efficient algorithms to solve the continuous relaxation of the coordination problem are described in [6, 18, 23, 4, 5, 8]. (The last three references address *design* problems, but the algorithms presented can also be used to solve the continuous coordination problem.)

It will be shown that for the voltage-unconstrained case, the *optimal* switched losses are not significantly worse than the *optimal* continuous losses obtained with the relaxed problem; however, the continuous values of capacitor and regulator settings that solve the relaxed problem are not necessarily a good guide to the discrete configuration that solves the switched problem. Instead, we utilize the solution to the relaxed problem *indirectly* as a basis for *approximating* the losses and voltages. In the next subsection, we outline the use of the relaxed problem and describe the organization of the rest of the chapter.

C   Organization

We use the solution to the continuous problem described in the last subsection as a *base-case* in the construction of a very accurate analytic *approximation* to the dependence of losses and voltages on the capacitor and regulator settings. The base-case conditions are defined in section II, computational and communication issues are addressed, and the approximation results are presented. It is shown that a quadratic formula is adequate for representing the losses, while a linear formula suffices for the voltages.

Using the approximation formulas, the optimal coordination problem is formulated in section III as an integer quadratic program with linear constraints. In section IV we discuss this problem and introduce *another* relaxed problem where:

- the switched nature of capacitor control is incorporated, but,

- the voltage constraints are still neglected.

In section V we describe algorithms for this relaxed problem and discuss their complexity as a function of the problem size [22, chapter 1]. It is shown that if:

1. voltage constraints are neglected, and,

2. all the capacitors have equal increments between adjacent settings,

then the problem of minimizing losses over switched capacitor settings can be solved in time that is polynomial in the problem size. This result obviates the need for non-polynomially bounded techniques such as branch and bound for this problem.

In section VI we incorporate fast algorithms for the relaxed problem into algorithms for the switched voltage-constrained case and in section VII report the test results. The computation times as a function of problem size are measured and compared to a benchmark procedure. The losses obtained with the proposed algorithms are compared to the losses obtained with a local control procedure. The results indicate the advantages of the proposed algorithms over the benchmark procedures. In section VIII we discuss the relationship between the design and coordination problems. In section IX we conclude the chapter.

II The Approximation Formulas

For the capacitor and regulator coordination problem we are interested in the dependence of losses and voltages on the settings of capacitors and regulators. The dependence is non-linear and involves the solution of the loadflow equations. Capacitor and regulator settings are *parameters* in these equations so that if the settings are changed then the direct approach is to recalculate the loadflow.

To avoid repeated loadflow calculations, we construct approximation formulas for the losses and voltages that are analogous to transmission system loss formulas [24, §9.2][25, 26]. The formulas are Taylor expansions about a base-case loadflow.

In general, the coefficients of the Taylor expansions are complicated so we further approximate them under the assumption that the the line impedances are small and that the tap ratios are near to unity; however, we do not assume that the capacitive susceptance is kept near the base-case value.

In the next subsection, we define the conditions for the base-case precisely. In subsection B, we discuss computational and communication issues. In subsection C, the formulas are presented and in subsection D they are compared to loadflow analysis using a 70 node test system. The formulas have

1. precisely defined asymptotic properties, and,

2. coefficients that are independent of the loads in the system.

The terms in the approximations are extremely simple to construct: the coefficients in the loss formula are the sums of line resistances along particular paths in the system, so that they depend only on the topology of the system and the line resistances. Similarly, the coefficients in the voltage formula depend only on the topology and the line reactances. We show that the loss function is convex, characterize the approximation error, and find a structure in the loss approximation formula that facilitates the construction of algorithms for the optimal switched coordination problem.

A   The base-case loadflow

In principle, the capacitive susceptance and regulator tap ratios for the base-case loadflow are arbitrary, but we choose the following conditions:

- regulator taps set to nominal, and,

- capacitive susceptances set to minimize losses.

This base-case requires the solution of a continuous optimization problem. There are three reasons for this choice of base-case:

1. the capacitor and regulator coordination problem described in section III seeks the *switched* capacitor and regulator settings that minimize losses. It will be demonstrated in section VI that it is rare for the minimum switched losses to be significantly larger than the minimum continuous losses so that the optimal switching schedule is relatively close to the chosen base-case conditions;

2. for this base-case, the first derivative of losses with respect to capacitor settings is zero so that the terms in the Taylor expansion that are linear in capacitance vanish; and,

3. there are fast algorithms to perform the required continuous optimization and software is available [4, 5, 6, 8, 18, 23].

To specify the base-case conditions precisely we introduce some nomenclature. Let P_{l0} denote the losses in the system; $\mathbf{B}_C$ the vector of capacitive susceptances; and, $\mathbf{V}$ the vector of voltage magnitudes at the nodes in the distribution system. We develop the analysis in terms of the capacitive susceptance in the system, although the capacitor current or reactive power could be used instead: the concentration on capacitive susceptance will be justified in section III. The analysis

$\mathcal{N}_N$	the set of nodes;
$\mathcal{N}_C$	the set of nodes with capacitors;
$\mathcal{N}_R$	the set of nodes immediately downstream of regulators;
$\mathbf{z}$	the vector of line impedances;
$\mathbf{t}$	the vector of off-nominal regulator tap ratio settings;
$\mathbf{B}_C$	the vector of capacitive susceptances;
$P_{l0}(\mathbf{z},\mathbf{t},\mathbf{B}_C)$	the losses in the system as a function of the line impedances, the regulator settings, and the capacitive susceptances;
$\mathbf{V}^2(\mathbf{z},\mathbf{t},\mathbf{B}_C)$	the vector of squares of the voltages in the system as a function of the line impedances, the regulator settings, and the capacitive susceptances;
$P_{l0}^{opt}(\mathbf{z},\mathbf{t})$	the minimum losses in the system, over values of the vector of capacitive susceptances, as a function of the line impedances and the regulator settings;
$\mathbf{B}_C^{opt}(\mathbf{z},\mathbf{t})$	the vector of capacitive susceptances for minimum losses, as a function of the line impedances and the regulator settings;
$\mathbf{V}^{2opt}(\mathbf{z},\mathbf{t})$	the vector of squares of system voltages at the minimum loss condition, as a function of the line impedances and the regulator settings.

Table 1: Glossary of symbols.

makes use of the *DistFlow* formulation to express the load flows in the system [8, §II]; therefore, it is convenient to refer to the squares of the voltages and we let $\mathbf{V}^2$ denote the vector of squares of the node voltage magnitudes.

To analyze the electrical properties of the distribution system we think of it as a collection of nodes, with loads, connected together by resistive-inductive branches. We assume that the branches form a tree rooted at the substation, so that the distribution system is 'radial.' We denote the nodes by $\mathcal{N}_N$, the nodes with capacitors by $\mathcal{N}_C$, and the nodes immediately downstream of regulators by $\mathcal{N}_R$. These definitions are summarized in table 1 and illustrated in figure 1.

The losses P_{l0} and the voltages $\mathbf{V}^2$ are functions of the line impedances and the regulator tap settings as well as of the capacitive susceptances $\mathbf{B}_C$. We express this dependency explicitly by introducing $\mathbf{z}$ as the vector of line impedances and

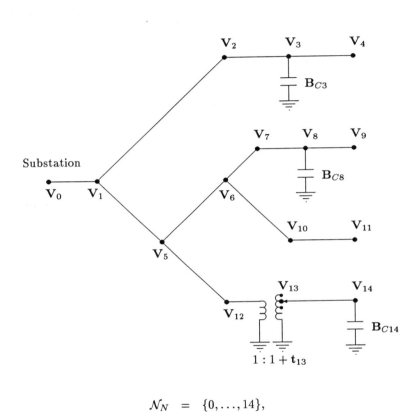

$$\mathcal{N}_N = \{0, \ldots, 14\},$$
$$\mathcal{N}_C = \{3, 8, 14\},$$
$$\mathcal{N}_R = \{13\}.$$

Figure 1: A skeleton distribution system, with loads and line impedances omitted, illustrating the definition of the vectors $\mathbf{B}_C, \mathbf{V}$, and $\mathbf{t}$ and the sets $\mathcal{N}_N, \mathcal{N}_C$, and $\mathcal{N}_R$.

Substation

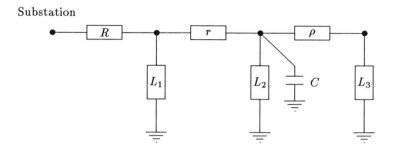

Figure 2: A simple system illustrating reactive flows for minimum losses.

$\mathbf{t}$ as the vector of off-nominal regulator tap ratios and write $P_{l0}(\mathbf{z}, \mathbf{t}, \mathbf{B}_C)$ and $\mathbf{V}^2(\mathbf{z}, \mathbf{t}, \mathbf{B}_C)$. We define the optimal continuous capacitive susceptances $\mathbf{B}_C^{opt}$ and the corresponding losses P_{l0}^{opt} and voltages $\mathbf{V}^{2opt}$ as follows:

$$P_{l0}^{opt}(\mathbf{z}, \mathbf{t}) = \min\{P_{l0}(\mathbf{z}, \mathbf{t}, \mathbf{B}_C) : \mathbf{B}_C \in \mathsf{R}^{\mathcal{N}_C}\}, \tag{1}$$

$$\mathbf{B}_C^{opt}(\mathbf{z}, \mathbf{t}) = \operatorname{argmin}\{P_{l0}(\mathbf{z}, \mathbf{t}, \mathbf{B}_C) : \mathbf{B}_C \in \mathsf{R}^{\mathcal{N}_C}\}, \tag{2}$$

$$\mathbf{V}^{2opt}(\mathbf{z}, \mathbf{t}) = \mathbf{V}^2(\mathbf{z}, \mathbf{t}, \mathbf{B}_C^{opt}(\mathbf{z}, \mathbf{t})), \tag{3}$$

where $\mathsf{R}^{\mathcal{N}_C}$ is the continuum of all possible values of capacitive susceptance. Equations (1)–(3) specify the base-case conditions. In the next subsection, we will discuss the practical calculation of the base-case.

B Computational and communication issues

Consider the reactive power flows that would result from application of the optimal continuous capacitive susceptances. The flows will not only compensate for local and downstream reactive loads, but, in general, also partially compensate the *upstream* loads. For example, consider the simple distribution system in figure 2. The three lines are resistive, having resistances R, r, and ρ, and it is assumed that R is bigger than r. To minimize losses, the capacitor, C, should supply some of the reactive power for L_1, as well as compensating for its local load L_2 and the 'downstream' load L_3. In contrast, compensation based *only* on information about the local and downstream load will not in general minimize the losses.

The optimal continuous capacitive susceptances required for the base-case can be calculated with continuous-valued optimization software, but despite the need

for non-local information, it is not necessary to *completely* centralize the calculation. For example, Neyer [23, §4.3] has proposed a *distributed* algorithm for
solving the continuous loss minimization problem that takes advantage of a distributed computing environment matching the topology of the electric distribution
system. With Neyer's algorithm, the minimum loss capacitive susceptances can
be calculated without telemetering the complete state of the system to a central
computer. Furthermore, the algorithm can be carried out in parallel with other
algorithms such as state estimation, therefore minimizing the total computation
costs.

The base-case susceptances, together with

1. the base-case losses and voltages, and,

2. the formulas presented in the next subsection,

constitute a *concise representation* of the approximate loss and voltage functions.
This representation can be efficiently transmitted to a single point if the approximation formulas are required by a centralized algorithm, such as the coordination
algorithms described in sections V and VI. Potentially, this would require much
less communication than transmitting the entire state of the system to a single
node for processing.

C   Formulas

The points $\mathbf{B}_C = \mathbf{B}_C^{opt}(\mathbf{z}, \mathbf{0})$ and $\mathbf{t} = \mathbf{0}$ are chosen as base-case points for the Taylor
expansion. The main result is that for sufficiently small values of line impedance
and for tap settings sufficiently close to nominal, the loss function and the voltages
in the system can be approximated by the following Taylor expansions, which we
refer to as the first order analytic approximations:

$$P_{l0}(\mathbf{z}, \mathbf{t}, \mathbf{B}_C) \approx P_{l0}^{opt}(\mathbf{z}, \mathbf{0}) + V_0^2 (\mathbf{B}_C - \mathbf{B}_C^{opt}(\mathbf{z}, \mathbf{0}))^T R^{CC} (\mathbf{B}_C - \mathbf{B}_C^{opt}(\mathbf{z}, \mathbf{0})), \quad (4)$$

$$\mathbf{V}^2(\mathbf{z}, \mathbf{t}, \mathbf{B}_C) \approx \mathbf{V}^{2opt}(\mathbf{z}, \mathbf{0}) + V_0^2 X^{NC} (\mathbf{B}_C - \mathbf{B}_C^{opt}(\mathbf{z}, \mathbf{0})) + 2V_0^2 K^{NR} \mathbf{t}, \quad (5)$$

where V_0^2 is the substation voltage and where the matrices $R^{CC} \in \mathbf{R}_+^{\mathcal{N}_C \times \mathcal{N}_C}$,
$X^{NC} \in \mathbf{R}_+^{\mathcal{N}_N \times \mathcal{N}_C}$, and $K^{NR} \in \mathbf{R}_+^{\mathcal{N}_N \times \mathcal{N}_R}$ are defined as follows. The ij-th element
of R^{CC} (X^{NC}) is the total resistance (inductive reactance) in the common part
of the paths from the substation to nodes i and j. The ij-th element of K^{NR}
is either zero or one: it is one if the regulator at node j is in the path from the

substation to node i. The formulas are derived in [27], where a general analysis of distribution system properties is carried out.

The matrices R^{CC}, X^{NC}, and K^{NR} are simple to construct; the structure of the matrix R^{CC} is simplified if we augment the capacitive susceptance vector $\mathbf{B}_C$ with 'dummy' capacitors as follows. First define the set of *distinguished*[1] nodes, $\mathcal{N}_D$, by considering the union of the paths from the nodes in $\mathcal{N}_C$ to node 0. Let $\mathcal{N}_B$ be the set of *branching nodes* [28, §2.2] in this union and let $\mathcal{N}_D = \mathcal{N}_C \cup \mathcal{N}_B$. The definition of these sets is illustrated in figure 3. The set of distinguished nodes satisfies $|\mathcal{N}_D| \le 2|\mathcal{N}_C|$: if the capacitors are all on the main feeder then $\mathcal{N}_D = \mathcal{N}_C$.

Secondly, augment $\mathbf{B}_C$ and $\mathbf{B}_C^{opt}$ to $\mathbf{B}_D \in \mathbf{R}^{\mathcal{N}_D}$ and $\mathbf{B}_D^{opt} \in \mathbf{R}^{\mathcal{N}_D}$, respectively, by defining:

$$\mathbf{B}_{Di} = \begin{cases} \mathbf{B}_{Ci}, & \text{if } i \in \mathcal{N}_C, \\ 0, & \text{if } i \in \mathcal{N}_D \setminus \mathcal{N}_C, \end{cases} \tag{6}$$

and defining $\mathbf{B}_D^{opt}$ analogously. If we define $R^{DD} \in \mathbf{R}^{\mathcal{N}_D \times \mathcal{N}_D}$ analogously to R^{CC} then [27]:

$$P_{l0}(\mathbf{z}, \mathbf{t}, \mathbf{B}_C) \approx P_{l0}^{opt}(\mathbf{z}, 0) + V_0^2 (\mathbf{B}_D - \mathbf{B}_D^{opt}(\mathbf{z}, 0))^T R^{DD} (\mathbf{B}_D - \mathbf{B}_D^{opt}(\mathbf{z}, 0)), \tag{7}$$

where

$$R^{DD} = U^{-T} D^{DD} U^{-1}, \tag{8}$$

and

1. U is non-singular and totally unimodular [29, §4.12], and,

2. D^{DD} is diagonal with non-negative entries.

The detailed definitions of U and D^{DD} are in [27].

An immediate corollary is that R^{DD} and R^{CC} are positive semi-definite so that for sufficiently small values of line impedance and for tap settings sufficiently close to nominal, the loss function is convex in $\mathbf{B}_C$. Consequently, continuous optimization algorithms can be guaranteed to converge to the minimum loss capacitive susceptance $\mathbf{B}_C^{opt}$. The assumption of convexity is implicit in our use of continuous optimization algorithms to calculate $\mathbf{B}_C^{opt}$ and has not been proved rigorously before in the literature. In the next subsection we test the validity of these approximation formulas in a model distribution system.

---

[1]I would like to thank Professor Dorit Hochbaum of the School of Business Administration and IEOR Department, University of California, Berkeley for suggesting this nomenclature.

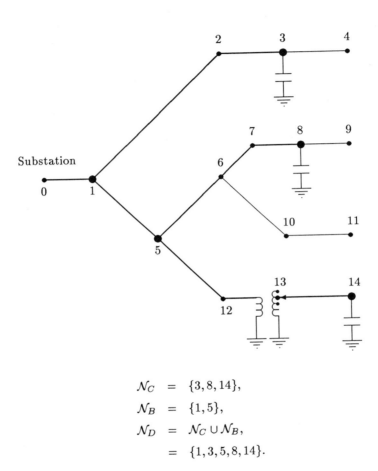

$$\mathcal{N}_C = \{3, 8, 14\},$$
$$\mathcal{N}_B = \{1, 5\},$$
$$\mathcal{N}_D = \mathcal{N}_C \cup \mathcal{N}_B,$$
$$= \{1, 3, 5, 8, 14\}.$$

Figure 3: A skeleton distribution system, with loads and line impedances omitted, illustrating the definition of the sets $\mathcal{N}_B$ and $\mathcal{N}_D$. The paths from the nodes in the set $\mathcal{N}_C$ to node 0 are shown thicker; the nodes in the set $\mathcal{N}_D$ are shown enlarged.

D   Numerical tests

In this subsection, the approximations are compared to exact loadflow solutions performed on a model test system. The test system used is the 70 node system described in [7]. Five capacitors were modeled at nodes 11, 18, 47, 52, and 69 and we obtained the optimal voltage-unconstrained continuous solution using the technique presented in [8]. The capacitor at node 52 was then varied about its optimal value and the reactive power generated by the other capacitors kept constant.[2] Figure 4 shows the comparison between the actual loadflow calculated values of losses and the theoretical values predicted by the first order analytic approximation. The maximum error between loadflow and analytic approximation is 1.5% expressed as a percentage of the loadflow calculated losses. The analytic approximation is slightly conservative in that the actual losses are slightly overestimated. Figure 5 shows the comparison between the loadflow calculated values of V_{52} and the analytic approximation: agreement is better than 0.01 per unit.

As a further test, the 32 possibilities corresponding to on-off values of the five capacitors were tested. 'On' values for C11, C18, C47, C52, and C69 were, respectively, 500, 500, 500, 1500, and 500 kVAr: these are large increments in reactive power chosen to thoroughly test the validity of the approximations. For comparison, the total reactive load in the test system is approximately 2700 kVAr. The results are shown in figure 6 where the analytic approximation in terms of B_C and the actual values of P_{l0} are plotted against the on-off choices for the capacitors. In all but one of the switch configurations, (that is, except for C11, C18, C47, and C69 'off' and C52 'on'), the analytic approximation in terms of B_C is conservative. The maximum error was less than 2.5%, again expressed as a percentage of the loadflow calculated losses.

The results demonstrate that a quadratic function can be used to represent losses, while a linear function suffices for the voltages, once the base-case loadflow has been solved. The formulas will be utilized in the next section.

---

[2]Due to software limitations, the loadflows were performed with loads and capacitors that were modeled as constant power and reactive power elements. The actual susceptance was then calculated using the actual voltage level. The results reflect the accuracy of the analytic approximations, as distinct from the load modeling accuracy.

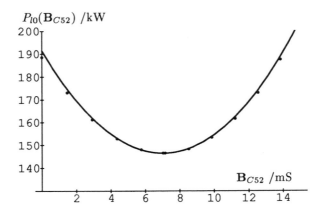

Figure 4: Analytic approximation to loss function (solid curve) and loadflow calculated values of loss function (points) versus capacitive susceptance B_{C52}.
Source: Ross Baldick and Felix F. Wu. Approximation Formulas for the Distribution System: the Loss Function and Voltage Dependence. To appear in *IEEE
Transactions on Power Delivery*. ©1990 IEEE

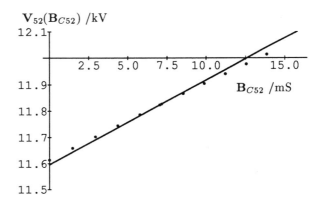

Figure 5: Analytic approximation to voltage at node 52 (solid curve) and loadflow
calculated values of voltage (points) versus capacitive susceptance B_{C52}.
Source: Ross Baldick and Felix F. Wu. Approximation Formulas for the Distribution System: the Loss Function and Voltage Dependence. To appear in *IEEE
Transactions on Power Delivery*. ©1990 IEEE

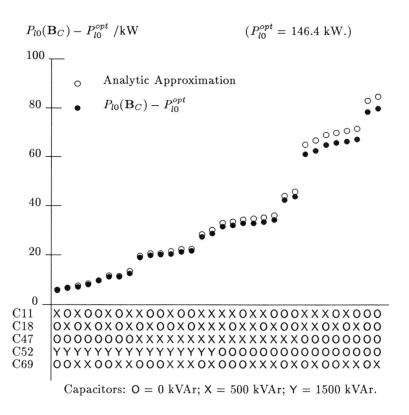

$P_{l0}(\mathbf{B}_C) - P_{l0}^{opt} / \text{kW}$ $(P_{l0}^{opt} = 146.4 \text{ kW.})$

Capacitors: O = 0 kVAr; X = 500 kVAr; Y = 1500 kVAr.

Figure 6: Analytic approximation in terms of $\mathbf{B}_C$ and loadflow calculated values versus on-off states of capacitors.

Source: Ross Baldick and Felix F. Wu. Approximation Formulas for the Distribution System: the Loss Function and Voltage Dependence. To appear in *IEEE Transactions on Power Delivery.* ©1990 IEEE

III   Formulation of the Coordination Problem

In this section we explicitly formulate the coordination problem using the approximation formulas presented in the last section. We first consider the control variable and then formulate the approximate problem.

A   Control variable

In the last section, we tacitly assumed that capacitive susceptance was used as the control variable for the capacitors. Other choices have been used in previous formulations: capacitor current in [4, 5, 6, 16, 18, 19, 30]; reactive power in [7, 8, 17, 31, 32]. These choices may be reasonable for the *continuous* capacitor coordination problem where the capacitors can be controlled to produce the desired reactive power or capacitor current. However, the only variable that is directly controlled by the *switching* of a capacitor is the capacitive susceptance presented to the system. For this reason, we formulate the problem in terms of capacitive susceptance. As noted in section II, the approximation in terms of the capacitive susceptance tends to give a conservative estimate of the losses so that the actual losses are consistently slightly lower than predicted by the analysis.

The approximation formulas can actually be expressed in terms of any one of the three variables: reactive power, capacitor current, or capacitive susceptance [27]. If, for example, static VAr compensators [33] are used instead of capacitors to generate reactive power, then a control variable other than capacitive susceptance may be appropriate to match the control scheme of the compensator. The analysis in the next subsection can be adapted to that case by rewriting the approximations in terms of the desired variable.

B   The approximate problem

We approximate problem 1 using the approximation formulas from section II:

Problem 2 (Approximate Switched Voltage-Constrained) *Given* $\mathbf{B}_C^{opt}$ *and* $\mathbf{V}^{2opt}$, *select the switching schedule for the capacitors and regulators so as to* minimize

$$V_0^2(\mathbf{B}_C - \mathbf{B}_C^{opt}(\mathbf{z}, \mathbf{0}))^T R^{CC}(\mathbf{B}_C - \mathbf{B}_C^{opt}(\mathbf{z}, \mathbf{0})), \qquad (9)$$

subject to *limits on*

$$\mathbf{V}^{2opt}(\mathbf{z}, \mathbf{0}) + V_0^2 X^{NC}(\mathbf{B}_C - \mathbf{B}_C^{opt}(\mathbf{z}, \mathbf{0})) + 2V_0^2 K^{NR}\mathbf{t}. \qquad (10)$$

As in [16], by using a continuous optimization as the basis of our objective function we are able to take advantage of the characteristics of continuous optimization algorithms while not neglecting the combinatorial aspects of the problem. In contrast to [16], we make use of the exact continuous solution, instead of approximating it, and solve the coordination problem for a more general class of distribution systems where capacitors can be placed on laterals as well as on the main feeder. As noted in section II, Neyer [23, §4.3] has proposed distributed algorithms for solving the continuous optimization problem required for the base-case.

We will transcribe problem 2 into a canonical integer quadratic programming problem with linear constraints by normalizing the capacitor and regulator variables. First consider the voltage constraints: let $\mathbf{V}^{2\max}$ and $\mathbf{V}^{2\min}$ be the vectors of the squares of the maximum and minimum allowed voltages, then the linearized voltage constraints can be expressed as

$$
\begin{aligned}
\Delta\mathbf{V}^{2\min} &= \mathbf{V}^{2\min} - \mathbf{V}^{2opt}, \\
&\leq V_0^2 X^{NC}(\mathbf{B}_C - \mathbf{B}_C^{opt}) + 2V_0^2 K^{NR}\mathbf{t}, \\
&\leq \mathbf{V}^{2\max} - \mathbf{V}^{2opt}, \\
&= \Delta\mathbf{V}^{2\max},
\end{aligned}
\tag{11}
$$

where we have omitted the arguments of $\mathbf{V}^{2opt}$ and $\mathbf{B}_C^{opt}$ for clarity.

Remark 2 In practice we may choose to enforce voltage constraints only at critical points such as at capacitors, regulators, and at the terminal nodes of the distribution system. In that case many of the rows of the matrices X^{NC} and K^{NR} can be discarded.

We assume that for each capacitor and each regulator, the *discrete* settings are separated by equal *increments*. We can express this with the following constraints:

$$
\forall \alpha \in \mathcal{N}_C, \mathbf{B}_{C\alpha}^{\min} \leq \mathbf{B}_{C\alpha}^{inc} x_\alpha \leq \mathbf{B}_{C\alpha}^{\max}, \tag{12}
$$

$$
\forall \alpha \in \mathcal{N}_C, x_\alpha \in \mathbf{Z}, \tag{13}
$$

$$
\forall \beta \in \mathcal{N}_R, \mathbf{t}_\beta^{\min} \leq \mathbf{t}_\beta^{inc} u_\beta \leq \mathbf{t}_\beta^{\max}, \tag{14}
$$

$$
\forall \beta \in \mathcal{N}_R, u_\beta \in \mathbf{Z}, \tag{15}
$$

where $\mathbf{Z}$ is the set of integers and where we have used x and u to represent normalized capacitor and regulator tap settings. For simplicity, we assume that for each capacitor α, $\mathbf{B}_{C\alpha}$ is a multiple of $\mathbf{B}_{C\alpha}^{inc}$ and for each regulator β, $\mathbf{t}_\beta$ is a multiple of $\mathbf{t}_\beta^{inc}$, but these assumptions can be easily removed. Note that we do not assume here that the increments are the same for all capacitors and for all regulators; however, we will make further assumptions about the increments in the discussion of particular algorithms in sections V and VI. Let

$$x_\alpha^{\min} = \mathbf{B}_{C\alpha}^{\min}/\mathbf{B}_{C\alpha}^{inc}, \tag{16}$$

$$x_\alpha^{\max} = \mathbf{B}_{C\alpha}^{\max}/\mathbf{B}_{C\alpha}^{inc}, \tag{17}$$

$$u_\beta^{\min} = \mathbf{t}_\beta^{\min}/\mathbf{t}_\beta^{inc}, \tag{18}$$

$$u_\beta^{\max} = \mathbf{t}_\beta^{\max}/\mathbf{t}_\beta^{inc}, \tag{19}$$

then the linearized constraints can be written as:

$$\begin{aligned}\Delta\mathbf{V}^{2\,\min} &\le X(x-b)+Ku,\\ &\le \Delta\mathbf{V}^{2\,\max},\end{aligned} \tag{20}$$

$$\begin{aligned}x^{\min} &\le x,\\ &\le x^{\max},\end{aligned} \tag{21}$$

$$\begin{aligned}u^{\min} &\le u,\\ &\le u^{\max},\end{aligned} \tag{22}$$

where the matrices X, K, and the vector b are defined by

$$\forall i \in \mathcal{N}_N, \forall j \in \mathcal{N}_C, X_{ij} = V_0^2 X_{ij}^{NC}\mathbf{B}_{Cj}^{inc}, \tag{23}$$

$$\forall i \in \mathcal{N}_N, \forall j \in \mathcal{N}_R, K_{ij} = 2V_0^2 K_{ij}^{NR}\mathbf{t}_j^{inc}, \tag{24}$$

$$\forall i \in \mathcal{N}_C, b_i = \mathbf{B}_{Ci}^{opt}/\mathbf{B}_{Ci}^{inc}. \tag{25}$$

We can rewrite the objective (9) of problem 2 as

$$f(x) = (x-b)^T R(x-b), \tag{26}$$

where the matrix $R \in \mathbf{R}^{\mathcal{N}_C} \times \mathbf{R}^{\mathcal{N}_C}$ is defined by

$$\forall i, j \in \mathcal{N}_C, R_{ij} = V_0^2 R_{ij}^{CC}\mathbf{B}_{Ci}^{inc}\mathbf{B}_{Cj}^{inc}. \tag{27}$$

The capacitor coordination problem can then be formulated as the following problem:

Problem 3 (Approximate Integer Voltage-Constrained) Instance:

   1. Sets of nodes $\mathcal{N}_N$, capacitor nodes $\mathcal{N}_C$, and regulator nodes $\mathcal{N}_R$;

2. vectors $\Delta V^{2\min}, \Delta V^{2\max} \in \mathbf{R}^{\mathcal{N}_N}$ defined in (11);

3. vectors $x^{\min}, x^{\max} \in \mathbf{Z}^{\mathcal{N}_C}, u^{\min}, u^{\max} \in \mathbf{Z}^{\mathcal{N}_R}$ defined in (16)–(19);

4. matrices $X \in \mathbf{R}^{\mathcal{N}_N \times \mathcal{N}_C}$, $K \in \mathbf{R}^{\mathcal{N}_N \times \mathcal{N}_R}$, and $R \in \mathbf{R}^{\mathcal{N}_C \times \mathcal{N}_C}$ defined in (23), (24), and (27), respectively; and,

5. a vector $b \in \mathbf{R}^{\mathcal{N}_C}$ defined in (25).

Question: *Solve*

$$
\begin{aligned}
\min f(x), \text{ subject to} \quad x &\in \mathbf{Z}^{\mathcal{N}_C}, \\
u &\in \mathbf{Z}^{\mathcal{N}_R}, \\
x^{\min} \leq \ x &\leq x^{\max}, \\
u^{\min} \leq \ u &\leq u^{\max}, \\
\Delta V^{2\min} &\leq X(x-b) + Ku, \\
&\leq \Delta V^{2\max},
\end{aligned}
\tag{28}
$$

where

$$
f(x) = (x-b)^T R(x-b).
\tag{29}
$$

□

In the next section, the characteristics of problem 3 will be discussed and a strategy for solving it will be introduced.

IV Discussion

Direct solution of problem 3 is difficult and no algorithms have been found that can treat the voltage constraints directly and that have run times polynomially bounded in the number of

1. capacitors, and,

2. voltage constraints.

Based on results concerning similar problems in [34, appendix C], we make the following conjecture:

Conjecture 1 Problem 3 is NP-hard in the strong sense [22, chapter 5].[3]

In the next subsection a strategy will be outlined that involves solving an integer problem with relaxed voltage constraints. Then in subsection B a bound will be presented for the relaxed problem and in subsection C a special case of the relaxed problem solved.

A   Solution strategy

An indirect strategy for problem 3 is motivated by the following observations:

1. The approximate quadratic objective in problem 3 is independent of the regulator tap settings so that the regulator taps can be adjusted to satisfy the voltage constraints without affecting the objective. This has been observed previously by Civanlar and Grainger [6, part I].

2. The effect of the capacitor settings on voltage is small in comparison to the effect of a typical regulator with 8 or 10 tap settings. For example, for the 70 node network described in section II, typical values of entries in the X matrix are 6×10^5 V^2, while entries in the K matrix are of the order of 1.5×10^6 V^2. Consequently, it is unlikely that a given capacitor switching schedule will be voltage infeasible while a slightly different schedule is voltage feasible.

3. As will be confirmed in section VII, low losses are correlated with good voltage profile since large losses *typically* mean heavy line flows that cause large voltage drops.

4. If voltage constraints are to be enforced at all or a large proportion of the nodes, then for any candidate capacitor switching schedule, more work is required to check the linearized voltage constraints for feasibility than is required to evaluate the quadratic objective function. Since the voltage-constrained problem resists solution by polynomially bounded algorithms, then some technique such as explicit search or branch and bound must be used if voltage constraints are to be treated directly. For sufficiently many capacitors and regulators, such a procedure will be hopelessly slow. This will be demonstrated for an explicit search procedure in section VII and in this case, the cost of computation may outweigh the loss savings.

---

[3]Strictly speaking, we mean that the subproblem of problem 3 where we restrict all numbers in the instance to being rationals is NP-hard in the strong sense.

Based on these considerations, the proposed algorithms proceed in three main steps. The first is to seek a solution or an approximate solution to the following problem:

Problem 4 (Approximate Integer Voltage-Unconstrained) Instance:

1. *A set of capacitor nodes $\mathcal{N}_C$;*

2. *vectors $x^{\min}, x^{\max} \in \mathbf{Z}^{\mathcal{N}_C}$ defined in (16)–(17);*

3. *a matrix $R \in \mathbf{R}^{\mathcal{N}_C \times \mathcal{N}_C}$ defined in (27); and,*

4. *a vector $b \in \mathbf{R}^{\mathcal{N}_C}$ defined in (25).*

Question: *Solve*

$$\min f(x), \text{ subject to } \quad x \; \in \; \mathbf{Z}^{\mathcal{N}_C},$$
$$x^{\min} \leq \; x \; \leq x^{\max}, \tag{30}$$

where

$$f(x) = (x - b)^T R (x - b). \tag{31}$$

□

The second step is to check if the regulators can be adjusted so that the solution to problem 4 is also voltage feasible for problem 3. That is, we generate tentative capacitor switching schedules without reference to voltage constraints and then check if the regulators can be adjusted to meet the voltage constraints. For reasons 1 to 3 above, a capacitor schedule that solves or approximately solves problem 4 will usually be voltage feasible for problem 3 for some value of the regulator settings.

It is possible, however, for a solution to problem 4 to be voltage infeasible for problem 3. In the third step of the algorithm we either output a feasible solution or in some cases resort to another technique such as enumeration, branch and bound, or local search to find an optimal or near optimal schedule.

In the following subsections, a bound is derived for problem 4 and a special case is solved.

B   A bounding result

Consider a random variable $z \in \mathbf{Z}^{\mathcal{N}_C}$ with independent components. Suppose that it is distributed on the vertices of the unit hypercube surrounding b in such a way that the expectation of $f(z)$ is minimized. The required distribution is [34, appendix C]:

$$\forall i \in \mathcal{N}_C, z_i = \begin{cases} \lfloor b_i \rfloor, & \text{with probability } \lceil b_i \rceil - b_i, \\ \lceil b_i \rceil, & \text{with probability } 1 + b_i - \lceil b_i \rceil, \end{cases} \tag{32}$$

where $\lfloor \bullet \rfloor$ and $\lceil \bullet \rceil$ are the floor and ceiling operators, respectively. For this distribution,

$$\begin{aligned} \mathcal{E}f(z) & \leq \frac{1}{4} \sum_{k \in \mathcal{N}_C} R_{kk}, \\ & = \frac{1}{4} V_0^2 \sum_{k \in \mathcal{N}_C} R_{kk}^{CC} (\mathbf{B}_{Ck}^{inc})^2, \end{aligned} \tag{33}$$

where $\mathcal{E}$ is the expectation operator. Moreover, there exists some x at one of these vertices such that

$$f(x) \leq \frac{1}{4} V_0^2 \sum_{k \in \mathcal{N}_C} R_{kk}^{CC} (\mathbf{B}_{Ck}^{inc})^2. \tag{34}$$

Now suppose that the optimal continuous voltage-unconstrained capacitive susceptance happens to lie within the range of installed capacitive susceptance, so that $x^{\min} \leq b \leq x^{\max}$. Then a feasible x exists satisfying (34). This bound generalizes that in [16], which only considered capacitors placed along the main feeder. It is possible to generalize the bound further to the case where $x^{\min} \leq b \leq x^{\max}$ does not hold [34, appendix C].

For the 70 node test system considered in section VII, with its five capacitors having tap increments of 300 kVAr, the bound (34) is about 3kW, which is only about 2% of the system losses under nominal loading conditions. In section V, we describe a 'randomized algorithm' [35] based on the bound that delivers a switching schedule that satisfies the bound in *expectation*.

C   A special case

A special case of problem 4 admits a simple exact solution based on 'rounding off' the continuous solution. Consider an instance of problem 4 that satisfies the following assumptions:

1. All the capacitors lie on the main feeder, and,

2. $x^{\min} \leq b \leq x^{\max}$.

To find the solution, note that since the capacitors all lie on the main feeder, we have [27]:

$$R^{CC} = U^{-T} D^{CC} U^{-1}, \qquad (35)$$

where

1. U is totally unimodular [29, §4.12], with:

   - 1's on the diagonal,

   - −1's on the superdiagonal, and,

   - zeros elsewhere;

   and,

2. D^{CC} is diagonal.

By the change of variables $y = U^{-1}x$ and $c = U^{-1}b$ we have

$$f(x) = g(y) = V_0^2 \sum_{i \in \mathcal{N}_C} D_{ii}^{CC} (\mathbf{B}_{Ci}^{inc})^2 (y_i - c_i)^2. \qquad (36)$$

If $y \in \mathbf{Z}^{\mathcal{N}_C}$, then $g(y)$ is minimized by setting

$$\forall i \in \mathcal{N}_C, y_i = [c_i], \qquad (37)$$

where $[\bullet]$ is the rounding function. By theorem 12.2, page 161 of [29], since U is totally unimodular,

$$y \in \mathbf{Z}^{\mathcal{N}_C} \Leftrightarrow x \in \mathbf{Z}^{\mathcal{N}_C}. \qquad (38)$$

If we temporarily neglect the constraints $x^{\min} \leq x \leq x^{\max}$ and let

$$x = Uy, \qquad (39)$$

then the given assignment to x yields the minimum objective for problem 4. Now note that if $[\bullet]$ rounds consistently for half-integers (that is, either always up or always down), then the given assignment to x also satisfies the constraints $x^{\min} \leq x \leq x^{\max}$ because of:

1. the special form of U, and,

2. the assumption on b.

To summarize, an optimal solution to the problem is given by:

$$c = U^{-1}b, \tag{40}$$

$$\forall i \in \mathcal{N}_C, y_i = [c_i], \tag{41}$$

$$x = Uy. \tag{42}$$

Similar results were obtained previously by Grainger et al. [16] using capacitor current as the control variable. We will use the transformation $y = U^{-1}x$ again in section V for the 'deterministic algorithm.'

If capacitors are arbitrarily placed on a radial distribution system then x calculated according to (40)–(42) will not, in general, satisfy $x^{\min} \leq x \leq x^{\max}$. However, if $x^{\min} \leq b \leq x^{\max}$, then we can think of setting

$$x = [b]. \tag{43}$$

This idea will be used in section V in the definition of the 'rounding algorithm.'

In section V, algorithms are proposed for problem 4 in the case that we do not restrict b. In section VI, the use of these algorithms in solving problem 3 will be discussed.

V   Algorithms for Problem 4

The bound (34) in the previous section indicates that low losses are achievable for problem 4. Furthermore, the actual minimum of this problem can be smaller than that implied by the bound. Conversely, in typical systems, many switching schedules yield losses that do not satisfy the bound. Therefore we must be careful to pick a schedule having low losses; however, there is a tradeoff between losses and computational speed.

We will describe three algorithms in this section for solving problem 4: a *rounding algorithm*, a *randomized algorithm*, and a *deterministic algorithm*. All have particular advantages and drawbacks in relation to the coordination problem. The rounding and randomized algorithms are very fast and can handle any size of capacitors and capacitor tap increments; however, they are not guaranteed to provide an optimal solution to problem 4, but instead generate *bounds* for the problem.

The deterministic algorithm, on the other hand, is slower and can only be applied to systems in which all increments in switched capacitance are equal, but for such systems obtains an optimal solution to problem 4. When the solution delivered by the deterministic algorithm is voltage feasible for problem 3, then it is also optimal for that problem. In the following subsections, the algorithms are described in detail.

A   The rounding algorithm

Following the discussion in section IV, we set

$$\forall i \in \mathcal{N}_C, x_i = \begin{cases} x_i^{\min}, & \text{if } b_i < x_i^{\min}, \\ [b_i], & \text{if } x_i^{\min} \leq b_i \leq x_i^{\max}, \\ x_i^{\max}, & \text{if } b_i > x_i^{\max}. \end{cases} \tag{44}$$

In general this strategy will produce suboptimal switching schedules; however, the average performance of this simple procedure is good [34, appendix C]. In the next subsection we discuss an analogous randomized algorithm.

B   The randomized algorithm

Again, following the discussion in section IV, we consider sampling a random vector z having independent components with the following distribution:

$$\forall i \in \mathcal{N}_C, z_i = \begin{cases} x_i^{\min}, & \text{if } b_i < x_i^{\min}, \\ \lfloor b_i \rfloor, & \text{with probability } \lceil b_i \rceil - b_i, \text{ if } x_i^{\min} \leq b_i \leq x_i^{\max}, \\ \lceil b_i \rceil, & \text{with probability } 1 + b_i - \lceil b_i \rceil, \text{ if } x_i^{\min} \leq b_i \leq x_i^{\max}, \\ x_i^{\max}, & \text{if } b_i > x_i^{\max}. \end{cases} \tag{45}$$

If $x^{\min} \leq b \leq x^{\max}$, then the expectation of $f(z)$ is relatively small. By minimizing $f(z)$ over several samples of the random variable, we expect to obtain a switching schedule with losses even smaller than those implied by the bound (34). We perform s trials, where s is a parameter chosen to be polynomially bounded in the problem size. In each trial we generate a random integer vector $z \in \mathbf{Z}^{\mathcal{N}_C}$ having independent components and with the above probability distribution.

C   The deterministic algorithm

This algorithm requires that the increments between adjacent settings be equal for all capacitors. To show the existence of a fast algorithm to perform the minimization we define another problem involving the distinguished nodes $\mathcal{N}_D$ introduced in section II. The problem is essentially the same as problem 4, but with extra 'capacitors' placed at the branching nodes in the set $\mathcal{N}_D \setminus \mathcal{N}_C$. For technical reasons involving the representation of numbers [22, §4.2] we restrict all vectors and matrices to having rational valued components. The problem is:

Problem 5 (Integer Voltage-Unconstrained Distinguished) Instance:

1. *A set of distinguished nodes $\mathcal{N}_D$;*

2. *vectors $x'^{\min}, x'^{\max} \in \mathbf{Z}^{\mathcal{N}_D}$;*

3. *a matrix $R' \in \mathbf{Q}^{\mathcal{N}_D \times \mathcal{N}_D}$ of the form*

$$R' = U^{-T}D'U^{-1}, \tag{46}$$

where

- *$D' \in \mathbf{Q}^{\mathcal{N}_D \times \mathcal{N}_D}$ is diagonal with non-negative entries, and,*
- *$U \in \mathbf{Q}^{\mathcal{N}_D \times \mathcal{N}_D}$ is non-singular and totally unimodular;*

and,

4. *a vector $b' \in \mathbf{Q}^{\mathcal{N}_D}$.*

Question: *Solve*

$$\min f(x'), \text{ subject to } x' \in \mathbf{Z}^{\mathcal{N}_D},$$
$$x'^{\min} \leq x' \leq x'^{\max}, \tag{47}$$

where

$$f(x') = (x' - b')^T R'(x' - b'). \tag{48}$$

□

Problem 5 can be solved in time that is polynomial in the size of its instance [22, §§4.2.1]; that is, polynomial in the parameters: $|\mathcal{N}_D|$, $\log_2 \|x'^{\max} - x'^{\min}\|_\infty$, and m', the maximum number of bits needed to represent any component of D' and b'. We use the form of R' to transform the objective into a separable quadratic function; the resulting problem has linear constraints with a totally unimodular constraint matrix; algorithm 4.4 of Hochbaum and Shanthikumar [36] is then used to solve this problem.

We now transform an instance of problem 4 into an instance of problem 5,

1. via a procedure taking time that is polynomial in $|\mathcal{N}_N|$ and the size of the instance of problem 4, and,

2. such that the size of the instance of problem 5 is polynomially bounded in the size of the instance of problem 4.

Consider an instance of problem 4 in which:

- the capacitor increments $\mathbf{B}_{C\alpha}^{inc}$ are the same for all capacitors, and,

- all matrices and vectors have rational valued components.

It can be transformed into an instance of problem 5 by defining

- $R^{DD} \in \mathbf{Q}^{\mathcal{N}_D \times \mathcal{N}_D}$ as in section II, and,

- the capacitor increment for 'capacitors' in the set $\mathcal{N}_D \setminus \mathcal{N}_C$ to be the same as that of the other capacitors,

and then setting

$$
\forall i \in \mathcal{N}_D, x_i'^{\min} = \begin{cases} x_i^{\min}, & \text{if } i \in \mathcal{N}_C, \\ 0, & \text{if } i \in \mathcal{N}_D \setminus \mathcal{N}_C, \end{cases} \tag{49}
$$

$$
\forall i \in \mathcal{N}_D, x_i'^{\max} = \begin{cases} x_i^{\max}, & \text{if } i \in \mathcal{N}_C, \\ 0, & \text{if } i \in \mathcal{N}_D \setminus \mathcal{N}_C, \end{cases} \tag{50}
$$

$$
\forall i \in \mathcal{N}_D, b_i' = \begin{cases} b_i, & \text{if } i \in \mathcal{N}_C, \\ 0, & \text{if } i \in \mathcal{N}_D \setminus \mathcal{N}_C, \end{cases} \tag{51}
$$

$$
\forall i, j \in \mathcal{N}_D, R_{ij}' = V_0^2 R_{ij}^{DD} \mathbf{B}_{Di}^{inc} \mathbf{B}_{Dj}^{inc}, \tag{52}
$$

where from section II,
$$
R^{DD} = U^{-T} D^{DD} U^{-1}, \tag{53}
$$

with U totally unimodular and D^{DD} diagonal. So, since $\mathbf{B}_{D\alpha}^{inc}$ is the same for all capacitors we have

$$
R' = U^{-T} D' U^{-1}, \tag{54}
$$

as required for problem 5, where $D' \in \mathbf{Q}^{\mathcal{N}_D \times \mathcal{N}_D}$ is diagonal with entries $D_{kk}' = V_0^2 D_{kk}^{DD} (\mathbf{B}_{Dk}^{inc})^2$.

To consider the size of the instance of problem 5 in relation to the instance of problem 4, note that we have added dummy capacitors of zero value to each node in the set $\mathcal{N}_D \setminus \mathcal{N}_C$, but that $|\mathcal{N}_D| \leq 2|\mathcal{N}_C|$. Furthermore, U and D' can be calculated in time that is polynomial in $|\mathcal{N}_N|$ and the size of the instance of problem 4, so that the transformation from problem 4 into problem 5 takes time that is polynomial in $|\mathcal{N}_N|$ and the size of the instance of problem 4. Therefore, if the capacitor tap increment is the same for all capacitors, then problem 4 can be solved in time that is polynomial in $|\mathcal{N}_N|, |\mathcal{N}_C|, \log_2 \|x^{\max} - x^{\min}\|_\infty$, and m, the maximum number of bits needed to represent any component of R and b.

Remark 3 If the topology of the system remains fixed then R' and D' need only be calculated once, so that each subsequent minimization can then be performed in time that is polynomial in $|\mathcal{N}_C|$, $\log_2 \|x^{\max} - x^{\min}\|_\infty$, and m.

The details of the deterministic algorithm for solving problem 5 are contained in figure 7. The main loop of the algorithm is transcribed from algorithm 4.4 in [36]. A proof of the correctness of the main loop is contained in [36]. The algorithm assumes that $x = 0$ corresponds to a feasible switch configuration. If it is not, then a suitable change of coordinates can be performed.

The first step in the algorithm is to transform the variables in the instance of problem 5 into a reference frame where the objective is separable. As in section IV,

$$
\begin{aligned}
f(x') &= g(y'), \\
&= \sum_{i \in \mathcal{N}_D} D'_{ii}(y'_i - c'_i)^2,
\end{aligned}
\tag{55}
$$

where $y' = U^{-1}x'$. Then, following algorithm 4.4 of [36], the y' search space is divided into a grid. We use the symbol $\overline{y}'$ to denote the value of y' normalized by the current grid spacing.

As described in [36], $\gamma = \max(1, \lceil \log_2 \|x^{\max} - x^{\min}\|_\infty \rceil - 1)$ iterations of the main loop are performed. The grid gets finer by a factor of two at each iteration, while the total number of grid divisions remains constant. Each new iteration involves a search conducted on a box that centers on the result of the previous iteration; the search in the first iteration centers on 0. For each capacitor, the search box extends $2n_D$ grid steps above and $2n_D$ below the solution to the previous iteration, where $n_D = |\mathcal{N}_D|$.

The objective in each iteration of the main loop of algorithm 4.4 of [36] is

$$
\sum_{i \in \mathcal{N}_D} \sum_{j=1}^{4n_D} [g_i^{L:s}(2s(\overline{y}'_i - n_D) + j) - g_i^{L:s}(2s(\overline{y}'_i - n_D) + j - 1)]\zeta_{ij}.
\tag{56}
$$

The objective is a linear function of the variable ζ, with coefficients specified in terms of the auxiliary function $g_i^{L:s}(\bullet)$, defined by

$$
\begin{aligned}
g_i^{L:s}(y'_i) &= (\lfloor y'_i/s + 1 \rfloor - y'_i/s)D'_{ii}(s\lfloor y'_i/s \rfloor - c'_i)^2 \\
&\quad + (y'_i/s - \lfloor y'_i/s \rfloor)D'_{ii}(s\lfloor y'_i/s + 1 \rfloor - c'_i)^2.
\end{aligned}
\tag{57}
$$

The linear constraints are

$$
\begin{aligned}
\lceil x'^{\min}/s - 2U(\overline{y}' - n_D 1) \rceil &\leq \sum_{i \in \mathcal{N}_D} \sum_{j=1}^{4n_D} U_i \zeta_{ij}, \\
&\leq \lfloor x'^{\max}/s - 2U(\overline{y}' - n_D 1) \rfloor,
\end{aligned}
\tag{58}
$$

```
begin
/* transform to y' coordinates */
```
$c' = U^{-1}b';$
$\overline{y}' = \mathbf{0};$
```
/* calculate number of iterations */
```
$\gamma = \max(1, \lceil \log_2 \|x^{\max} - x^{\min}\|_\infty \rceil - 1);$
```
/* main loop */
for  (k = γ - 1, ..., 0) {
```
$\quad s = 2^k;$
```
    /* solve linear program */
```
$\quad \zeta = \operatorname{argmin} \sum_{i \in \mathcal{N}_D} \sum_{j=1}^{4n_D} [g_i^{L:s}(2s(\overline{y}_i' - n_D) + j) - g_i^{L:s}(2s(\overline{y}_i' - n_D) + j - 1)]\zeta_{ij}$
```
        subject to
```
$\quad\quad\quad [x'^{\min}/s - 2U(\overline{y}' - n_D\mathbf{1})] \leq \sum_{i \in \mathcal{N}_D} \sum_{j=1}^{4n_D} U_i \zeta_{ij}$
$$\leq \lfloor x'^{\max}/s - 2U(\overline{y}' - n_D\mathbf{1}) \rfloor$$
```
        and subject to
```
$\quad\quad \zeta_{ij} \in [0, 1];$
```
    /* update ȳ' for new grid spacing */
    for  (i ∈ N_D)
```
$\quad\quad \overline{y}_i' = 2(\overline{y}_i' - n_D) + \sum_{j=1}^{4n_D} \zeta_{ij};$
```
}
/* transform back to x' coordinates */
```
$y' = \overline{y}';$
$x' = Uy';$
```
end.
```

Figure 7: The deterministic algorithm.

Source: The main loop of the algorithm is transcribed from algorithm 4.4 of [36].

where

- U_i is the i-th column of U, and,

- $\mathbf{1}$ is the vector of all ones.

Since U is totally unimodular, the constraints of the linear program are also totally unimodular, so that any basic feasible solution of the linear program is integer valued. The solution is used to update the value of $\overline{y}'$ for the next finer grid spacing. At the final iteration, the grid spacing is unity and the linear objective coincides with g at all integer points. After this iteration, the problem solution in the y' coordinates is transformed back to the original capacitor switching coordinates.

Remark 4 Because the constraint matrix is totally unimodular, the linear programs are in principle solvable in time that is polynomially bounded in the problem size [13, chapter I.6][37, chapter 8]. However, in practice we use a fast but non-polynomially bounded algorithm such as simplex to solve the linear program.

VI Algorithms for Problem 3

As discussed in section IV, our approach is to start with a solution or approximate solutions to problem 4 provided by the algorithms in the last section. We then check for voltage feasibility for problem 3. In the next subsection we describe how to check for voltage feasibility given a candidate capacitor switching schedule. Then in subsection B we discuss the options if the capacitor switching schedule is voltage infeasible.

A Voltage feasibility for problem 3

A vector x, representing a capacitor switching schedule, is voltage feasible if the following constraints can be satisfied:

$$\Delta \mathbf{V}^{2\min} - X(x - b) \leq Ku,$$
$$\leq \Delta \mathbf{V}^{2\max} - X(x - b), \tag{59}$$
$$u^{\min} \leq u,$$
$$\leq u^{\max}, \tag{60}$$
$$u \in \mathbf{Z}^{N_R}. \tag{61}$$

We first consider the case where the regulator tap increment is the same for all regulators. With this assumption, the constraint matrix K is a multiple of a totally unimodular matrix [29, §4.12]. Therefore, by corollary 12.3, page 161 of [29], the constraints can be satisfied if and only if the following continuous linear constraints can also be satisfied:

$$\lceil (\Delta \mathbf{V}^{2\min} - X(x - b))/K^{inc} \rceil \leq (K/K^{inc})u,$$
$$\leq \lfloor (\Delta \mathbf{V}^{2\max} - X(x - b))/K^{inc} \rfloor, \quad (62)$$
$$u^{\min} \leq u,$$
$$\leq u^{\max}, \quad (63)$$

where $K^{inc} = 2V_0^2 t_\beta^{inc}$ is equal to the magnitude of the non-zero elements in K and where $\lceil \bullet \rceil$ and $\lfloor \bullet \rfloor$ are the coordinate-wise ceiling and floor operators. The satisfiability of linear constraints can be checked efficiently with linear programming software. Moreover, every basic feasible vector satisfying (62) and (63) is integer valued, yielding a vector u corresponding to a set of feasible tap settings.

If the assumption of equal tap increments for the regulators does not hold then we must make subsidiary assumptions in order to efficiently check for feasibility. Examples of such conditions are:

1. If there are a large number of closely spaced tap settings on each regulator, then K^{inc} is small and the discrete nature of the taps can usually be ignored. In this case, (59)–(60) can be used to check for feasibility.

2. If it assumed that the voltage in the system is monotonically decreasing away from the substation except at the regulators, then the regulators can be adjusted in a breadth first search from the substation downstream. As each regulator β is reached by the search, it is set to the *highest* tap position that is voltage feasible for the voltage constraints of all nodes that are:

   - downstream of β, but,
   - upstream of all regulators not yet reached by the search.

   Under low load conditions, a similar procedure applies if the voltage is monotonically increasing except at the regulators.

If the tap increments for the regulators are unequal and none of the subsidiary assumptions hold, then it is much more difficult to check for feasibility: even if (62) and (63) can be satisfied, we cannot easily tell whether the switching schedule is voltage feasible unless we perform the time consuming task of enumerating the transformer tap combinations. If (59) and (60) cannot be satisfied, however, then the switching schedule is voltage infeasible.

Remark 5 Extra constraints such as line flow limits might be linearized and incorporated into this formulation generating an augmented coefficient matrix K'. If K' is a multiple of a totally unimodular matrix then feasibility can be checked easily; if not, then once again the determination of feasibility relies on subsidiary assumptions.

B   The switching schedule

For the rounding and deterministic algorithms, the candidate capacitor schedule is checked for voltage feasibility. In the case of the randomized algorithm, we actually perform s trials. Given the s random vectors from the s trials, we take the minimum value of the objective over those that are voltage feasible. In the case of the deterministic algorithm, if the capacitor switching schedule is voltage feasible, then it is optimal. No such guarantee is possible for the rounding and randomized algorithms.

A practical strategy is to run the deterministic algorithm first. Then, the minimum voltage-unconstrained switched losses are obtained if the corresponding capacitor configuration is voltage feasible. If not, a suboptimal but good solution can be sought from the rounding and randomized algorithms.

In the case that no voltage feasible vector is generated by any of the algorithms, then we must resort to an enumerative or branch and bound technique to find the optimal schedule. If only the lower voltage constraints are violated, a possible heuristic is to locally search in the neighborhood of a solution to problem 4 by switching in successively more capacitors until voltage constraints can be satisfied.

In the next section we report test results with these algorithms for the model system and compare to two benchmark procedures.

VII   Test Results

We coded the algorithms in the C language using calls to the FORTRAN NAG library to perform the linear programming and constraint checking [38, chapter E04]. The test system used is the 70 node system introduced in section II. As in section II, five capacitors were modeled at nodes 11, 18, 47, 52, and 69: nominal

values for C11, C18, C47, C52, and C69 of, respectively, 600, 600, 600, 900, and
600 kVAr were chosen with the nominal capacitor increment for all capacitors being 300 kVAr. Two regulators were modeled: the first was placed just before node
8 with tap increments of 0.5% and tap range of ±5.0%; the second was placed
just before node 47 with tap increments of 0.5% and tap range of ±2.5%. Voltage
constraints of 0.95 and 1.05 pu were enforced at each distinguished node, at the
nodes on either side of both regulators, and at all terminal nodes: that is, at nodes
2, 7, 8, 11, 18, 26, 34, 38, 41, 46, 47, 52, 54, 56, 58, 69, and 90. Two load levels
were examined. We will refer to them as medium and high and they correspond,
respectively, to loads of 1.0 and 1.5 times the loads in the nominal system in [7,
table 1].

The parameter s in the randomized algorithm was set to 10. In the next
subsection, two benchmark algorithms will be described. Then in subsection B
the performance of the algorithms described in sections V and VI will be evaluated.
In subsection C, the execution times of the algorithms are measured.

A   Benchmarks

We define two procedures to evaluate the performance of the algorithms described
in sections V and VI:

1. a simple *explicit search* procedure, to serve as a benchmark for the execution
   times of the algorithms, and,

2. a *local control* procedure that compensates only for local and downstream
   reactive loads, to serve as a benchmark for evaluating the gains achieved
   through the use of the *global* information contained in the loss formula.

The explicit search was conducted by systematically checking each possible
capacitor switching schedule for voltage feasibility and then taking the minimum
of the losses over all the voltage feasible switching schedules. To speed up the
checking of voltage feasibility, each successive linear constraint run was initialized
with the final regulator setting from the previous run. The explicit search procedure is a convenient benchmark, although an adaptation of a branch and bound
procedure such as in [16] may be much faster in practice.

The local control procedure chooses capacitor settings to compensate only
for local and downstream loads. In practice, a variety of schemes are used to
locally control capacitors [3, §§8-5-2], and it is difficult to evaluate the performance

without specifying the control scheme in detail. We will use one possible scheme as a benchmark: each capacitor controller chooses the capacitor setting that is closest to the sum of the local and downstream loads as measured locally at the capacitor. We will neglect voltage considerations and it is assumed that the downstream capacitor controllers operate before the upstream ones, so that the procedure is well-defined.

B   Losses and voltage profile

1   Medium load

For the medium load case, all 324 possible capacitor switching schedules were voltage feasible after adjustment of the regulator taps. The minimum switched losses were 0.7 kW in excess of the optimal voltage-unconstrained continuous losses of 146.4 kW.

The rounding and randomized algorithms both delivered switching schedules with losses that were 2.2 kW in excess of the optimal continuous losses, reflecting the fact that they are not guaranteed to obtain the minimum voltage-unconstrained losses. The losses are, however, very close to the best achievable. The voltage-unconstrained optimal switch configuration was voltage feasible so that the deterministic algorithm delivered the optimal voltage-constrained configuration. These results are summarized in figure 8 and the switching schedules determined by the algorithms are shown in table 2.

The losses obtained with the algorithms are not significantly worse than those obtainable with continuous control of capacitors. For comparison, the losses corresponding to no installed capacitors were about 80 kW in excess of the optimal continuous losses. With the local control procedure, the losses were 49.8 kW in excess of the optimal continuous losses, indicating the considerable gains achievable through the global information provided by the loss formulas.

With the capacitors set to $\mathbf{B}_C^{opt}$ and regulator taps at nominal, the minimum voltage was 11.79 kV (0.93 pu). After finding the optimal switched capacitor settings and adjusting the taps, the voltage improved to 12.06 kV (0.95 pu) as shown in figure 9. The voltages under the optimal switched schedules were calculated using the voltage formula: unfortunately, no distribution system loadflow was available that incorporated regulators, so that the accuracy of these figures has not been confirmed.

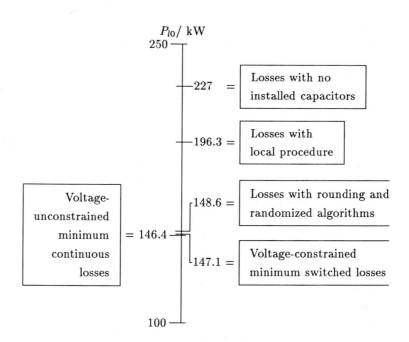

Figure 8: Losses under medium load.

node	in-stalled	optimal cont.	optimal switched	local control	round-ing	random-ized	determin-istic
11	0.60	0.32	0.30	0.30	0.30	0.30	0.30
18	0.60	0.22	0.30	0.00	0.30	0.30	0.30
47	0.60	0.02	0.00	0.60	0.00	0.00	0.00
52	0.90	0.98	0.90	0.30	0.90	0.90	0.90
69	0.60	0.29	0.60	0.00	0.30	0.30	0.60

Table 2: Capacitor switching schedules delivered by the benchmark procedures and the algorithms for medium load level with the optimal continuous levels for comparison.

All switching schedules are indicated in terms of the nominal MVAr ratings of the capacitors. The optimal continuous values are shown for comparison and were calculated as described in section II. The optimal switched values are from the explicit search procedure, while the local control procedure delivered the schedule in the fifth column. The rounding, randomized, and deterministic algorithms delivered the schedules shown in the last three columns.

Table 3: Capacitor switching schedules delivered by the benchmark procedures and the algorithms for high load level with the optimal continuous levels for comparison.

node	in-stalled	optimal cont.	optimal switched	local control	round-ing	random-ized	determin-istic
11	0.60	0.49	0.60	0.30	0.60	0.60	0.60
18	0.60	0.33	0.60	0.30	0.30	0.60	0.60
47	0.60	0.04	0.00	0.60	0.00	0.00	0.00
52	0.90	1.48	0.90	0.30	0.90	0.90	0.90
69	0.60	0.47	0.60	0.00	0.60	0.60	0.60

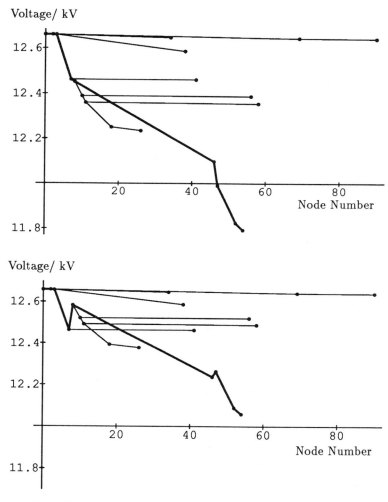

Figure 9: Top: Voltage profile of system under medium load with optimal continuous capacitive susceptance and with regulator taps set to nominal. Bottom: Voltage profile under medium load with optimal switched capacitive susceptance and with taps adjusted for voltage feasibility. In both cases the lateral with the lowest voltage profile is shown thicker: the voltage increases in the bottom profile occur at the regulators. Points represent the voltage at a node while the lines joining them indicate the topology of the system.

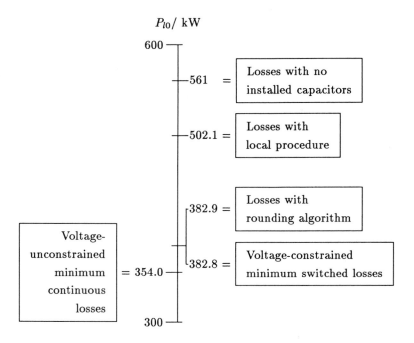

Figure 10: Losses under high load.

2   High load

For the high load case, out of 324 possible capacitor switching schedules, 135 were voltage feasible after adjustment of the regulator taps. The minimum switched losses were 28.8 kW in excess of the optimal continuous losses of 354.0 kW.

The rounding algorithm delivered losses that were 29.0 kW in excess of the optimal continuous losses. Of the 10 random switch configurations considered, all were voltage feasible and both the randomized and the deterministic algorithm delivered a switching schedule with the minimum switched losses of 28.8 kW in excess of the optimal continuous losses. These results are summarized in figure 10 and the capacitor switching schedules determined by the algorithms are shown in table 3.

Again, the losses obtained by the algorithms are only marginally higher than the losses obtainable with continuous control of capacitors: the losses corresponding to no installed capacitors were 207 kW in excess of the optimal continuous losses.

The losses obtained with the local control procedure were 148.1 kW in excess of the optimal continuous losses, again indicating the considerable gains achievable through the global information provided by the loss formulas. As discussed in section II, local information is insufficient to globally minimize the losses. In this case, much of the critical reactive load was *upstream* of the capacitors and so the local control procedure did not detect the need for reactive compensation. A local procedure based on *voltage* may have been able to detect this situation and compensate appropriately; however, in general, global information is required to globally minimize the losses.

With the capacitors set to $\mathbf{B}_C^{opt}$ and regulator taps at nominal, the minimum voltage was 11.31 kV (0.89 pu). After finding the optimal switched capacitor settings and adjusting the taps, the voltage improved to 12.03 kV (0.95 pu) as shown in figure 11. With no installed capacitors, the high load level is voltage infeasible, even with regulators. *Both* the capacitors and regulators are necessary for voltage feasibility in the high load case; however, the regulators play a larger role in voltage feasibility than the capacitors.

Remark 6 The high load case indicates that voltage feasibility is correlated with proximity in capacitive susceptance space to the minimum continuous loss condition, confirming the claim made in section IV. In this example the switched capacitor schedules surrounding the optimal continuous capacitive susceptance are more likely to be voltage feasible than an arbitrarily chosen schedule.

C   Execution times

The run time of the explicit search procedure for the five capacitor and two regulator system, excluding the calculation of the base-case loadflow and all input-output processing, ranged between 2.6 and 2.9 seconds of VAX 6420 cpu time. The rounding algorithm took 0.08 second; the randomized algorithm took between 0.15 and 0.19 seconds; and, the deterministic algorithm took between 0.5 and 0.6 seconds. Increasing the number of voltage constraints can be expected to increase the run time of the explicit search procedure and the rounding and randomized algorithms in approximately the same ratio: there would be a smaller effect on the deterministic algorithm.

A ten capacitor and four regulator system was simulated by making two copies of the test system. The explicit search procedure took between 1700 and 1800 seconds of cpu time, while the rounding, randomized, and deterministic algorithms

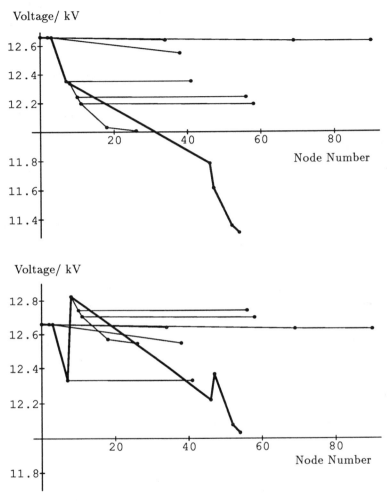

Figure 11: Top: Voltage profile of system under high load with optimal continuous capacitive susceptance and with regulator taps set to nominal. Bottom: Voltage profile under high load with optimal switched capacitive susceptance and with taps adjusted for voltage feasibility. In both cases the lateral with the lowest voltage profile is shown thicker: the voltage increases in the bottom profile occur at the regulators. Points represent the voltage at a node while the lines joining them indicate the topology of the system.

case	explicit	rounding	randomized	deterministic
5 capacitors, 2 regulators	2.6–2.9	0.08	0.15–0.19	0.5–0.6
10 capacitors, 4 regulators	1700–1800	0.09–0.12	0.27–0.28	2.4–3.1

Table 4: Cpu time for algorithms in VAX 6420 cpu seconds.

took 0.09–0.12, 0.27–0.28, and 2.4–3.1 seconds, respectively. This demonstrates the explosive growth in processing time required by non-polynomially bounded procedures such as explicit search and branch and bound: for the explicit search procedure, the cost of cpu time *alone* would probably be larger than the value of the expected energy savings. A branch and bound procedure would almost certainly improve on this situation; however, it is unlikely that it could compete with the three proposed algorithms, all of which have relatively modest computational requirements. The cpu times are summarized in table 4.

The low losses obtained with the algorithms justify the effort involved in the calculation of the base-case, particularly if, as mentioned in section II, the base-case calculation can be bundled with other distribution automation functions. The deterministic algorithm delivers switching schedules with losses that are on average lower than the schedules delivered by the randomized algorithm, which in turn delivers better schedules than the rounding algorithm; however, the improvement in losses of the deterministic over the rounding and randomized algorithms may not be warranted by its somewhat increased computation costs. *A fortiori* the occasional failure of the three algorithms to generate the optimal switching schedule would not justify the significantly increased cpu time required by the explicit search procedure. A similar conclusion could be expected to hold for a branch and bound procedure.

VIII   Relationship Between the Design and Coordination Problems

In section IV, using (34), the difference between the voltage-unconstrained minimum switched losses and the voltage-unconstrained minimum continuous losses was bounded by about 3 kW. In the medium load case, $\mathbf{B}_C^{opt}$ satisfies

$$\mathbf{B}_C^{\min} \leq \mathbf{B}_C^{opt} \leq \mathbf{B}_C^{\max}, \tag{64}$$

and all algorithms deliver schedules that satisfy the bound. In the high load case, the difference between the minimum voltage-unconstrained switched losses and the minimum voltage-unconstrained continuous losses is much greater than the bound, since $\mathbf{B}_C^{opt}$ does not satisfy (64); however, in this case, (34) is a valid bound for the difference between:

1. the optimal voltage-unconstrained continuous losses, subject to the *additional* constraint (64), and,

2. the optimal voltage-unconstrained switched losses.

The minimum continuous losses, subject to (64), are therefore only slightly lower than the minimum switched losses. This suggests that continuously controllable sources of reactive power will not generally be cost-effective over switched capacitors, unless other issues such as voltage constraints, voltage flicker, or fast load swings constrain the capacitor design [39]. Furthermore, the placement and sizing of *switched* capacitors can be performed *assuming* continuous control, subject to (64), so long as optimal or near optimal switched coordination is actually implemented: the bound, (34), can then be used to evaluate the slightly increased average losses under optimal switched control of capacitor placement and sizing designs that are based on the assumption of continuous control. In comparison, if a *local* strategy, such as the local control benchmark, is used to control the capacitors, then the design algorithms in the literature may give misleadingly optimistic results.

IX   Conclusion

We have analyzed the loss and voltage functions in a distribution system and given approximate formulas for them based on knowledge of the unconstrained minimum loss capacitive susceptance. The approximate formulas are accurate and the loss formula is conservative in the sense that the estimated losses are slightly larger than the actual losses. A useful byproduct of the analysis is a proof that the losses in the system is a convex function, so that steepest descent algorithms for the voltage-unconstrained minimum continuous losses problem can be guaranteed to converge to the global minimum. Distributed algorithms can be used to perform this base-case calculation. It was shown that the loss approximation has a special structure.

Algorithms, based on the approximations, have been presented that seek the voltage-constrained minimum loss switching schedule. The deterministic algorithm relies on the special structure of the loss formula. The algorithms are efficient for large systems in the sense that their run times are polynomially bounded in the problem size. Test results on a 70 node system confirm the theoretical predictions.

It was shown that the optimal continuous losses are typically only slightly lower than the optimal switched losses, therefore vindicating an approach to design that ignores the discrete nature of capacitors. This allows fast continuous algorithms to be used in the design problem with the confidence that combinatorial algorithms used for actual coordination will yield essentially the same loss levels. Further work includes:

- combining capacitor and regulator coordination with switch reconfiguration of the distribution system [40, 41];

- considering the tradeoff between

  - losses, and,
  - the communication costs involved with the several algorithms we have proposed for coordination,

  to optimize the choice of algorithm for coordination; and,

- considering the tradeoff between

  - losses,
  - the maintenance and operation cost of switched elements, and
  - the variability of distribution system load,

  to optimize the frequency with which capacitor and regulator schedules are updated.

The optimal coordination of capacitors and regulators promises to reduce losses and increase the efficiency of the electric distribution system.

References

[1] Edward Kahn. *Electric Utility Planning and Regulation.* American Council for an Energy-Efficient Economy, Washington, DC, 1988.

[2] T. J. E. Miller. The theory of load compensation. In T. J. E. Miller, editor, *Reactive Power Control in Electric Systems*, chapter 1, pages 1–48. John Wiley and Sons, New York, 1982.

[3] Turan Gönen. *Electric Power Distribution System Engineering.* McGraw-Hill series in electrical engineering. McGraw-Hill, New York, 1986.

[4] S. H. Lee and J. J. Grainger. Optimum placement of fixed and switched capacitors on primary distribution feeders. *IEEE Transactions on Power Apparatus and Systems*, PAS-100(1):345–352, January 1981.

[5] J. J. Grainger, S. Civanlar, and S. H. Lee. Optimal design and control scheme for continuous capacitive compensation of distribution feeders. *IEEE Transactions on Power Apparatus and Systems*, PAS-102(10):3271–3278, October 1983.

[6] S. Civanlar and J. J. Grainger. Volt/VAr control on distribution systems with lateral branches using shunt capacitors and voltage regulators: Parts I–III. *IEEE Transactions on Power Apparatus and Systems*, PAS-104(11):3278–3297, November 1985.

[7] Mesut E. Baran and Felix F. Wu. Optimal capacitor placement on radial distribution systems. *IEEE Transactions on Power Delivery*, 4(1):725–734, January 1989.

[8] Mesut E. Baran and Felix F. Wu. Optimal sizing of capacitors placed on a radial distribution system. *IEEE Transactions on Power Delivery*, 4(1):735–743, January 1989.

[9] K. A. Wirgau. Reactive power coordination. In T. J. E. Miller, editor, *Reactive Power Control in Electric Systems*, chapter 11, pages 353–363. John Wiley and Sons, New York, 1982.

[10] J. C. Picard and H. D. Ratliff. A graph-theoretic equivalence for integer programs. *Operations Research*, 21:261–269, 1973.

[11] J. C. Picard and H. D. Ratliff. Minimum cuts and related problems. *Networks*, 5(4):357–370, 1975.

[12] Jean-Claude Picard and Maurice Queyranne. Selected applications of minimum cuts in networks. *INFOR*, 20(4):394–422, November 1982.

[13] George L. Nemhauser and Laurence A. Wolsey. *Integer and Combinatorial Optimization*. John Wiley and Sons, New York, 1988.

[14] R. F. Cook. Optimizing the application of shunt capacitors for reactive-volt-ampere control and loss reduction. *AIEE Transactions, part III*, 80:430–444, August 1961.

[15] J. J. Grainger, S. H. Lee, and A. A. El-Kib. Design of a real-time switching control scheme for capacitive compensation of distribution feeders. *IEEE Transactions on Power Apparatus and Systems*, PAS-101(8):2420–2428, August 1982.

[16] J. J. Grainger, S. Civanlar, K. N. Clinard, and L. J. Gale. Discrete-tap control scheme for capacitive compensation of distribution feeders. *IEEE Transactions on Power Apparatus and Systems*, PAS-103(8):2098–2107, August 1984.

[17] Ross Baldick. Optimal on-off control of capacitors in a distribution system. Master's thesis, University of California, Berkeley, May 1988.

[18] J. J. Grainger, S. Civanlar, K. N. Clinard, and L. J. Gale. Optimal voltage dependent continuous-time control of reactive power on primary feeders. *IEEE Transactions on Power Apparatus and Systems*, PAS-103(9):2714–2722, September 1984.

[19] D. R. Brown. Performance analysis of a variable source of reactive power on distribution system primary feeders. *IEEE Transactions on Power Apparatus and Systems*, PAS-100(11):4364–4372, November 1981.

[20] A. A. El-Kib, J. J. Grainger, K. N. Clinard, and L. J. Gale. Placement of fixed and/or non-simultaneously switched capacitors on unbalanced three-phase feeders involving laterals. *IEEE Transactions on Power Apparatus and Systems*, PAS-104(11):3298–3305, November 1985.

[21] W. R. Owens. Save dollars with power-factor correction. *Electrical World*, pages 109–111, March 1983.

[22] Michael R. Garey and David S. Johnson. *Computers and Intractability: A Guide to the Theory of NP-Completeness.* W. H. Freeman and Company, San Francisco, 1979.

[23] Andreas Felix Neyer. *Distributed Algorithms for Monitoring and Control of Electric Power Transmission and Distribution Systems.* PhD thesis, University of California, Berkeley, 1989.

[24] William D. Stevenson, Jr. *Elements of Power System Analysis.* McGraw-Hill. 1982.

[25] F. D. Galiana and M. Banakar. Approximation formulae for dependent load flow variables. *IEEE Transactions on Power Apparatus and Systems,* PAS-100(3):1128–1137, March 1981.

[26] H. Glavitsch and M. Spoerry. Quadratic loss formula for reactive dispatch. In *1983 Power Industry Computer Application Conference—PICA-83,* pages 27–33. IEEE Power Engineering Society, 1983.

[27] Ross Baldick and Felix F. Wu. Approximation formulas for the distribution system: the loss function and voltage dependence. Paper 90 SM 286-5 PWRD presented at the IEEE Power Engineering Society 1990 Summer Meeting, Minneapolis, Minnesota, July 15–19, 1990. To appear in *IEEE Transactions on Power Delivery,* 1991.

[28] Hsiao-Dong Chiang and Mesut E. Baran. On the existence and uniqueness of load flow solution for radial distribution power networks. *IEEE Transactions on Circuits and Systems,* CAS-37(3):410–416, March 1990.

[29] Eugene L. Lawler. *Combinatorial Optimization: Networks and Matroids.* Holt, Rinehart and Winston, 1976.

[30] J. J. Grainger and S. H. Lee. Optimum size and location of shunt capacitors for reduction of losses on distribution feeders. *IEEE Transactions on Power Apparatus and Systems,* PAS-100(3):1105–1118, March 1981.

[31] Hsiao-Dong Chiang, Jin-Cheng Wang, Orville Cockings, and Hyoun-Duck Shin. Optimal capacitor placements in distribution systems: Part 1: A new formulation and the overall problem. *IEEE Transactions on Power Delivery,* 5(2):634–642, April 1990.

[32] Hsiao-Dong Chiang, Jin-Cheng Wang, Orville Cockings, and Hyoun-Duck Shin. Optimal capacitor placements in distribution systems: Part 2: Solution algorithms and numerical results. *IEEE Transactions on Power Delivery*, 5(2):643–649, April 1990.

[33] T. J. E. Miller and R. W. Lye. Principles of static compensators. In T. J. E. Miller, editor, *Reactive Power Control in Electric Systems*, chapter 4, pages 181–222. John Wiley and Sons, New York, 1982.

[34] Ross Baldick. *Improving Power System Operating Efficiency Through Real-Time Control and Tariffs*. PhD thesis, University of California, Berkeley, 1990.

[35] Michael O. Rabin. Probabilistic algorithms. In J. F. Traub, editor, *Algorithms and Complexity: New Directions and Recent Results*, pages 21–39. Academic Press, New York, 1976.

[36] Dorit S. Hochbaum and J. George Shanthikumar. Convex separable optimization is not much harder than linear optimization. Manuscript, April 1988, Revised January 1989, To appear in the Journal of the ACM, 1990.

[37] Christos H. Papadimitriou and Kenneth Steiglitz. *Combinatorial Optimization: Algorithms and Complexity*. Prentice-Hall, Inc., Englewood Cliffs, New Jersey, 1982.

[38] The Numerical Algorithms Group. *The NAG Fortran Library Manual*, 1987.

[39] T. J. E. Miller and A. R. Oltrogge. Reactive compensation and the electric arc furnace. In T. J. E. Miller, editor, *Reactive Power Control in Electric Systems*, chapter 9, pages 299–330. John Wiley and Sons, New York, 1982.

[40] Hsiao-Dong Chiang and René Jean-Jumeau. Optimal network reconfigurations in distribution systems: Part 1: A new formulation and a solution methodology. *IEEE Transactions on Power Delivery*, 5(4):1902–1909, November 1990.

[41] Hsiao-Dong Chiang and René Jean-Jumeau. Optimal network reconfigurations in distribution systems: Part 2: Solution algorithms and numerical results. *IEEE Transactions on Power Delivery*, 5(3):1568–1574, July 1990.

OPTIMAL OPERATIONAL PLANNING: A UNIFIED APPROACH TO REAL AND REACTIVE POWER DISPATCHES

KWANG Y. LEE

Department of Electrical and Computer Engineering
The Pennsylvania State University
University Park, Pennsylvania 16802

YOUNG MOON PARK

Department of Electrical Engineering
Seoul National University
Seoul 151, Korea

I.   INTRODUCTION

The optimal operation of a power system requires a judicious planning of available resources and facilities to their maximum potential before investing for additional facilities. This leads to the operational planning problem. The purpose of the operational planning problem is to minimize the fuel costs, system losses or some other appropriate objective functions, while maintaining an acceptable system performance in terms of voltage profile, contingencies, or system security. The operational planning problem was first formulated as an optimal power flow problem [1-4] by selecting the fuel cost as the objective function and the network or load-flow equations as constraints. The problem was solved for an optimal allocation of real power generation to units, resulting in an economic dispatch [5-8]. Recently, voltage stability, or voltage collapse [9] has been an increasingly important issue to utility as the power system is approaching its limit of operation due to economical and environmental constraints. Generators alone can no longer supply the reactive power that is needed to maintain the voltage profile within the allowed range throughout the power system. Additional reactive power or var sources need to be introduced and coordinated with generators. This has motivated many researchers to formulate optimal reactive power problems [10-12], wherein the system loss is used as an objective function, resulting in an economic reactive power dispatch [6,13].

To handle the large-scale problems of this nature, the idea of P-Q decomposition was applied to the optimal power flow [14-16], where the problem is decomposed into real-power (P) optimization problem and the reactive-power (Q) optimization problem. The P-optimization problem is to minimize the production cost under the assumption that system voltages are held constant, and the Q-optimization problem is to minimize the transmission loss under the assumption that real-power generation is held constant. Owing to the loose coupling between two problems, the sequential optimization of the these provided a considerable advantage over the simultaneous optimization of all control variables. It also should be noted that a number of articles are devoted to the Q-problem alone for the optimal control of reactive power flow [10-13].

The above two problems, optimal real and reactive power problems, were mostly formulated using two different objective functions; the fuel cost for the P-optimization problem and the system loss for the Q-optimization problem. However, it was found that unless all generating units are of the same efficiency, the minimization of the system loss does not guarantee the minimization of the operation (fuel) cost [6]. Lee and his co-workers [6-8] developed a unified approach which uses the one objective function, the fuel cost, for both P- and Q- optimization problems and avoids the switching of objective functions from one to another. Compared to the usual approach of loss minimization [12,13,15], this fuel cost minimization approach gives the least operation cost compared to the usual system loss minimization approaches [6].

The fuel cost minimization was also used for the long-term var planning problem [7] and for the operation of a large-scale power system [8]. In these works [6-8],

the network equations were linearized and the Jacobian matrix was partitioned d-
ifferently for the P- and Q- optimization problems. Although the fuel cost was the
objective function for the Q- optimization problem, it was expressed as a quadratic
function of reactive power supplies. These two factors sometimes cause oscillations
in the iterative optimization procedure.

The augmented Jacobian matrix can be partitioned in a unified way so that the
resulting sensitivity matrices can be made common to both P- and Q- optimizations.
Furthermore, the objective function for the Q- optimization problem can be defined
by augmenting the fuel cost function with the power balance equation, and by
using the voltage magnitudes and tap settings as the control variables. This allows
both the P- and Q- optimization problems to minimize the same objective function
iteratively without causing oscillations. Moreover, the Q- optimization becomes a
linear formulation because of the use of the bus voltage magnitudes as the control
variables.

Both the P- and Q-optimization problems are special cases of a general nonlinear programming problem with a nonlinear objective function and nonlinear functional equality and/or inequality constraints. A typical approach to solve this problem is to augment the constraints into an objective function by using the Lagrange
multipliers [5] and/or penalty functions, and to minimize the augmented objective
function by using an optimization scheme, such as the steepest   descent algorithms [1,15], or the sequential unconstrained minimization technique [14]. Other
approaches are the use of linear programming approximation to the nonlinear problem [10,12], or the use of quadratic approximation to the objective function in
order to apply the quadratic programming technique [17]. Owing to the size of the
problem as well as the large number of functional inequality constraints, improvement on computational efficiency has been the thrust of most works. Recently, Lee
and his co-workers [6-8] used a modified version of the gradient projection method
(GPM) [18], which introduces several advantages over traditional approaches. This
method avoids the use of penalty functions and/or calculations of Lagrange multipliers [1,14,15]; additionally, the step length is calculated without the need of a
linear search algorithm as in other methods.

In this development, a unified method is presented for the optimal operational planning of a power system. The method is based upon the three modules coupled with each other. First, the P-optimization module, which is equivalent to the economic dispatch, optimally allocates the real power generation among generators. The second module, Q-optimization module, optimally determines the reactive power output of generators and the var sources as well as tap-settings. Finally, the optimal loadflow module is used to make fine adjustments on the results of P- and Q- optimization modules.

II. OPERATIONAL PLANNING PROBLEM

The objective of the operational planning problem is to minimize the operation cost, which is the cost encountered in producing a certain amount of power to meet the demand within a system. The policy is to reduce the amount of fuel required to generate real power, while maintaining an acceptable system performance in terms of voltage profile and system security.

The process is decomposed into two modules, the real power (P) optimization and the reactive power (Q) optimization modules, in order to enhance the computation efficiency. In the P module, the real power generation is optimized by minimizing the incremental cost function, with the reactive power being adjusted as a dependent variable due to the coupling. In the Q module the optimal reactive power generation or bus voltages and transformer tap-settings are obtained by minimizing an equivalent cost function. A general optimization problem for the operational planning will be first formulated . Then the quadratic fuel cost function will be introduced and a procedure of obtaining the incremental cost function will be presented.

A.   Problem Formulation

The cost function is a function of real power generation and the objective is to minimize the fuel cost function:

Minimize

$$C(P_{sg}) = f(P_{sg}, Q_{sgc}, N), \tag{1}$$

subject to

$$g(P_{sg}, Q_{sgc}, N) = 0, \tag{2a}$$

and

$$\underline{P}_{sg} \le P_{sg} \le \overline{P}_{sg}$$

$$\underline{Q}_{sgc} \le Q_{sgc} \le \overline{Q}_{sgc}$$

$$\underline{V} \le V \le \overline{V} \tag{2b}$$

$$\underline{N} \le N \le \overline{N}$$

$$h(\delta, V) \le \overline{h},$$

where

s, g, c : indices for swing bus, other generator buses, and
            buses with compensators, respectively,

$C(P_{sg})$: total fuel cost for the generators,

P_{sg} : vector of real power generations,

Q_{sgc} : vector of reactive power supplies,

δ : vector of bus voltage angles,

V : vector of bus voltage magnitudes,

N : vector of tap-settings,

$h(.)$: vector of transmission line flows,

$g(P_{sg})$: vector of power supply-demand equations,

$\overline{(.)}, \underline{(.)}$: the upper and lower limits, respectively.

Here the objective function $f(.)$ is the total power production cost, or, more specifically, the total summation of generator fuel costs. It is expressed in terms of control variables P_{sg}, Q_{sgc}, and N as was in reference [6-8], where P_{sg} was the control variable for the P-module, and Q_{sgc} and N were the control variables for the Q-module. The use of Q_{sgc} as control variables in these works tends to create an oscillation problem and, thus, an improvement can be made. Since the reactive power Q_{sgc} and the bus voltage magnitude V are directly related to one another, V can be chosen as the control variable instead of Q_{sgc} in the Q-module. This converts the Q module to a linear objective function with linear constraints and, thus, the optimization is faster and without oscillations.

The problem defined by Eqs. (1) and (2) is a nonlinear programming problem and can be solved directly using an optimal power flow technique [1-4]. However the minimization of a nonlinear cost function (1) while satisfying the nonlinear power balance equation (2a) is not an easy job. An alternative way of handling the nonlinear power balance equation is to linearize the nonlinear equations and use the Jacobian matrix representation; a complete parallel way similar to the Newton-Raphson loadflow method. Since the variables in this method have to be small incremental values, the corresponding nonlinear cost function (1) also needs to be expressed in terms of the incremental variables.

B.   Incremental Cost Function

In order to decompose the operational planning problem into the P- and Q-optimization modules, the objective function and the network equations need to be expressed in terms of incremental variables. The cost function C can be expressed as a quadratic function of real power generations P_{sg} :

$$C(P_{sg}) = \sum_{k=1}^{m}(\alpha_k + \beta_k P_k + \gamma_k P_k^2), \tag{3}$$

where

$\alpha_k, \beta_k, \gamma_k$: the fuel cost coefficients,

m: the total number of generators,

P_k: the real power output of generator k.

The quadratic function (3) has been the most commonly used fuel cost function. Some utilities, however, traditionally may have been using other forms of nonlinear functions, such as a cubic function [5,7]. In these cases a least squares approximation can be used to find the fuel cost coefficients for an equivalent quadratic cost function [7].

Since the nonlinear power balance equation (2a) will be linearized with the incremental variables, the cost function (3) also needs to be expressed in terms of the incremental variables, leading to the incremental cost function. A procedure of deriving the incremental cost function is given below:

From Eq.(3), the cost for an increased production can be written as

$$C(P_{sg} + \Delta P_{sg}) = \sum_{k=1}^{m} [\alpha_k + \beta_k(P_k + \Delta P_k) + \gamma_k(P_k + \Delta P_k)^2]$$

$$= \sum_{k=1}^{m} [\alpha_k + \beta_k P_k + \gamma_k P_k^2 + (\beta_k + 2\gamma_k P_k)\Delta P_k + \gamma_k \Delta P_k^2]. \qquad (4)$$

Subtracting Eq. (3) from Eq. (4), the incremental cost function is obtained as

$$\Delta C(\Delta P_{sg}) \triangleq C(P_{sg} + \Delta P_{sg}) - C(P_{sg})$$

$$= \sum_{k=1}^{m} (\beta_k + 2\gamma_k P_k)\Delta P_k + \gamma_k \Delta P_k^2,$$

where

ΔP_{sg} : vector changes in real power generation P_{sg}.

$\Delta C(.)$: incremental cost function

This incremental cost is also in a quadratic form, and can be expressed in a matrix form:

$$\Delta C(\Delta P_{sg}) = C_p^T \Delta P_{sg} + \Delta P_{sg}^T D_p \Delta P_{sg}, \qquad (5)$$

where $(.)^T$ denotes the transpose, and

$C_p^T \triangleq [\beta_1 + 2\gamma_1 P_1, \beta_2 + 2\gamma_2 P_2, \dots, \beta_m + 2\gamma_m P_m]$

$D_p \triangleq diag[\gamma_1, \gamma_2, \dots, \gamma_m].$

It should be noted that the usual linearization approach to the nonlinear cost function (3) will yield only a linear incremental cost function, which is not as accurate as the quadratic incremental cost function (5). This cost function is directly usable in the P-optimization module since ΔP_{sg} itself is the control variable in that module. In the Q-optimization module, however, ΔP_{sg} should be expressed as a function of the Q-optimization control variables ΔQ_{sgc} or ΔV, and ΔN. The main operational problem defined in Eqs. (1) and (2) is now decomposed into P-optimization and Q-optimization problems.

III.   THE P-Q DECOMPOSITION

The concept of P-Q decomposition was found to be very useful in solving the loadflow problem; the fast decoupled loadflow technique is based on the decomposition principle using the fact the real and reactive power controls are loosely coupled [4]. In a similar way, the optimization problem can be solved more efficiently by using the decomposition principle. The incremental cost function (5) is to be minimized while satisfying the power balance equation (2a), which is a set of nonlinear algebraic equations in terms variables P_{sg}, Q_{sgc} or V, and N. In order to decompose the operational planning problem, the power balance equation needs to be linearized and expressed in terms of the incremental variables $\Delta P_{sg}, \Delta Q_{sgc}$ or ΔV, and ΔN. This will make it possible to develop sensitivity relationships among the incremental variables; the relationships will then serve as constraints in the P- and Q- optimization modules.

A.   Linearization of Power Balance Equation

The power balance equation (2a), can be linearized and expressed in the following matrix form [19]:

$$\begin{bmatrix} \Delta P \\ \Delta Q \end{bmatrix} = [J] \begin{bmatrix} \Delta \delta \\ \Delta V \\ \Delta N \end{bmatrix}, \tag{6}$$

where

    J : Jacobian matrix,
    ΔP : vector of changes in bus real power,
    ΔQ : vector of changes in bus reactive power,
    $\Delta \delta$: vector of changes in bus voltage angles,
    ΔV : vector of changes in bus voltage magnitudes,
    ΔN : vector of changes in transformer tap settings.

The Jacobian J in Eq. (6) can be partitioned as

$$
\begin{bmatrix}
\Delta P_s \\
-- \\
\Delta P_g \\
\Delta P_l \\
-- \\
\Delta Q_{sgc} \\
-- \\
\Delta Q_{l'}
\end{bmatrix}
=
\begin{bmatrix}
J_{10} & J_{11} & J_{12} & J_{13} \\
-- & -- & -- & -- \\
 & & & \\
J_{20} & J_{21} & J_{22} & J_{23} \\
-- & -- & -- & -- \\
J_{30} & J_{31} & J_{32} & J_{33} \\
-- & -- & -- & -- \\
J_{40} & J_{41} & J_{42} & J_{43}
\end{bmatrix}
\begin{bmatrix}
\Delta \delta_s \\
-- \\
\Delta \delta_g \\
\Delta \delta_l \\
-- \\
\Delta V \\
-- \\
\Delta N
\end{bmatrix}
\tag{7}
$$

where

s, g, c : indices for swing bus, generators, and reactive power
compensating devices, respectively

l, l' : indices for all load buses and load buses which do not
have reactive compensating devices, respectively.
(These symbols are also used for the maximum number
of corresponding buses for notational brevity.)

It should be noted that the Jacobian matrix contains entries for all buses, including the swing bus. This is in contrast with the usual Newton-Raphson loadflow, where the entries corresponding to the swing bus are deleted in the Jacobian. The entries in the block matrices can be derived from the standard power balance equations [19]. The last column block can be obtained by differentiating the nodal power equations with respect to the off-nominal tap-setting values N. Note that the Jacobian matrix in Eq. (7) unifies the two Jacobian matrices used for the P- and Q-optimization modules in references [6-8]. This simplifies the decomposition and leads to a consistent set of sensitivity matrices and a smooth transition from the P-optimization module to the Q- optimization module.

B.   Sensitivity Equations

In the usual decoupled loadflow the off-diagonal blocks in the Jacobian are assumed to be very small and, thus, neglected. This yields a complete decoupling between the real and reactive powers. In reality, however, the real and reactive powers are coupled; although the degree of coupling may be small. When the real power is changed or controlled, the reactive power is also changed. Similarly, when the reactive power is controlled, the real power is also changed. This implies that the reactive power control also affects the fuel cost. This is the motivation behind the fuel cost minimization for the Q-optimization problem.

The relationship between the real and reactive powers can be represented by sensitivity equations. These sensitivity relationships not only reflect the control variables in the fuel cost, but also provide linear constraint equations for each optimization problem.

From Eq. (7) the following sensitivity relationship is obtained:

By applying the condition $\Delta \delta_s = 0$,

$$
\begin{aligned}
\Delta P_s &= \frac{\partial P_s}{\partial \delta_{gl}} \Delta \delta_{gl} + \frac{\partial P_s}{\partial V} \Delta V + \frac{\partial P_s}{\partial N} \Delta N \\
&= [J_{11}] \begin{bmatrix} \Delta \delta_g \\ \Delta \delta_l \end{bmatrix} + J_{12} \Delta V + J_{13} \Delta N,
\end{aligned} \tag{8}
$$

where

$J_{11} \triangleq \frac{\partial P_s}{\partial \delta_{gl}}$: 1x $(m + l - 1)$ row vector

$J_{12} \triangleq \frac{\partial P_s}{\partial V}$: 1x$(m + l)$ row vector

$J_{13} \triangleq \frac{\partial P_s}{\partial N}$: 1x N_t row vector

N_t : the total number of tap-setting transformers.

Similarly, from Eq. (7),

$$
\begin{aligned}
\begin{bmatrix} \Delta P_g \\ \Delta P_l \end{bmatrix} &= \frac{\partial P_{gl}}{\partial \delta_{gl}} \Delta \delta_{gl} + \frac{\partial P_{gl}}{\partial V} \Delta V + \frac{\partial P_{gl}}{\partial N} \Delta N \\
&= J_{21} \begin{bmatrix} \Delta \delta_g \\ \Delta \delta_l \end{bmatrix} + J_{22} \Delta V + J_{23} \Delta N,
\end{aligned} \tag{9}
$$

where

$J_{21} \triangleq \frac{\partial P_{gl}}{\partial \delta_{gl}}$: $(m + l - 1)$ x $(m + l - 1)$ matrix

$J_{22} \triangleq \frac{\partial P_{gl}}{\partial V}$: $(m + l - 1)$ x $(m + l)$ matrix

$J_{23} \triangleq \frac{\partial P_{gl}}{\partial N} : (m + l - 1) \times N_t$ matrix.

Solving Eq. (9) for voltage angles,

$$\begin{bmatrix} \Delta \delta_g \\ \Delta \delta_l \end{bmatrix} = J_{21}^{-1} \left[\begin{bmatrix} \Delta P_g \\ \Delta P_l \end{bmatrix} - J_{22} \Delta V - J_{23} \Delta N \right], \tag{10}$$

and substituting this in Eq. (10),

$$\Delta P_s = [J_{11} J_{21}^{-1}] \begin{bmatrix} \Delta P_g \\ \Delta P_l \end{bmatrix} + [J_{12} - J_{11} J_{21}^{-1} J_{22}] \Delta V$$

$$+ [J_{12} - J_{11} J_{21}^{-1} J_{23}] \Delta N.$$

Since $\Delta P_l = 0$ for normal load buses,

$$\Delta P_s = J_A \Delta P_g + J_B \Delta V + J_C \Delta N, \tag{11}$$

where

$J_A \triangleq$ the $1^{st}(m - 1)$ elements of $[J_{11} J_{21}^{-1}]$

$J_B \triangleq [J_{12} - J_{11} J_{21}^{-1} J_{22}]$

$J_C \triangleq [J_{13} - J_{11} J_{21}^{-1} J_{23}]$.

Equation (11) can also be written as

$$[1 : -J_A] \Delta P_{sg} = \Delta S, \tag{12}$$

where

$$\Delta S \triangleq J_B \Delta V + J_C \Delta N.$$

This is a power balance or sensitivity equation, which is a linearized version of the general power balance equation (2a). This equation shows the relationship between the real power and the control variables, ΔV and ΔN, for the Q-optimization problem. In the P-optimization module the sensitivity ΔS is set to zero.

It should be noted that the use of power balance equation (12) eliminates the use of usual B-coefficients. Since the B-coefficients are used in a very crude way to approximate the system loss and, furthermore, hard to compute for a large system, the conventional economic dispatch which is based upon B-coefficients is not optimal in reality. Thus, even the P-optimization alone can provide better operational planning solution.

Similarly from Eq. (7), the reactive power is expanded as

$$\Delta Q_{sgc} = \frac{\partial Q_{sgc}}{\partial \delta_{gl}} \Delta \delta_{gl} + \frac{\partial Q_{sgc}}{\partial V} \Delta V + \frac{\partial Q_{sgc}}{\partial N} \Delta N$$
$$= J_{31} \Delta \delta_{gl} + J_{32} \Delta V + J_{33} \Delta N, \qquad (13)$$

where

$$J_{31} \overset{\Delta}{=} \frac{\partial Q_{sgc}}{\partial \delta_{gl}} : (m + l - l') \times (m + l - 1) \text{ matrix}$$

$$J_{32} \overset{\Delta}{=} \frac{\partial Q_{sgc}}{\partial V} : (m + l - l') \times (m + l) \text{ matrix}$$

$$J_{33} \overset{\Delta}{=} \frac{\partial Q_{sgc}}{\partial N} : (m + l - l') \times N_t \text{ matrix.}$$

Substituting $\Delta \delta_{gl}$ of Eqs. (10), Eq. (13) becomes

$$\Delta Q_{sgc} = [J_{31} J_{21}^{-1}] \begin{bmatrix} \Delta P_g \\ \Delta P_l \end{bmatrix} + [J_{32} - J_{31} J_{21}^{-1} J_{22}] \Delta V$$
$$+ [J_{33} - J_{31} J_{21}^{-1} J_{23}] \Delta N,$$

which then becomes

$$\Delta Q_{sgc} = J_D \Delta P_g + J_E \Delta V + J_F \Delta N, \qquad (14)$$

where

$$J_D \overset{\Delta}{=} \text{ the first (m-1) columns of } [J_{31} J_{21}^{-1}]$$

$$J_E \overset{\Delta}{=} [J_{32} - J_{31} J_{21}^{-1} J_{22}]$$

$$J_F \overset{\Delta}{=} [J_{33} - J_{31} J_{21}^{-1} J_{23}].$$

The sensitivity equation (14) gives the values of the reactive power supplies whenever the control variables for the P- and Q- optimization modules are determined.

Finally, the reactive power change in load buses is from Eq. (7),

$$\Delta Q_{l'} = \frac{\partial Q_{l'}}{\partial \delta_{gl}} \Delta \delta_{gl} + \frac{\partial Q_{l'}}{\partial V} \Delta V + \frac{\partial Q_{l'}}{\partial N} \Delta N$$
$$= J_{41} \Delta \delta_{gl} + J_{42} \Delta V + J_{43} \Delta N, \qquad (15)$$

where

$J_{41} \triangleq \frac{\partial Q_{l'}}{\partial \delta_{gl}} : l' \times (m + l - 1)$ matrix

$J_{42} \triangleq \frac{\partial Q_{l'}}{\partial V} : l' \times (m + l)$ matrix

$J_{43} \triangleq \frac{\partial Q_{l'}}{\partial N} : l' \times N_t$ matrix.

Substituting $\Delta \delta_{gl}$ of Eq. (10), Eq. (15) becomes

$$\Delta Q_{l'} = [J_{41} J_{21}^{-1}] \begin{bmatrix} \Delta P_g \\ \Delta P_l \end{bmatrix} + [J_{42} - J_{41} J_{21}^{-1} J_{22}] \Delta V$$
$$+ [J_{43} - J_{41} J_{21}^{-1} J_{23}] \Delta N,$$

which becomes

$$\Delta Q_{l'} = J_G \Delta P_g + J_H \Delta V + J_I \Delta N, \qquad (16)$$

where

$J_G \triangleq$ the first $(m - 1)$ columns of $J_{41} J_{21}^{-1}$

$J_H \triangleq J_{42} - J_{41} J_{21}^{-1} J_{22}$

$J_I \triangleq J_{43} - J_{41} J_{21}^{-1} J_{23}$.

The sensitivity equation (16) reflects the changes in the control variables, ΔP_g, ΔV, and ΔN, on the reactive power in load buses.

C.   The P-Optimization Module

In this module, the P-optimization problem is solved by minimizing the incremental cost function (5) as a function of the incremental change in real power generation. For the P module, using the decoupling condition

$$\Delta P_l = \Delta V = \Delta \delta_s = \Delta N = 0, \qquad (17)$$

the linearized constraints, Eqs. (12) and (14), are simplified as

$$\Delta P_s = J_A \Delta P_g \tag{18}$$

and

$$\Delta Q_{sgc} = J_D \Delta P_g. \tag{19}$$

In the general case of iteration, using the incremental cost function (5) and constraint (12), the P- optimization module becomes as follows:

Minimize

$$\Delta C_p(\Delta P_{sg}) = C_p^T \Delta P_{sg} + \Delta P_{sg}^T D_p \Delta P_{sg}, \tag{20}$$

subject to

$$[1 \ : \ -J_A]\Delta P_{sg} = \Delta S \tag{21}$$

$$\underline{\Delta P_{sg}} \leq \Delta P_{sg} \leq \Delta \overline{P_{sg}},$$

where

$$\underline{\Delta P_{sg}} = \underline{P_{sg}} - P_{sg}$$

$$\Delta \overline{P_{sg}} = \overline{P_{sg}} - P_{sg}$$

$$SS \stackrel{\Delta}{=} J_B \Delta V + J_C \Delta N.$$

In the P-module, the sensitivity ΔS is set to zero. This quadratic optimization problem is solved by the gradient projection method (GPM) [6] to find the optimal solution ΔP_{sg}. Another benefit of this P-optimization module is that, as a by-product, it computes the reactive power changes Q_{sgc} caused by the P-optimization as given by the sensitivity equation (19). Finally, the solution of the P-optimization module is used to update P and Q as follows:

$$P_{sg}^{k+1} = P_{sg}^k + \Delta P_{sg}^k$$

$$Q_{sgc}^{k+1} = Q_{sgc}^k + \Delta Q_{sgc}^k. \tag{22}$$

D.   Q - Optimization Module: Linear Form

After solving the P-optimization module, the load flow is run to fine tune the system operating condition, and then followed by the Q optimization. In this Q-optimization module, the incremental cost function (20) is to be minimized by the incremental changes in voltage magnitude and tap-settings. A constraint for this module is, from Eq. (21),

$$[1 \; : -J_A]\Delta P_{sg} = J_B \Delta V + J_C \Delta N. \tag{23}$$

Additional constraints are obtained from Eqs. (13) and (15), by using decoupling conditions $\Delta \delta_{gl} = 0$ and $\Delta Q_{l'} = 0$ for the Q module,

$$\Delta Q_{sgc} = J_{32}\Delta V + J_{33}\Delta N \tag{24a}$$

$$0 = J_{42}\Delta V + J_{43}\Delta N. \tag{24b}$$

The constraint equation (23) implies that when ΔV and ΔN are adjusted in the Q-optimization module, ΔP_{sg} will also be changed, which will then change the incremental cost (20). Therefore, the cost function for the Q module can be obtained by augmenting the real cost function (20) with constraint (23):

$$\Delta C_q = \Delta C_p + \lambda\{[1 \; : -J_A]\Delta P_{sg} - J_B \Delta V - J_C \Delta N\}, \tag{25}$$

where λ is the Lagrange multiplier.

Note in Eq.(25) that the variation

$$\Delta C_p + \lambda\{[1 \; : -J_A]\Delta P_{sg}\}$$

is assumed already optimized in the P-optimization module, and, thus, fixed constant in this Q-optimization module. Additional minimization, then, can be achieved by the changes in ΔV and ΔN.

Thus, from the constraints (23)-(24) and the cost function (25), the Q - optimization module becomes as follows:

Minimize

$$\Delta C_q(\Delta V, \Delta N) = -\lambda[J_B \Delta V + J_C \Delta N], \qquad (26)$$

subject to

$$J_{42}\Delta V + J_{43}\Delta N = 0$$

$$\underline{\Delta Q}_{sgc} \leq J_{32}\Delta V + J_{33}\Delta N \leq \overline{\Delta Q}_{sgc}$$

$$\underline{\Delta N} \leq \Delta N \leq \Delta \overline{N} \qquad (27)$$

$$\underline{\Delta V} \leq \Delta V \leq \Delta \overline{V}$$

$$h(V) \leq \overline{h},$$

where

$\Delta C_q(.)$: the incremental operational cost for the
    reactive compensation

$\overline{\Delta V} \triangleq \overline{V} - V$

$\underline{\Delta V} \triangleq \underline{V} - V$

$\overline{\Delta Q}_{sgc} \triangleq \overline{Q}_{sgc} - Q_{sgc}$

$\underline{\Delta Q}_{sgc} \triangleq \underline{Q}_{sgc} - Q_{sgc}$

$\overline{\Delta N} \triangleq \overline{N} - N$

$\underline{\Delta N} \triangleq \underline{N} - N.$

The Q-optimization problem can also be solved by the same gradient projection method (GPM) to find the optimal solution (ΔV and ΔN), which then gives optimum reactive power supply ΔQ_{sgc} by the sensitivity equation (24a). It may be noted, however, that the cost function (26) is a linear function. Thus, computation will be faster and a linear programming method can be used instead of the GPM. Finally, the solution of the Q-optimization module is used to update Q, V, and N as follows:

$$Q_{sgc}^{k+1} = Q_{sgc}^k + \Delta Q_{sgc}^k$$

$$V^{k+1} = V^k + \Delta V^k, \tag{28}$$

$$N^{k+1} = N^k + \Delta N^k.$$

E.   Q-Optimization Module: Quadratic Form

The control variables for the Q-optimization module were ΔV and ΔN in the above formulation. The voltage magnitude ΔV can be replaced by the reactive power ΔQ_{sgc} as the control variable, which was the case for references [6-8]. In this case the sensitivity of the real power ΔP_{sg} with respect to the reactive power ΔQ_{sgc} can be obtained, and the fuel cost function is represented by a quadratic function in ΔQ_{sgc}.

For this formulation it is necessary to eliminate the dependent variable ΔV and use ΔQ_{sgc} and ΔN as the control variables.

Using the same decoupling conditions $\Delta \delta_{sgl} = 0$, Eqs. (8) and (9) can be combined as

$$\Delta P_{sgl} = \begin{bmatrix} \Delta P_s \\ \Delta P_g \\ \Delta P_l \end{bmatrix} = \begin{bmatrix} J_{12} \\ J_{22} \end{bmatrix} \Delta V + \begin{bmatrix} J_{13} \\ J_{23} \end{bmatrix} \Delta N,$$

or

$$\Delta P_{sg} = \begin{bmatrix} \Delta P_s \\ \Delta P_g \end{bmatrix} \triangleq J_{p2} \Delta V + J_{p3} \Delta N, \tag{29}$$

where

$$J_{p2} \triangleq \text{the first } m \text{ columns of } \begin{bmatrix} J_{12} \\ J_{22} \end{bmatrix}$$

$$J_{p2} \triangleq \text{the first } m \text{ columns of } \begin{bmatrix} J_{12} \\ J_{22} \end{bmatrix}.$$

Similarly Eqs. (13) and (15) can be combined as

$$\Delta Q = \begin{bmatrix} \Delta Q_{sgc} \\ \Delta Q_{l'} \end{bmatrix} = \begin{bmatrix} J_{32} \\ J_{42} \end{bmatrix} \Delta V + \begin{bmatrix} J_{33} \\ J_{43} \end{bmatrix} \Delta N, \tag{30}$$

$$\triangleq J_{q2} \Delta V + J_{q3} \Delta N.$$

Accordingly, the dependent variables ΔV can be expressed as a function of control variables, ΔQ_{sgc} and ΔN, as

$$\Delta V = J_{q2}^{-1} \begin{bmatrix} \Delta Q_{sgc} \\ \Delta Q_{l'} \end{bmatrix} - J_{q2}^{-1} J_{q3} \Delta N$$

$$\triangleq J_J \Delta Q_{sgc} - J_{q2}^{-1} J_{q3} \Delta N. \tag{31}$$

where

$J_J \triangleq$ the first $(m + l - l')$ columns of J_{q2}^{-1}.

Here the decoupling condition $Q_{l'} = 0$ is used. Then, rearranging Eq (31),

$$\Delta V = [J_J \ : \ -J_{q2}^{-1} J_{q3}] \begin{bmatrix} \Delta Q_{sgc} \\ \Delta N \end{bmatrix}$$

$$\triangleq J_K \begin{bmatrix} \Delta Q_{sgc} \\ \Delta N \end{bmatrix}, \tag{32}$$

where

$J_K \triangleq [J_J \ : \ -J_{q2}^{-1} J_{q3}].$

Also, substitution of Eq. (31) into Eq.(29) results in

$$\Delta P_{sg} = J_{p2} J_J \Delta Q_{sgc} + [J_{p3} - J_{p2} J_{q2}^{-1} J_{q3}] \Delta N$$

$$= [J_{p2} J_J \ : \ J_{p3} - J_{p2} J_{q2}^{-1} J_{q3}] \begin{bmatrix} \Delta Q_{sgc} \\ \Delta N \end{bmatrix}$$

$$\triangleq J_L \begin{bmatrix} \Delta Q_{sgc} \\ \Delta N \end{bmatrix}, \tag{33}$$

where

$J_K \triangleq [J_{p2} J_J \ : \ J_{p3} - J_{p2} J_{q2}^{-1} J_{q3}].$

Equation (32) represents the sensitivity of the voltage magnitude ΔV on the control variables ΔQ_{sgc} and ΔN. Equation (33) represents the sensitivity relationship which reflects the changes in ΔQ_{sgc} and ΔN on the real power generation ΔP_{sg}; this, then, is reflected in the changes in the fuel cost.

Consequently from Eqs. (5), (32) and (33), the Q-optimization becomes as follows:

Minimize

$$\Delta C_q(\Delta Q_{sgc}, \Delta N) = C_q^T \begin{bmatrix} \Delta Q_{sgc} \\ \Delta N \end{bmatrix} + \begin{bmatrix} \Delta Q_{sgc} \\ \Delta N \end{bmatrix}^T D_q \begin{bmatrix} \Delta Q_{sgc} \\ \Delta N \end{bmatrix}, \qquad (34)$$

subject to

$$\Delta \underline{V} \leq J_K \begin{bmatrix} \Delta Q_{sgc} \\ \Delta N \end{bmatrix} \leq \Delta \overline{V}$$

$$\Delta \underline{Q}_{sgc} \leq \Delta Q_{sgc} \leq \overline{\Delta Q}_{sgc} \qquad (35)$$

$$\Delta \underline{N} \leq \Delta N \leq \Delta \overline{N},$$

where

$$\overline{\Delta V} \triangleq \overline{V} - V$$

$$\underline{\Delta V} \triangleq \underline{V} - V$$

$$\overline{\Delta Q}_{sgc} \triangleq \overline{Q}_{sgc} - Q_{sgc}$$

$$\underline{\Delta Q}_{sgc} \triangleq \underline{Q}_{sgc} - Q_{sgc}$$

$$\overline{\Delta N} \triangleq \overline{N} - N$$

$$\underline{\Delta N} \triangleq \underline{N} - N$$

$$C_q^T \triangleq C_p^T J_L$$

$$D_q \triangleq J_L^T D_p J_L.$$

It should be noted that the cost function (34) is a quadratic function, and, thus, slower in speed than the linear formulation in Eq.(26). This quadratic form was used in references [6-8].

IV. GRADIENT PROJECTION METHOD

The P- and Q-optimization problems can be solved by any nonlinear programming method. The gradient projection method (GPM) was found to be very efficient algorithm for these optimization problems. The GPM, originally introduced by Rosen [18], is an iteratative numerical procedure to find the minimum (or maximum) of a nonlinear function which is constrained by linear constraints.

Assuming $f(x)$ is a convex function defined in R^n, the mathematical programming problem with linear constraints can be defined as follows:

Minimize

$$f(x),\tag{36}$$

subject to

$$M^T x - b \geq 0,\tag{37}$$

where

$M \triangleq [m_1, ..., m_E, m_{E+1}, ..., m_{E+I}]$

$B \triangleq [b_1, ..., b_E, b_{E+1}, ..., b_{E+I}]$

E = number of equality constraints

I = number of inequality constraints.

The vector m_j is the $n-$ dimensional inward normal vector of the constraints j, in the form

$$m_j x - b_j \geq 0.\tag{38}$$

The method consists of projecting the negative of the gradient of the objective function (36) along the boundaries of the *active* constraints contained in Eq.(37). Assuming x^i is an initial feasible point, the new point x^{i+1} will be defined as

$$x^{i+1} = x^i + \tau z^i\tag{39}$$

where τ is the step length, and z^i is the unit vector in the direction of the projected gradient defined by

$$z^i = \frac{P_q \left[\frac{-\partial f^{(i)}}{\partial x} \right]}{\left|\left| P_q \left[\frac{-\partial f^{(i)}}{\partial x} \right] \right|\right|}.\tag{40}$$

Here P_q is called the projection matrix which is calculated using the matrix M_q that contains the normal vectors of q *active* constraints:

$$P_q = I - M_q [M_q^T M_q]^{-1} M_q^T.\tag{41}$$

In this method the step length τ is calculated to be the maximum step that can be taken without violating any constraint. It is the smallest positive value of τ_j found by evaluating

$$\tau_j = \frac{b_j - m_j^T x^i}{m_j^T z^i} \qquad (42)$$

for all nonactive constraints.

One of the problems of this method for computer applications is the computation of an initial feasible point. This problem can be overcome by projecting the initial point perpendicular to the nearest violated constraint, then the new point is projected along this constraint until it hits another violated constraint. This procedure is repeated until a feasible point is found.

This optimization method has been shown to be very useful in the solution of the optimal power dispatch problem. The followings are some of the most important characteristics of this method:

(a) fast convergence,

(b) capacity of handling equality and inequality constraints without the use of penalty functions or Lagrange multipliers,

(c) step length τ is calculated without the need of linear searches, as in other methods such as the steepest descent method,

(d) inversion of matrix $[M_q^T M_q]^{-1}$ in Eq. (41) is evaluated using a recursive formula, which avoids the need of inverting the matrix each time a constraint is dropped or added from M_q, thus saving a considerable amount of time.

V.   SIMULATION RESULTS

The proposed algorithm was applied in three test systems. These are the 6-bus system [6], the IEEE 14-bus system, and the modified IEEE 30-bus system [4].

A.   The 6-Bus System

Figure 1 shows the one line diagram of the 6-bus system with 7 lines, 2 generator banks, and 2 tap-changing transformers. The line data and generator data are

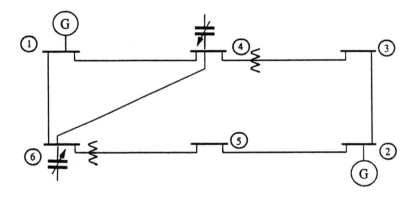

Fig. 1.   One Line Diagram of the 6-Bus System.

shown in Table I. Table II shows the initial loadflow, which also includes specified generator and load data. No control is made on the capacitor banks or tap-changing transformers. It can be noted that the voltages at load buses are very low; all below the lower limits of 0.90 per unit (p.u.). Generation cost and system loss for this case are 619.53 $/h and 9.8 MW, respectively.

Three case studies are made for comparison: (1) Q-optimization, (2) P- optimization, and (3) P-Q optimization, with results summarized in Table III.

(1) Q-optimization

In order to see the impact of the reactive power dispatch, only the Q - optimization module was run, while keeping the real power generation constant. In this case all means of reactive power control, generator reactive power, (buses 1 and 2), capacitor banks at (buses 4 and 6), and tap-settings (lines 4 and 7), are fully utilized. This has resulted in the increase of bus voltage magnitudes throughout the system, with the minimum value of 0.978 p.u. at bus 3.

It should be noted that, besides the voltage improvement, the swing bus power is reduced due to the Q-optimization; consequently, the generation cost and system loss are reduced to 586.33 $/h and 6.62 MW, respectively.

(2) P-optimization

For comparison purpose the usual economic dispatch was run with the P-optimization module only. In this case much generation is shifted from generator 1 to generator 2, which is more efficient unit among the two. Consequently, the generation cost is reduced to 454.10 $/h. However, the system loss is increased considerably to 26.99 MW.

It can be noted that the voltage magnitudes are decreased from the initial case since the P module does not include the voltage constraint.

(3) P-Q optimization

Finally, both P- and Q-optimization modules are run iteratively. In this case, the improvements made in the first two cases are both observed. The reactive power controls are fully utilized to improve the voltage profile and the real power is adjusted to decrease the generation cost. The generation cost is significantly

Table I. Data For 6 Bus System

(a) Line Data

Line Number	From bus	To bus	Line R[p.u.]	Imped. X[p.u.]	Tap setting	Line Charg.
1	1	6	0.123	0.518	0.0000	0.000
2	1	4	0.080	0.370	0.0000	0.000
3	4	6	0.097	0.407	0.0000	0.000
4	6	5	0.000	0.300	1.0250	0.000
5	5	2	0.282	0.640	0.0000	0.000
6	2	3	0.723	1.050	0.0000	0.000
7	4	3	0.000	0.133	1.1000	0.000

(b) Generator Data

number	Cost α	Coeffs. β	γ
1	0.0000	1.0000	0.005
2	0.0000	1.0000	0.010

Table II.   Initial Loadflow For 6-Bus System

BUS POWER INFORMATION

BUS	V[P.U]	ANG[DEG]	PG[MW]	QG[Mvar]	PD[MW]	QD[Mvar]	P[MW]	Q[Mvar]
1	1.000	0.000	94.836	53.845	0.000	0.000	94.836	53.845
2	1.000	-5.046	50.000	25.193	0.000	0.000	50.005	25.193
3	0.784	-15.916	0.000	0.000	55.000	13.000	-54.993	-12.990
4	0.883	-11.242	0.000	0.000	0.000	0.000	-0.009	-0.009
5	0.820	-15.284	0.000	0.000	30.000	18.000	-30.000	-18.001
6	0.873	-14.504	0.000	0.000	50.000	5.000	-49.998	-4.999

TOTAL LOSS POWER =   9.836+ j   43.039

INITIAL FUEL COST =   619.5305

Table III. Summary of Results For 6-Bus System

Variable	Lower	Limits Upper	Initial State	First Study	Second Study	Third Study
P_1(MW)	0.0	100.0	94.836	91.620	61.987	52.786
P_2(MW)	0.0	100.0	50.000	50.000	100.000	100.000
Q_1(Mvar)	-20.0	100.0	53.845	28.305	72.984	42.349
Q_2(Mvar)	-20.0	100.0	25.193	7.633	35.320	8.733
V_1(p.u.)	1.00	1.10	1.090	1.060	1.000	1.090
V_2(p.u.)	1.00	1.15	1.000	1.104	1.000	1.145
V_3(p.u.)	0.90	1.05	0.784	0.978	0.718	0.967
V_4(p.u.)	0.90	1.05	0.883	0.999	0.838	1.003
V_5(p.u.)	0.90	1.05	0.820	1.004	0.727	0.998
V_6(p.u.)	0.90	1.05	0.873	0.990	0.824	0.987
N_4	0.90	1.10	1.025	0.938	1.025	0.913
N_7	0.90	1.10	1.100	1.000	1.100	1.000
Q_{c4}(Mvar)	0.0	15.0	0.0	14.772	0.000	14.648
Q_{c6}(Mvar)	0.0	15.5	0.0	15.291	0.000	15.092
Generation	Cost [$/h]		619.531	586.329	454.103	392.106
System	Loss [MW]		9.836	6.620	26.987	17.786

First Study - with Q-optimization module only

Second Study - with P-optimization module only

Third Study - with both P- and Q-optimization modules

reduced to 392.11 \$/h, which is a considerable savings from 455.11 \$/h of case 2. The result is summarized in Table IV.

Figure 2 shows the convergence characteristics of case 3, where both P and Q modules were used iteratively in each iteration. It should be noted that the economic dispatch alone will stop at the first iteration with generation cost of 454.10 \$/h (case 2). However, the iterative use of P and Q modules was able to reduce the cost to 392.11 \$/h after 10 iterations. It should also be noted that, except at the initial iteration, the role of Q module is more significant than the P module; but the Q module alone will not bring the cost down.

B.   IEEE 14-Bus System

Figure 3 shows the one line diagram of the IEEE 14-bus with 20 lines, 5 generators (at buses 1,2,3,6 and 8), 4 capacitor banks (at buses 10,11,13 and 14), and 3 tap-changing transformers (at lines 8,9 and 11). The line data and generator data are shown in Tables V and VI, respectively.

The initial loadflow was run and summarized in Table VII. It can be noted that no control is made on the capacitor banks or tap-changing transformers. It can also be noted that the voltages at the load buses are very low; most below the lower limit of 0.95 p.u. Generation cost and system losses for this case are 1075.59 \$/h and 4.31 MW, respectively.

Three cases are made for comparison: (1) Q optimization, (2) P optimization, and (3) P-Q optimization, with results summarized in Table VIII.

(1) Q optimization

In order to study the impact of reactive power dispatch, only the Q - optimization module was run; while keeping the real power generation constant, except at the swing bus. In this case all means of reactive power control, generator reactive powers (buses 1,2,3,6 and 8), capacitive compensators (buses 10,11,13 and 14), and tap-settings (lines 8,9 and 11), are fully utilized. This has resulted in the increase of bus voltage magnitudes throughout the system, with the minimum value 0.974 p.u. at bus 12.

Table IV. Optimal Operational Planning For 6-Bus System

BUS POWER INFORMATION

BUS	V[P.U]	ANG[DEG]	PG[MW]	QG[Mvar]	PD[MW]	QD[Mvar]	P[MW]	Q[Mvar]
1	1.090	0.000	52.786	42.349	0.000	0.000	52.786	42.349
2	1.145	15.603	100.000	8.733	0.000	0.000	100.000	8.733
3	0.967	-6.865	0.000	0.000	55.000	13.000	-54.999	-12.998
4	1.003	-5.000	0.000	14.648	0.000	0.000	-0.005	14.648
5	0.998	-2.634	0.000	0.000	30.000	18.000	-29.996	-18.000
6	0.987	-6.134	0.000	15.092	50.000	5.000	-50.001	10.092

TOTAL LOSS POWER =   17.786+ j 44.824

OPERATION COST =   392.1062

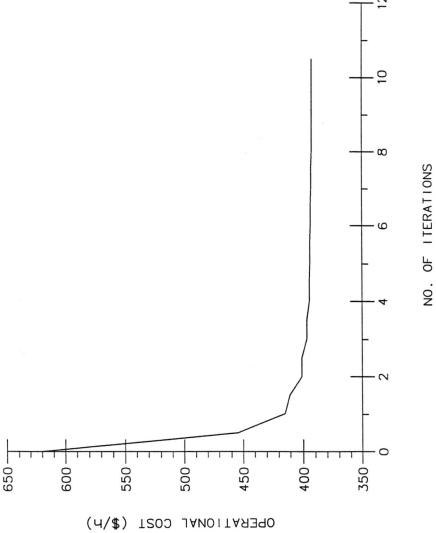

Fig. 2.   Convergence of Optimal Operational Planning for the 6-Bus System

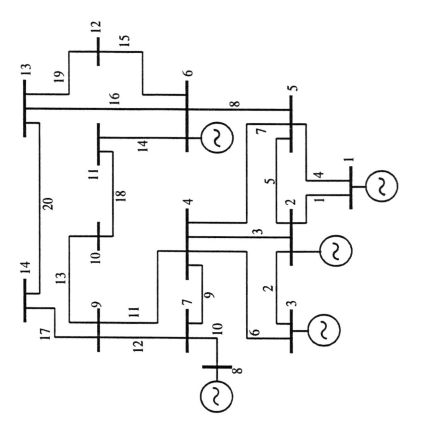

Fig. 3.   One Line Diagram of the 14-Bus System.

Table V. Line Data For 14-Bus System

Line Number	From bus	To bus	Line R[p.u.]	Imped. X[p.u.]	Tap setting	Line Charg.
1	1	2	0.0194	0.0592	0.0000	0.0264
2	2	3	0.0470	0.1980	0.0000	0.0219
3	2	4	0.0581	0.1763	0.0000	0.0187
4	1	5	0.0540	0.2230	0.0000	0.0246
5	2	5	0.0570	0.1739	0.0000	0.0170
6	3	4	0.0670	0.1710	0.0000	0.0173
7	4	5	0.0134	0.0421	0.0000	0.0064
8	5	6	0.0000	0.2520	1.0000	0.0000
9	4	7	0.0000	0.2091	1.0000	0.0000
10	7	8	0.0000	0.1762	0.0000	0.0000
11	4	9	0.0000	0.5562	1.0000	0.0000
12	7	9	0.0000	0.1100	0.0000	0.0000
13	9	10	0.0318	0.0845	0.0000	0.0000
14	6	11	0.0950	0.1989	0.0000	0.0000
15	6	12	0.1229	0.2558	0.0000	0.0000
16	6	13	0.0662	0.1303	0.0000	0.0000
17	9	14	0.1271	0.2704	0.0000	0.0000
18	10	11	0.0821	0.1921	0.0000	0.0000
19	12	13	0.2209	0.1999	0.0000	0.0000
20	13	14	0.1709	0.3480	0.0000	0.0000

Table VI. Generator Data For 14 Bus System

Bus number	Cost α	Coeffs. β	γ
1	0.0000	2.2500	0.0083
2	0.0000	1.0000	0.0625
3	0.0000	1.7500	0.0175
6	0.0000	3.0000	0.0250
8	0.0000	3.0000	0.0260

Table VII.   Initial Loadflow For 14-Bus System

BUS POWER INFORMATION

BUS	V[P.U]	ANG[DEG]	PG[MW]	QG[Mvar]	PD[MW]	QD[Mvar]	P[MW]	Q[Mvar]
1	1.000	0.000	95.309	0.000	0.000	0.000	95.309	-20.517
2	1.000	-2.238	34.000	28.429	21.700	12.700	12.300	15.729
3	1.000	-4.676	80.000	29.878	94.200	19.000	-14.200	10.878
4	0.983	-4.982	0.000	0.000	47.800	-3.900	-47.800	3.900
5	0.983	-4.587	0.000	0.000	27.600	11.600	-27.600	-11.600
6	1.000	-6.963	50.000	53.498	11.200	7.500	38.800	45.998
7	0.974	-4.132	0.000	0.000	0.000	0.000	0.000	0.000
8	1.000	1.058	50.000	17.111	0.000	0.000	50.000	17.111
9	0.957	-7.057	0.000	0.000	20.500	10.600	-20.500	-10.600
10	0.943	-8.061	0.000	0.000	20.000	10.800	-20.000	-10.800
11	0.950	-8.425	0.000	0.000	20.500	10.800	-20.500	-10.800
12	0.955	-8.521	0.000	0.000	20.100	10.600	-20.095	-10.597
13	0.970	-8.010	0.000	0.000	10.500	5.800	-10.506	-5.805
14	0.946	-8.286	0.000	0.000	10.900	5.000	-10.900	-5.000

TOTAL LOSS POWER =   4.309+ j   7.897

INITIAL FUEL COST =   1075.5894 $/h

Table VIII. Summary of Results For 14-Bus System

Variable	Lower	Limits Upper	Initial State	First Study	Second Study	Third Study
P_1(MW)	0.0	340.00	95.309	94.484	133.140	133.357
P_2(MW)	0.0	70.00	34.000	34.000	28.840	28.688
P_3(MW)	0.0	80.00	80.000	80.000	80.000	80.000
P_6(MW)	0.0	90.00	50.000	50.000	80.000	80.000
P_8(MW)	0.0	70.00	50.000	50.000	35.248	34.599
Q_1(Mvar)	-50.0	200.0	0.000	0.5222	34.111	33.923
Q_2(Mvar)	-40.0	50.0	28.429	6.019	0.000	23.978
Q_3(Mvar)	0.0	40.0	29.878	19.265	37.425	0.000
Q_6(Mvar)	0.0	50.0	53.498	26.717	31.848	14.680
Q_8(Mvar)	0.0	50.0	17.111	16.221	56.686	21.823
V_1(p.u.)	0.95	1.05	1.000	1.039	16.158	6.377
V_2(p.u.)	0.95	1.05	1.000	1.028	1.000	1.036
V_3(p.u.)	0.95	1.05	1.000	1.021	1.000	1.011
V_4(p.u.)	0.95	1.05	0.983	1.017	1.000	0.996
V_5(p.u.)	0.95	1.05	0.983	1.017	0.979	0.992
V_6(p.u.)	0.95	1.05	1.000	1.011	0.979	0.997
V_7(p.u.)	0.95	1.05	0.974	1.007	1.000	0.998
V_8(p.u.)	0.95	1.05	1.000	1.031	0.973	1.010
V_9(p.u.)	0.95	1.05	0.957	0.998	1.000	1.020
V_{10}(p.u.)	0.95	1.05	0.943	0.992	0.956	1.003
V_{11}(p.u.)	0.95	1.05	0.950	0.991	0.943	0.995
V_{12}(p.u.)	0.95	1.05	0.955	0.974	0.950	0.993
V_{13}(p.u.)	0.95	1.05	0.970	0.995	0.955	0.966
V_{14}(p.u.)	0.95	1.05	0.946	0.987	0.970	0.990
N_8	0.90	1.10	1.00	1.000	0.946	0.999
N_9	0.90	1.10	1.00	1.000	1.0250	1.0375
N_{11}	0.90	1.10	1.00	1.000	1.0125	0.9750
Q_{c10}(Mvar)	0.0	150.0	0.0	12.6566	1.0125	0.9750
Q_{c11}(Mvar)	0.0	150.0	0.0	10.3078	0.000	10.7137
Q_{c13}(Mvar)	0.0	150.0	0.0	8.6371	0.000	16.3161
Q_{c14}(Mvar)	0.0	150.0	0.0	4.2165	0.000	11.3648
					0.000	11.0420
Generation	Cost [$/h]		1075.599	1072.433	1048.910	1045.201
System	Loss [MW]		4.309	3.484	6.340	5.567

First Study - with Q-optimization module only

Second Study - with P-optimization module only

third Study - with both P and Q-optimization modules

326 KWANG Y. LEE AND YOUNG MOON PARK

It can be noted that not only the voltage profile is improved considerably due to the Q-optimization, but also the swing bus power is reduced; consequently the generation cost and system loss are reduced to 1072.43 $/h and 3.48 MW, respectively, from the initial values.

(2) P optimization

The economic dispatch is simulated by using the P module alone. In this case the voltage magnitude profile remains about the same as in the initial loadflow. The real power generation is re-distributed so that the generation cost is reduced to 1048.91 $/h; but the system loss is increased to 6.34 MW.

(3) P-Q optimization

Finally, both P- and Q-optimization modules are used for complete optimization. In this case, the real power generations are optimized in the P module, and the tap-settings and bus voltages, or reactive power generations, of generators and capacitor banks are optimized in the Q-module. It can be noted that the bus voltage magnitudes are decreased somewhat, compared to case 2 and, therefore, the system loss is increased somewhat from 3.48 MW to 5.57 MW. However, the generation cost is further reduced to 1045.20 $/h from the previous values of 1072.43 $/h in case 1, or 1048.91 $/h in case 2.

This final optimal operation results are summarized in Table IX. The security constraints can be examined for voltage magnitudes and angles. All voltage magnitudes are, of course, within the limits, with the minimum value of 0.966 p.u. at bus 12. The angles are from the minimum of -11.649^0 to the maximum of 0.0^0. The convergence characteristics is shown in Fig. 4. The convergence is achieved in 3 iterations, where each iteration has both P- and Q- optimization steps; the half iteration is after the P optimization and the full iteration is after the Q optimization.

C.   IEEE 30-Bus System

The proposed algorithm is also applied to a larger system of the IEEE 30-bus model [4,6]. The line data and generator data are shown in Tables X and XI, respectively.

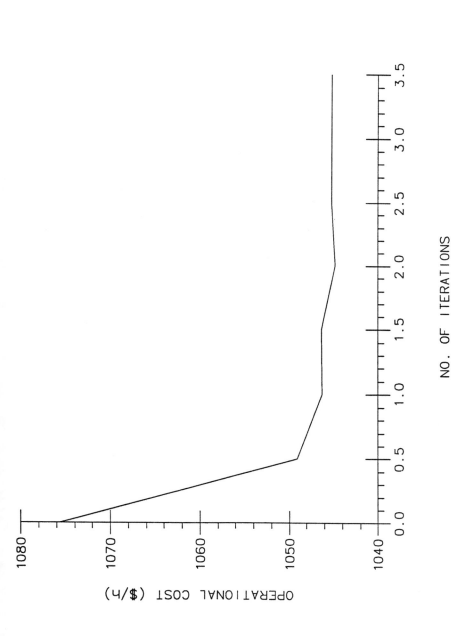

Fig. 4. Convergence of Optimal Operational Planning for the 14-Bus System

Table IX.   Optimal Operational planning For 14-Bus System

BUS POWER INFORMATION

BUS	V[P.U]	ANG[DEG]	PG[MW]	QG[Mvar]	PD[MW]	QD[Mvar]	P[MW]	Q[Mvar]
1	1.036	0.000	133.357	23.978	0.000	0.000	133.357	23.978
2	1.011	-2.575	28.688	0.000	21.700	12.700	6.988	-17.416
3	0.996	-5.223	80.000	14.680	94.200	19.000	-14.200	-4.320
4	0.992	-6.219	0.000	0.000	47.800	-3.900	-47.800	3.900
5	0.997	-5.674	0.000	0.000	27.600	11.600	-27.600	-11.600
6	0.998	-9.526	34.599	21.823	11.200	7.500	23.399	14.323
7	1.010	-7.268	0.000	0.000	0.000	0.000	0.000	0.000
8	1.020	-3.942	33.923	6.377	0.000	0.000	33.923	6.377
9	1.003	-9.938	0.000	0.000	20.500	10.600	-20.500	-10.600
10	0.995	-11.130	0.000	10.714	20.000	10.800	-20.000	-0.086
11	0.993	-11.649	0.000	16.316	20.500	10.800	-20.500	5.516
12	0.966	-11.183	0.000	0.000	20.100	10.600	-20.100	-10.605
13	0.990	-11.057	0.000	11.365	10.500	5.800	-10.499	5.565
14	0.999	-11.632	0.000	11.042	10.900	5.000	-10.900	6.042

TOTAL LOSS POWER =   5.567+ j   11.074

OPERATION COST =   1045.2007 $/h

Table X. Line Data For 30-Bus System

Line Number	From bus	To bus	Line R[p.u.]	Imped. X[p.u.]	Tap setting	Line Charg.
1	1	2	0.0192	0.0575	0.0000	0.0000
2	1	3	0.0452	0.1852	0.0000	0.0000
3	2	4	0.0570	0.1737	0.0000	0.0000
4	3	4	0.0132	0.0379	0.0000	0.0000
5	2	5	0.0472	0.1983	0.0000	0.0000
6	2	6	0.0581	0.1763	0.0000	0.0000
7	4	6	0.0119	0.0414	0.0000	0.0000
8	5	7	0.0460	0.1160	0.0000	0.0000
9	6	7	0.0267	0.0820	0.0000	0.0000
10	6	8	0.0120	0.0420	0.0000	0.0000
11	6	9	0.0000	0.2080	1.0780	0.0000
12	6	10	0.0000	0.5560	1.0690	0.0000
13	9	11	0.0000	0.2080	0.0000	0.0000
14	9	10	0.0000	0.1100	0.0000	0.0000
15	4	12	0.0000	0.2560	1.0320	0.0000
16	12	13	0.0000	0.1400	0.0000	0.0000
17	12	14	0.1231	0.2559	0.0000	0.0000
18	12	15	0.0662	0.1304	0.0000	0.0000
19	12	16	0.0945	0.1987	0.0000	0.0000
20	14	15	0.2210	0.1997	0.0000	0.0000
21	16	17	0.0824	0.1932	0.0000	0.0000
22	15	18	0.1070	0.2185	0.0000	0.0000
23	18	19	0.0639	0.1292	0.0000	0.0000
24	19	20	0.0340	0.0680	0.0000	0.0000
25	10	20	0.0936	0.2090	0.0000	0.0000
26	10	17	0.0324	0.0845	0.0000	0.0000
27	10	21	0.0348	0.0749	0.0000	0.0000
28	10	22	0.0727	0.1499	0.0000	0.0000
29	21	22	0.0116	0.0236	0.0000	0.0000
30	15	23	0.1000	0.2020	0.0000	0.0000
31	22	24	0.1150	0.1790	0.0000	0.0000
32	23	24	0.1320	0.2700	0.0000	0.0000
33	24	25	0.1885	0.3292	0.0000	0.0000
34	25	26	0.2544	0.3800	0.0000	0.0000
35	25	27	0.1093	0.2087	0.0000	0.0000
36	28	27	0.0000	0.3960	1.0680	0.0000
37	27	29	0.2198	0.4153	0.0000	0.0000
38	27	30	0.3202	0.6027	0.0000	0.0000
39	29	30	0.2399	0.4533	0.0000	0.0000
40	8	28	0.0636	0.2000	0.0000	0.0000

Table XI. Generator Data For 30-Bus System

Bus number	Cost Coeffs. α	β	γ
1	0.0000	2.0000	0.0037
2	0.0000	1.7500	0.0175
5	0.0000	1.0000	0.0625
8	0.0000	3.2500	0.0083
11	0.0000	3.0000	0.0250
13	0.0000	3.0000	0.0250

The initial loadflow along with load data are shown in Table XII. Since no attempt is made to optimize the operation the capacitor banks are all set to zero. For the given load data, generations and voltage magnitudes for the PV buses are chosen rather arbitrarily. The generation cost and system loss for this case are 903.31 $/h and 6.51 MW, respectively. The voltages at load buses are mostly below 1.00 p.u., with the minimum of 0.858 p.u. at bus 30.

In a way similar to the 6-bus and 14-bus systems, three different studies are made: (1) Q optimization only, (2) P optimization only, and (3) P-Q optimization, with results compared in Table XIII.

(1) Q optimization

The reactive power dispatch is simulated by using the Q module alone, while keeping the real power generation constant. In this case all means of reactive power control are fully utilized. This has resulted in an increase of bus voltage magnitudes throughout the system, with the minimum value of 0.984 p.u. at bus 26.

The Q optimization also reduced the swing bus generation, and the generation cost and system loss are reduced to 899.37 $/h and 5.06 MW, respectively.

(2) P optimization

Again using the P module alone, the usual economic dispatch is simulated. As before for the 14 bus system, the voltage magnitude profile remains about the same as in the initial loadflow. Because of the economic allocation of generation among the six generators the generation cost is reduced to 807.82 $/h, but the system loss is increased to 11.34 MW.

(3) P-Q optimization

The final result is summarized in Table XIV, where both P- and Q- optimization modules are used iterative for the optimal operation. The reduction in the generation cost (800.14 $/h, compared to 903.31 $/h for case 1 and 807.82 $/h for case 2) clearly shows the advantage of the unified minimization approach using both modules. Although the system loss is increased somewhat (9.43 MW, compared to the initial value of 5.06 MW), the generation cost is decreased significantly. The security constraints are also checked for voltage magnitudes and angles. All voltage

Table XII.   Initial Loadflow For 30-Bus System

BUS POWER INFORMATION

BUS	V[P.U]	ANG[DEG]	PG[MW]	QG[Mvar]	PD[MW]	QD[Mvar]	P[MW]	Q[Mvar]
1	1.000	0.000	99.905	0.000	0.000	0.000	99.905	-21.531
2	1.000	-2.142	80.000	15.973	21.700	12.700	58.300	3.273
3	0.990	-4.486	0.000	0.000	2.400	1.200	-2.400	-1.200
4	0.988	-5.384	0.000	0.000	7.600	1.600	-7.600	-1.600
5	1.000	-7.637	50.000	44.132	94.200	19.000	-44.200	25.132
6	0.991	-6.312	0.000	0.000	0.000	0.000	0.000	0.000
7	0.986	-7.380	0.000	0.000	22.800	10.900	-22.800	-10.900
8	1.000	-6.783	20.000	56.127	30.000	30.000	-10.000	26.127
9	0.938	-7.989	0.000	0.000	0.000	0.000	0.000	0.000
10	0.917	-10.344	0.000	0.000	5.800	2.000	-5.800	-2.000
11	1.000	-5.447	20.000	30.238	0.000	0.000	20.000	30.238
12	0.955	-9.747	0.000	0.000	11.200	7.500	-11.200	-7.500
13	1.000	-8.067	20.000	32.499	0.000	0.000	20.000	32.499
14	0.935	-10.855	0.000	0.000	6.200	1.600	-6.205	-1.596
15	0.926	-10.884	0.000	0.000	8.200	2.500	-8.195	-2.505
16	0.931	-10.322	0.000	0.000	3.500	1.800	-3.500	-1.800
17	0.915	-10.599	0.000	0.000	9.000	5.800	-9.000	-5.800
18	0.909	-11.582	0.000	0.000	3.200	0.900	-3.200	-0.900
19	0.902	-11.754	0.000	0.000	9.500	3.400	-9.500	-3.400
20	0.905	-11.473	0.000	0.000	2.200	0.700	-2.200	-0.700
21	0.902	-10.943	0.000	0.000	17.500	11.200	-17.500	-11.200
22	0.903	-10.931	0.000	0.000	0.000	0.000	0.000	0.000
23	0.906	-11.333	0.000	0.000	3.200	1.600	-3.200	-1.600
24	0.889	-11.499	0.000	0.000	8.700	6.700	-8.700	-6.701
25	0.886	-11.572	0.000	0.000	0.000	0.000	0.000	0.000
26	0.866	-12.128	0.000	0.000	3.500	2.300	-3.500	-2.300
27	0.895	-11.272	0.000	0.000	0.000	0.000	0.000	0.000
28	0.985	-6.751	0.000	0.000	0.000	0.000	0.000	0.000
29	0.871	-12.892	0.000	0.000	2.400	0.900	-2.400	-0.900
30	0.858	-14.067	0.000	0.000	10.600	1.900	-10.600	-1.900

TOTAL LOSS POWER =   6.505+ j   31.238

INITIAL FUEL COST =   903.3090 $/h

Table XIII. Summary of Results For 30-Bus System

Variable	Limits Lower	Limits Upper	Initial State	First Study	Second Study	Third Study
P_1(MW)	50.0	200.00	99.905	98.463	176.136	177.923
P_2(MW)	80.0	70.00	80.000	80.000	48.840	48.568
P_5(MW)	15.0	50.00	50.000	50.000	21.776	21.296
P_8(MW)	10.0	35.00	20.000	20.000	23.545	21.143
P_{11}(MW)	10.0	30.00	20.000	20.000	12.501	11.896
P_{13}(MW)	12.0	40.00	20.000	20.000	12.000	12.000
Q_1(Mvar)	-20.0	200.0	0.000	5.019	0.000	27.493
Q_2(Mvar)	-20.0	100.0	15.973	17.570	36.547	14.712
Q_5(Mvar)	-15.0	80.0	44.132	32.063	55.528	24.093
Q_8(Mvar)	-15.0	60.0	56.127	27.349	58.544	34.840
Q_{11}(Mvar)	-10.0	50.0	30.238	24.276	30.347	23.077
Q_{13}(Mvar)	-15.0	60.0	32.499	14.401	33.347	13.263
V_1(p.u.)	0.95	1.05	1.000	1.061	1.000	1.080
V_2(p.u.)	0.95	1.05	1.000	1.050	1.000	1.052
V_3(p.u.)	0.95	1.05	0.990	1.037	0.988	1.041
V_4(p.u.)	0.95	1.05	0.988	1.032	0.986	1.033
V_5(p.u.)	0.95	1.05	1.000	1.032	1.000	1.014
V_6(p.u.)	0.95	1.05	0.991	1.028	0.989	1.025
V_7(p.u.)	0.95	1.05	0.986	1.021	0.985	1.012
V_8(p.u.)	0.95	1.05	1.000	1.026	1.000	1.026
V_9(p.u.)	0.95	1.05	0.938	1.019	0.937	1.015
V_{10}(p.u.)	0.95	1.05	0.917	1.011	0.916	1.002
V_{11}(p.u.)	0.95	1.05	1.000	1.065	1.000	1.060
V_{12}(p.u.)	0.95	1.05	0.955	1.020	0.953	1.002
V_{13}(p.u.)	0.95	1.05	1.000	1.039	1.000	1.020
V_{14}(p.u.)	0.95	1.05	0.935	1.009	0.933	0.992
V_{15}(p.u.)	0.95	1.05	0.926	1.008	0.925	0.992
V_{16}(p.u.)	0.95	1.05	0.931	1.010	0.929	0.996
V_{17}(p.u.)	0.95	1.05	0.915	1.008	0.914	0.997
V_{18}(p.u.)	0.95	1.05	0.909	1.000	0.908	0.986
V_{19}(p.u.)	0.95	1.05	0.902	0.999	0.901	0.985
V_{20}(p.u.)	0.95	1.05	0.905	1.003	0.904	0.990

Table XIII. Summary of Results For 30-Bus System

(Continued)

Variable	Limits		Initial	First	Second	Third
	Lower	Upper	State	Study	Study	Study
V_{21}(p.u.)	0.95	1.05	0.902	1.002	0.901	0.992
V_{22}(p.u.)	0.95	1.05	0.903	1.003	0.902	0.993
V_{23}(p.u.)	0.95	1.05	0.906	1.005	0.905	0.990
V_{24}(p.u.)	0.95	1.05	0.889	0.997	0.888	0.983
V_{25}(p.u.)	0.95	1.05	0.886	1.002	0.886	0.984
V_{26}(p.u.)	0.95	1.05	0.866	0.984	0.865	0.966
V_{27}(p.u.)	0.95	1.05	0.895	1.014	0.894	0.993
V_{28}(p.u.)	0.95	1.05	0.985	1.022	0.983	1.021
V_{29}(p.u.)	0.95	1.05	0.871	1.005	0.871	0.984
V_{30}(p.u.)	0.95	1.05	0.858	0.989	0.858	0.968
N_{11}	0.90	1.10	1.00	1.0405	1.0250	1.0280
N_{12}	0.90	1.10	1.00	0.9940	1.0125	0.9740
N_{15}	0.90	1.10	1.00	1.0195	1.0125	1.0445
N_{36}	0.90	1.10	1.00	0.9930	1.0250	1.0180
Q_{c15}(Mvar)	0.0	5.0	0.0	3.9674	0.000	4.6161
Q_{c17}(Mvar)	0.0	5.0	0.0	4.2388	0.000	3.7857
Q_{c20}(Mvar)	0.0	5.0	0.0	4.7140	0.000	3.4805
Q_{c21}(Mvar)	0.0	5.0	0.0	4.4270	0.000	3.9138
Q_{c23}(Mvar)	0.0	5.0	0.0	4.1195	0.000	4.6377
Q_{c24}(Mvar)	0.0	5.0	0.0	4.6113	0.000	4.3726
Q_{c29}(Mvar)	0.0	5.0	0.0	3.4917	0.000	3.6659
Generation	Cost [$/h]		903.309	899.368	807.818	800.1415
System	Loss [MW]		6.505	5.063	11.398	9.426

First Study - with Q-optimization module only

Second Study - with P-optimization module only

third Study - with both P and Q-optimization modules

Table XIV. Optimal Operational Planning For 30-Bus System

BUS POWER INFORMATION

BUS	V[P.U]	ANG[DEG]	PG[MW]	QG[Mvar]	PD[MW]	QD[Mvar]	P[MW]	Q[Mvar]
1	1.080	0.000	177.923	27.493	0.000	0.000	177.923	27.493
2	1.052	-3.293	48.568	14.712	21.700	12.700	26.868	2.012
3	1.041	-5.320	0.000	0.000	2.400	1.200	-2.400	-1.200
4	1.033	-6.406	0.000	0.000	7.600	1.600	-7.600	-1.600
5	1.014	-9.861	21.296	24.093	94.200	19.000	-72.904	5.093
6	1.025	-7.513	0.000	0.000	0.000	0.000	0.001	0.000
7	1.012	-8.981	0.000	0.000	22.800	10.900	-22.800	-10.900
8	1.026	-7.775	21.143	34.840	30.000	30.000	-8.857	4.840
9	1.015	-9.834	0.000	0.000	0.000	0.000	0.000	0.000
10	1.002	-11.792	0.000	0.000	5.800	2.000	-5.800	-2.000
11	1.060	-8.517	11.896	23.077	0.000	0.000	11.896	23.077
12	1.002	-10.885	0.000	0.000	11.200	7.500	-11.200	-7.500
13	1.020	-9.942	12.000	13.263	0.000	0.000	12.000	13.263
14	0.992	-11.938	0.000	0.000	6.200	1.600	-6.200	-1.604
15	0.992	-12.234	0.000	4.616	8.200	2.500	-8.200	2.116
16	0.996	-11.612	0.000	0.000	3.500	1.800	-3.500	-1.800
17	0.997	-12.018	0.000	3.786	9.000	5.800	-9.000	-2.014
18	0.986	-12.902	0.000	0.000	3.200	0.900	-3.200	-0.900
19	0.985	-13.090	0.000	0.000	9.500	3.400	-9.500	-3.400
20	0.990	-12.877	0.000	3.481	2.200	0.700	-2.200	2.781
21	0.992	-12.363	0.000	3.914	17.500	11.200	-17.500	-7.286
22	0.993	-12.351	0.000	0.000	0.000	0.000	0.001	0.000
23	0.990	-12.845	0.000	4.638	3.200	1.600	-3.200	3.038
24	0.983	-12.923	0.000	4.373	8.700	6.700	-8.700	-2.327
25	0.984	-12.629	0.000	0.000	0.000	0.000	0.000	0.000
26	0.966	-13.078	0.000	0.000	3.500	2.300	-3.500	-2.300
27	0.993	-12.166	0.000	0.000	0.000	0.000	0.000	0.000
28	1.021	-8.041	0.000	0.000	0.000	0.000	0.000	0.000
29	0.984	-13.803	0.000	3.666	2.400	0.900	-2.400	2.766
30	0.968	-14.584	0.000	0.000	10.600	1.900	-10.600	-1.900

TOTAL LOSS POWER =   9.426+ j   39.746

OPERATION COST = 800.1405 $/h

magnitudes are above the lower limit, with the minimum value of 0.966 p.u. at bus 26. The voltage angles are from the minimum of -14.584^0 to the maximum of 0.0^0.

Figure 5 shows the convergence property of the optimal operation planning algorithms, where the P- and Q-optimization modules are run iteratively in each iteration. As was seen in case 2, the P module alone will stop after the first half iteration (807.82 \$/h). However, the iterative use of P and Q modules brought the cost down further to the final value of 800.14 \$/h.

The simulation was run with VAX 8550 in the Engineering Computer Laboratory at the Penn State University. Although the loadflow (L) module is optional, it was used twice per iteration; first after P module, and then after the Q module. With this procedure, i.e.; P-L-Q-L per each iteration, the CPU time was estimated for each system. The average CPU time was 0.120 sec, 0.261 sec, or 0.826 sec per iteration for the 6-bus, 14-bus, or 30-bus systems, respectively. When the loadflow module was used only once after the P module, i.e., P-L-Q per each iteration, then the corresponding CPU time was reduced to 0.088 sec, 0.21 sec, or 0.583 sec per iteration.

VI.   CONCLUSIONS

An optimal operational planning algorithm has been developed for real and reactive powers. The algorithm consists of the real power (P) and reactive power (Q) optimization modules. Both the P- and Q- optimization modules use the same fuel cost function as the objective function, unlike other conventional methods which use the transmission loss function for the Q module.

The proposed sensitivity relationships are developed which are common to both P and Q modules. The Q module in a linear form is developed by augmenting the fuel cost function with the power balance equation in order to provide a smooth transition between the two modules. An alternative Q module is also presented in a quadratic form, where the reactive power and tap-settings are used as control variables.

Through the simulation of test systems, the unified methodology has demonstrated that it can effectively provide an optimal set of real power generations,

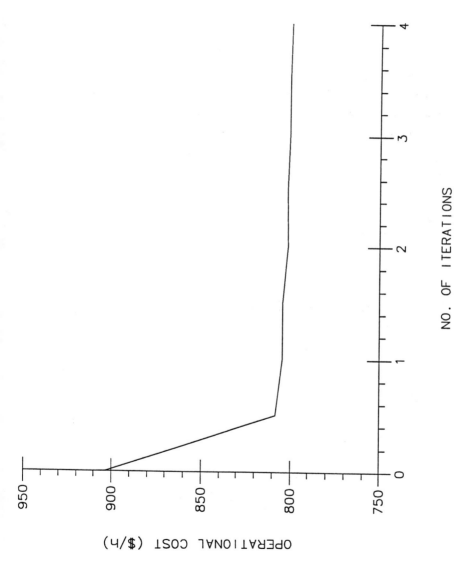

Fig. 5.   Convergence of Optimal Operational Planning for the 30-Bus System

transformer tap-settings, and reactive power generations of generators and capacitor banks. It has also improved the voltage profile throughout the system.

Finally, it has been shown that the usual loss minimization approach does not guarantee the optimal operation of power systems unless all the units have equal efficiency.

ACKNOWLEDGEMENT

This work has been supported in parts by the NSF under Grant INT-8617329, Allegheny Power System, and the Korea Electric Power Corporation. The authors wish to thank their former Ph.D. students, J. L. Ortiz, K. J. Kim, J. B. Kim, and M. K. W. Mangoli, for their significant contribution.

REFERENCES

1. H. W. Dommel and W.F. Tinney, *IEEE Trans. on Power Apparatus and Syst.*, *PAS-87*, 1866-87 (1968).

2. H. H. Happ, *IEEE Trans. on Power Apparatus and Syst., PAS-96*, 841-854 (1977).

3. J. Carpentier, Jr., *Int. J. of Electrical Power and Energy Systems, 1*, 13-15 (1979).

4. O. Alsac and B. Scott, *IEEE Trans. on Power Apparatus and Syst., PAS - 93*, 745 - 751 (1974).

5. F. J. Trefny and K. Y. Lee, *IEEE Trans. on Power Apparatus and Syst., PAS-100*, 3466-3477 (1981).

6. K. Y. Lee, Y.M. Park, and J.J. Ortiz, *IEEE Trans. on Power Apparatus and Syst., PAS-104*, 1147-1153 (1985).

7. K. Y. Lee, J. L. Ortiz, Y. M. Park, and M.A. Mohtadi, *IEEE Trans. on Power Systems, PWRS-3*, 413-420 (1988).

8. K. Y. Lee, J. L. Ortiz, Y. M. Park, and L. G. Pond, *IEEE Trans. on Power Systems, PWRS-1*, 153-159 (1986).

9. B. H. Lee and K. Y. Lee, *IEEE PES Winter Meeting, #91 WM 123-0 PWRS*, New York (1991).

10. J. Peschon, D. S. Piercy, W. F. Tinney, and O. J. Tveit, *IEEE Trans. on Power Apparatus and Syst., PAS-87,* 40-48 (1968).

11. R. A. Fernandes, H. H. Happ, and K. A. Wirgau, *Int. J. of Electrical Power and Energy Systems, 2,* 133-139 (1980).

12. K. R. C. Mamandur and R. O. Chenoweth, *IEEE Trans. on Power Apparatus and Syst., PAS-100,* 3185-3194 (1981).

13. H. Glavitsch and M. Spoerry, *Proc. 1983 Power Industry Computer Application Conference,* 27-33 (1983).

14. R. R. Shoults and D. T. Sun, *IEEE Trans. on Power Apparatus and Syst., PAS-101,* 397-405, (1982).

15. R. C. Burchett, H. H. Happ, D. R. Vierath, and K. A. Wirgau, *IEEE Trans. on Power Apparatus and Syst., PAS-101,* 406-414, (1982).

16. F. F. Wu, G. Gross, J. F. Luini, and P. M. Look, *Proc. 1979 PICA Conference,* 126-136 (1979).

17. M. C. Biggs and M. A. Laughton, *Math. Program., 13,* (167-182) 1981.

18. J. B. Rosen, *J. Soc. Indust. Appl. Math. 3,* 181-217 (1960)

19. G. W. Stagg and A. H. El-Abiad, " Computer Methods in Power Systems Analysis ", McGraw-Hill, New York, 1968.

MULTISTAGE LINEAR PROGRAMMING METHODS FOR OPTIMAL ENERGY PLANT OPERATION*

TAKUSHI NISHIYA

and

MOTOHISA FUNABASHI

Systems Development Laboratory

Hitachi, Ltd.

1099 Ozenji Asao-Ku

Kawasaki, 215

Japan

1 . INTRODUCTION

The development of optimization methods has a long history, however, algorithmic innovation is still required particularly for operating large-scale dynamic plants. Because of the high dimensionality of the plants, technological bases for the problem in operating such large plants are usually found in the area of linear programming.

---

*This paper is a revised version of [8].

CONTROL AND DYNAMIC SYSTEMS, VOL. 42

In applying linear programming for optimal dynamic-plant operation ,
the constraint appears in the form of a staircase structure. In
exploiting this structure, several attempts at devising efficient
algorithms have been made. Glassey [1] and Ho-Manne [2] applied the
Dantzig-Wolfe decomposition principle to iteratively generate a series
of subproblems. Perold-Dantzig [3] developed the basis factorization
method where the correction factor has a minimal number of structural
columns. Ho and Loute [4] have empirically compared the conventional
simplex method to these methods, and have shown that both special
techniques are approximately two or three times faster than the direct
simplex method. However, when applied to the area of automatic
control, more improvements in speed are required.

In this paper, a variant of the basis factorization method is
developed for solving multistage linear programming, in which some
outputs of one stage are the inputs to the next stage. This method is
devised to use the special staircase structure of the original basis. In
this method, a set of elementary column operations transform the
original basis into a working basis consisting of square and
nonsingular diagonal blocks.

Based on this fact, a new algorithm is proposed where the relative
cost factor for the original nonbasis is easily obtained by memorizing
the column vectors corresponding to the variables which are included
in the original basis but not in the working basis.

Operational scheduling for a plant is carried out a couple of times a
day responding to the change in production rate of the process.
However, in scheduling a plant operation by linear programming, the
optimal solution often becomes infeasible. When the solution becomes
infeasible, rescheduling becomes inevitable. The proposed algorithm
overcomes these problems. It generates the problem incrementing the
number of stages one by one and optimizes successively these
restricted problems. Therefore, if the optimal solution is infeasible,
the stage number and the name of the variable which caused
infeasibility are obtained in the optimization process. On the contrary,
in the case of the above stated special methods, feasibility is not

checked until the solution of the whole problem is obtained. This feature of the proposed algorithm enables plant operators to obtain optimal operational guidance quickly. When the problem is infeasible, the operators can easily detect and change the constraints which cause infeasible solutions and can obtain another operational scheduling quickly. In this case, the solution process can be restarted from the present state of the variables. This improvement in speed brings quick response between operators and the computer and makes it possible for operators to execute these scheduling tasks iteratively on minicomputers.

The problem is mathematically described and the idea of the basis factorization is shown in Section II. The simplex procedure in the multistage linear programming problems exploiting a newly defined correction factor is explained in Section III. The superiority of the algorithm over the conventional simplex method is demonstrated by its application to the in-plant energy operation problem, as described in Section IV.

II. PROBLEM AND THE BASIS FACTORIZATION

A. Problem

Let us consider an optimization problem in operating large-scale plants. In many cases, plant behavior and operating objectives are described by linear equations similar to those presented in Section IV. This optimization problem can be represented as follows:

$$(\text{P1}) \ \min. \quad z = \sum_{t=1}^{K} c^{\tau} \, x(t) \tag{1}$$

$$\text{subj. to} \ \begin{cases} Ux(1) = r^{1} \\ Tx(t-1) + Ux(t) = r^{\tau}, \quad t = 2,3,\cdots,K \end{cases} \tag{2}$$

$$a^{\tau} \leqq x(t) \leqq b^{\tau}, \quad t = 1,2,\cdots,K \tag{3}$$

where $x(t)$ is a vector of state and control variables, T and U are $m \times n$

matrices. The bounds a^t and b^t are n-dimensional column vectors, and r^t is an m-dimensional column vector. The objective coefficient c^t is an n-dimensional row vector. The plant equation (2) represents both static and dynamic mass and/or energy balance requirement at time stage t in the plant.

The problem (P1) has the following staircase coefficient matrix:

$$
A = \begin{bmatrix}
U & & & \\
T & U & & \\
 & T & U & \\
 & & \cdots\cdots & \\
 & & & T & U
\end{bmatrix}. \tag{4}
$$

Using a conventional simplex method to solve (P1), the large compound matrix A requires a large amount of computational effort. Since matrix A is characterized by a staircase structure, analytical manipulation significantly reduces the computational efforts. The development of an efficient manipulation algorithm is the central subject of this paper.

B. Factorization of the Basis

The following basis matrix B has a similar staircase structure as matrix A:

$$
B = \begin{bmatrix}
B_{11} & & & & \\
B_{12} & B_{22} & & & \\
 & B_{32} & B_{33} & & \\
 & & \cdots & \cdots & \\
 & & & B_{K,K-1} & B_{KK}
\end{bmatrix}, \quad B_{tt} : m_t \times m_t \tag{5}
$$

where m_t is the number of basis variables chosen from time stage t.

Using a conventional simplex method to solve (P1), basis matrix B becomes a large matrix and a large amount of computational effort is required to obtain its inversion. On the other hand, when m basis variables are chosen from each stage, its basis matrix $\bar{B}$ will have this staircase structure

$$\bar{B} = \begin{bmatrix} \bar{B}_{11} & & & & \\ \bar{B}_{12} & \bar{B}_{22} & & & \\ & \bar{B}_{32} & \bar{B}_{33} & & \\ & & \cdots & \cdots & \\ & & & \bar{B}_{K,K-1} & \bar{B}_{KK} \end{bmatrix} \; , \; \bar{B}_{tt} : m \times m \tag{6}$$

where the diagonal blocks $\bar{B}_{tt}$ are square and nonsingular. Therefore, the inversion of $\bar{B}$ is easily obtained by using inversions of diagonal blocks $\bar{B}_{tt}$.

The basis matrices B and $\bar{B}$ have a simple relation

$$\begin{aligned} B &= [B_2, B_1] \\ &= \bar{B}\bar{B}^{-1}[B_2, B_1] \\ &= \bar{B}F \end{aligned} \tag{7}$$

where B_1 is a common coefficient matrix in B and $\bar{B}$, and B_2 is another coefficient matrix. If B and $\bar{B}$ have $(mK-k)$ columns in common, the correction factor F will contain precisely $(mK-k)$ unit columns $\bar{B}^{-1}B_1$; see [3]. From the equation above, F takes the following form:

$$F = \begin{bmatrix} G & 0 \\ H & I \end{bmatrix} \tag{8}$$

where I is an identity matrix of order $(mK-k)$, and the other k columns in F, F_j, are obtained by a column vector P_j of the coefficient matrix A that is contained in B but not in $\bar{B}$, i.e.,

$$F_j = \bar{B}^{-1}P_j. \tag{9}$$

In other words, the relation (7) is equivalent to multiple pivoting among basis B and $\bar{B}$.

Using the relation (7), basis matrix B is factored into $\bar{B}$ and F. Since $\bar{B}^{-1}$ and F^{-1} are easily obtained by elementary manipulation of the column vectors, the development of an efficient manipulation algorithm reduces the required computational effort and this is the central subject in this paper.

In the following discussion, the notations used are defined as follows.

【 Definition 】

Vectors X_B and X_N are basis and nonbasis variables, respectively, associated with the original basis

$$X_B \equiv [x_B(1), x_B(2), \cdots, x_B(K)]^1$$
$$X_N \equiv [x_N(1), x_N(2), \cdots, x_N(K)]^1.$$

Vectors $X_{\bar{B}}$ and $X_{\bar{N}}$ are basis and nonbasis variables, respectively, associated with the working basis

$$X_{\bar{B}} \equiv [x_{\bar{B}}(1), x_{\bar{B}}(2), \cdots, x_{\bar{B}}(K)]^1$$
$$X_{\bar{N}} \equiv [x_{\bar{N}}(1), x_{\bar{N}}(2), \cdots, x_{\bar{N}}(K)]^1.$$

Lastly let X_S be a vector of the variables in $X_{\bar{B}}$ but not in X_B, X_H be a vector of the variables in X_B but not in $X_{\bar{B}}$, $X_{\bar{B}-S}$ be a vector of the variables in both X_B and $X_{\bar{B}}$, and $X_{\bar{N}-H}$ be a vector of the variables in both X_N and $X_{\bar{N}}$. They are expressed as follows

$$X_{\bar{B}} \equiv [X_S, X_{\bar{B}-S}]^1$$
$$X_{\bar{N}} \equiv [X_H, X_{\bar{N}-H}]^1$$
$$X_B \equiv [X_H, X_{\bar{B}-S}]^1$$
$$X_N \equiv [X_S, X_{\bar{N}-H}]^1.$$

---

[1]All partitioned vectors will be written as row vectors but are assumed to be column vectors.

III. MULTISTAGE LINEAR PROGRAMMING

A. Simplex Method for Multistage Linear Programming

The simplex procedure for solving (P1) is composed of the following steps.

1. Calculation of the simplex multipliers' vector.
2. Calculation of the relative cost factors.
3. Test for optimality.
4. Pivoting.
5. Updating factorization.

In this section, we will first describe the outline of the proposed algorithm followed by explanation of the above steps.

B. Outline of the Algorithm

The explanation of the algorithm assumes that the number of the last stage is K and the initial values of the variables are known (see Section III -D). The main idea is that the nonbasic variables considered are always confined to a stage. This restricted problem is handled step by step until the optimal condition is satisfied in all the nonbasic variables as shown in the flow diagram in Fig. 1.

1. Set $t = K$.
2. Calculate simplex multipliers' vector π and relative cost factor λ^t. In the initial pass of this step, F has no structural column and is equal to the identity matrix.
3. Check the optimality condition for λ^t. If it is satisfied, go to Step 4, otherwise go to Step 5.
4. If $t \neq 1$, set $t = t-1$ and return to Step 2. Otherwise ($t = 1$), two possibilities exist. The first is when no pivot operation is executed

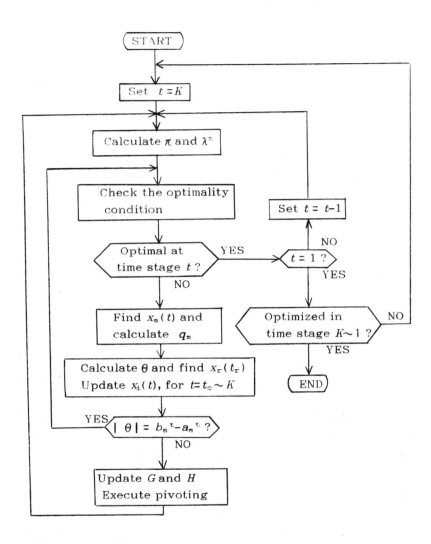

Fig. 1. Flow diagram of the proposed algorithm.

from $t = K$ to $t = 1$. Here, an optimal solution is obtained and the algorithm stops. The second is when at least one pivot operation is executed. In this case, return to Step 1 and repeat the procedure.

5. Find the variable, $x_s(t)$, newly entered into basis B and calculate q_s. Then, find the variable, $x_r(t_r)$, removed from basis B. Furthermore, update the basic variables $x_i(t)$ for $t = t_c, \cdots, K$ and go on to Step 6.

6. If $|\theta| = b_s^t - a_s^t$ in Step 5, then return to Step 3, otherwise update G and H, and return to Step 2.

C. Explanation of the Simplex Procedure

1. Simplex Multipliers' Vector

The original basis B and the artificial basis $\bar{B}$ have the relation (7), therefore simplex multipliers' vector π,

$$\pi = [\ \pi_1, \pi_2, \cdots, \pi_K\ ]$$

is calculated as:

$$\begin{aligned} \pi &= c_B B^{-1} \\ &= c_B F^{-1} \bar{B}^{-1} \end{aligned} \tag{10}$$

where c_B is the corresponding cost vector for the original basis.

From (10), we define

$$c_B' = c_B F^{-1} \tag{11}$$

both c_B and c_B' are partitioned into two parts corresponding to X_H and $X_{\bar{B}-s}$,

$$\begin{aligned} c_B &= [\ c_{B,H}\ |\ c_{B,\bar{B}-s}\ ] \\ c_B' &= [\ c_{B,H}'\ |\ c_{B,\bar{B}-s}'\ ]. \end{aligned}$$

Substituting (8) into (11), yields

$$[c'_{B,H} \mid c'_{B,\bar{B}-S}] = [c_{B,H} \mid c_{B,\bar{B}-S}] \begin{bmatrix} G^{-1} & 0 \\ -HG^{-1} & I \end{bmatrix}$$

$$= [(c_{B,H} - c_{B,\bar{B}-S} H) G^{-1} \mid c_{B,\bar{B}-S}] . \qquad (12)$$

Thus, if the correction factor F is maintained in the computer storage, c'_B can be calculated easily.

From (10), simplex multipliers' vector π and cost vector c_B have the relation:

$$[\pi_1, \pi_2, \cdots, \pi_K] \begin{bmatrix} \bar{B}_{11} & & & \\ \bar{B}_{21} & \bar{B}_{23} & & \\ & \cdots & \cdots & \\ & & \bar{B}_{K,K-1} & \bar{B}_{KK} \end{bmatrix} = [c'_{B1}, c'_{B2}, \cdots, c'_{BK}].$$

The above relation yields the following sequence of equations:

$$\begin{cases} \pi_K \bar{B}_{KK} = c'_{BK} \\ \pi_{K-1} \bar{B}_{K-1,K-1} + \pi_K \bar{B}_{K,K-1} = c'_{B,K-1} \\ \text{------------------------} \\ \pi_1 \bar{B}_{11} + \pi_2 \bar{B}_{21} = c'_{B1} \end{cases}$$

and solving for π we get

$$\begin{cases} \pi_K = c'_{BK} \bar{B}_{KK}{}^{-1} \\ \pi_{K-1} = (c'_{B,K-1} - \pi_K \bar{B}_{K,K-1}) \bar{B}_{K-1,K-1}{}^{-1} \\ \text{------------------------} \\ \pi_1 = (c'_{B1} - \pi_2 \bar{B}_{21}) \bar{B}_{11}{}^{-1}. \end{cases} \qquad (13)$$

Thus π can be obtained successively using small inverse matrices $\bar{B}_{\tau\tau}{}^{-1}$. Note: the diagonal blocks $\bar{B}_{\tau\tau}$ are inverted by the product form [6] and are stored in sparse form. As calculation of (13) is required for each simplex procedure, it is very important to devise an efficient new manipulation method to save computational time.

The following is a newly defined vector μ, i.e.,

$$\begin{cases} \mu = (\ \mu_1, \mu_2,\ \cdots\ , \mu_K\ )^1 \\ \mu_t = \pi_t\ \bar{B}_{tt}\ . \end{cases}$$

Solving for μ, (13) can be rewritten as follows:

$$\begin{cases} \mu_K = c'_{BK} \\ \mu_{K-1} = c'_{B,K-1} - \mu_K \bar{B}_{KK}^{-1} \bar{B}_{K,K-1} \\ \text{------------------------} \\ \mu_1 = c'_{B1} - \mu_2 \bar{B}_{22}^{-1} \bar{B}_{21}. \end{cases} \qquad (14)$$

The matrix $\bar{B}_{tt}^{-1}\bar{B}_{t,t-1}$ does not change unless the pivot operation is executed at time stage t or $t-1$. Furthermore, column size of $\bar{B}_{t,t-1}$ is usually small. Hence, the matrices $\bar{B}_{tt}^{-1}\bar{B}_{t,t-1}$ ($t=2,3,\ \cdots\ ,K$) are efficiently maintained in the computer storage and used in other calculations which will be mentioned later.

2. Relative Cost Factor

The relative cost factor for $x_N(t)$ ($= [x_s(t), x_{\bar{N}-H}(t)]$) is

$$\lambda_N{}^t = c_N{}^t - \pi P_N$$

where $c_N{}^t$ is a vector composed of objective coefficients for $x_N(t)$ and P_N is a matrix composed of column coefficients in matrix A corresponding to $x_N(t)$. P_N has its nonzero elements at the t-th and $(t+1)$th time stages, i.e.,

$$P_{Nj} = (\ 0,\ P_{Nj}{}^t,\ P_{Nj}{}^{t+1},\ 0\ )^1$$

and $\lambda_N{}^t$ is easily calculated as

$$\lambda_{Nj}{}^{t} = c_{Nj}{}^{t} - \pi_{t}P_{Nj}{}^{t} - \pi_{t+1}P_{Nj}{}^{t+1}. \tag{15}$$

We can localize the calculation and save computational time using this calculation.

3. *Test for Optimality*

In this procedure the upper bound method [6] is used. Optimality conditions for the nonbasic variables at time stage t are

$$\left. \begin{array}{l} x_{Nj}(t) = a_{Nj}{}^{t} \ \ \text{and} \ \ \lambda_{Nj}{}^{t} \geqq 0 \\ \text{or} \\ x_{Nj}(t) = b_{Nj}{}^{t} \ \ \text{and} \ \ \lambda_{Nj}{}^{t} \leqq 0 \end{array} \right\} \ \text{for the nonbasic variable } x_{Nj}(t).$$

When the above conditions are not satisfied, a nonbasic variable $x_{s}(t)$ with a maximum absolute value $|\lambda_{Ns}{}^{t}|$ can be found among the nonbasic variables not satisfying the above conditions.

4. *Pivoting*

As $x_{s}(t)$ changes from $x_{s}(t)$ to $x_{s}(t)+\theta$, the value of the objective function is reduced, and the values of the basic variables vary with the relation

$$x_{i}(t)^{*} = x_{i}(t) - \theta \cdot q_{si}{}^{t}, \quad \text{for } i = 1, 2, \cdots, m_{j};$$
$$t = t_{c}, t_{c}+1, \cdots, K \tag{16}$$

where t_{c} is the time stage given by the relation (21), and $q_{s}{}^{t}$ is a column vector defined by

$$q_{s} = B^{-1}P_{s}$$
$$= (q_{s}{}^{1}, q_{s}{}^{2}, \cdots, q_{s}{}^{K})^{1}. \tag{17}$$

P_{s} is a column vector of coefficient matrix A, corresponding to $x_{s}(t)$ and has the following form:

$$P_s = (0, P_s{}^t, P_s{}^{t+1}, 0)^1.$$

There are two cases for obtaining q_s. One case is $x_s(t) \in X_{\bar{N}-H}$ and the other is $x_s(t) \in X_s$.

Case 1. $x_s(t) \in X_{\bar{N}-H}$

Substituting (7) into (17), we obtain

$$q_s = F^{-1} \bar{B}^{-1} P_s. \tag{18}$$

From (18), we get

$$\bar{q}_s = \bar{B}^{-1} P_s$$
$$= (\bar{q}_s{}^1, \bar{q}_s{}^2, \cdots, \bar{q}_s{}^K)^1 \tag{19}$$

and obtain the following relation

$$
\begin{bmatrix}
\bar{B}_{1,1} & & & & & \\
\cdots & \cdots & & & & \\
& \bar{B}_{t-1,t} & \bar{B}_{tt} & & & \\
& & \bar{B}_{t+1,t} & \bar{B}_{t+1,t+1} & & \\
& & & \cdots & \cdots & \\
& & & & \bar{B}_{K,K-1} & \bar{B}_{KK}
\end{bmatrix}
\begin{bmatrix}
\bar{q}_s{}^1 \\
\vdots \\
\bar{q}_s{}^t \\
\bar{q}_s{}^{t+1} \\
\vdots \\
\bar{q}_s{}^K
\end{bmatrix}
=
\begin{bmatrix}
0 \\
\vdots \\
P_s{}^t \\
P_s{}^{t+1} \\
\vdots \\
0
\end{bmatrix}.
$$

The above relation yields the following sequence of equations:

$$
\begin{cases}
\bar{q}_s{}^1 = 0, & i = 1, 2, \cdots, t-1 \\
\bar{q}_s{}^t = \bar{B}_{tt}{}^{-1} P_s{}^t \\
\bar{q}_s{}^{t+1} = \bar{B}_{t+1,t+1}{}^{-1} (P_s{}^{t+1} - \bar{B}_{t+1,t} q_s{}^t) \\
\bar{q}_s{}^i = -\bar{B}_{ii}{}^{-1} \bar{B}_{i,i-1} q_s{}^{i-1}, & i = t+2, t+3, \cdots, K
\end{cases}
$$

In the above calculation, the same matrices $\bar{B}_{jj}{}^{-1} \bar{B}_{j,j-1}$ $(j = t+1, t+2, \cdots, K)$ are used and maintained as stated previously. Because of this, calculation efficiency is greatly improved.

Substituting (19) into (18), we obtain q_s as

$$
\begin{bmatrix} q_s^{\,1} \\ q_s^{\,2} \end{bmatrix} = \begin{bmatrix} G^{-1} & 0 \\ -HG^{-1} & I \end{bmatrix} \begin{bmatrix} \bar{q}_s^{\,1} \\ \bar{q}_s^{\,2} \end{bmatrix}
$$

$$
= \begin{bmatrix} G^{-1}\bar{q}_s^{\,1} \\ \\ \bar{q}_s^{\,2} - HG^{-1}\bar{q}_s^{\,1} \end{bmatrix} \tag{20}
$$

where q_s and $\bar{q}_s$ are permuted and partitioned as

$$q_s = [\, q_s^{\,1} \mid q_s^{\,2}\,]^1$$
$$\bar{q}_s = [\, \bar{q}_s^{\,1} \mid \bar{q}_s^{\,2}\,]^1$$

corresponding to G and H.

Case 2. $x_s(t) \in X_S$

When P_s is a column of $\bar{B}$, i.e., $x_s(t)$ is a variable in working basis $\bar{B}$ but not in the original basis B. In this case $\bar{B}^{-1}P_s$ becomes a unit vector, then from (18) q_s is obtained as a corresponding column of F^{-1}.

Using q_s, we can search for the variable $x_r(t_r)$ which is removed from the original basis. In the searching process, it is sufficient to consider the range from time stage t_c to K, where t_c is

$$t_c = \min(\; t, \text{ minimum stage number of}$$
$$\text{the variables in } X_H \;). \tag{21}$$

The reason is that the correction factor has the structure:

$$
F_j = \begin{bmatrix} 1 & & & & \\ & \ddots & & \boxed{} & \\ & & \boxed{} & & \\ & & \boxed{} & \ddots & \\ & & & & 1 \end{bmatrix} \begin{array}{l} \\ \leftarrow t_{hj} \\ \\ \end{array} \tag{22}
$$
$$\uparrow \atop t_{sj}$$

and each characteristic column has its nonzero element only after t_{hj}.

As $x_s(t)$ changes from $x_s(t)$ to $x_s(t)+\theta$, the value of the objective function is reduced, and the values of the basic variables vary with the relation (16). The next step is determined by the following two cases.

Case 1.

$x_s(t)$ reaches its upper bound $b_s{}^t$ or lower bound $a_s{}^t$, without any basic variable violating its upper or lower bound.

Case 2.

Some basic variables reach either of its bounds before $x_s(t)$ does.

In case 1, set $x_s(t)$ to its upper (lower) bound, record the new values of the basic variables, and repeat this procedure without changing the basis. After perhaps several repetitions of this case, either optimality or case 2 will be reached. In case 2, let $x_r(t_r)$ be one of the basic variables to reach its upper or lower bound first. Perform a basic change, with $x_r(t_r)$ entering and $x_s(t)$ leaving.

Let us assume that $x_s(t)$ is decreasing from its upper bound, $b_s{}^t$. The basic and nonbasic variables are related by

$$x_i(t)^* = x_i(t) + \theta \cdot q_{si}{}^t, \text{ for } x_i(t) \in X_B \qquad (23)$$
$$t = t_c, t_c+1, \cdots, K ; \quad i = 1, 2, \cdots, m_t.$$

Since we want

$$a_i{}^t \leq x_i(t)^* \leq b_i{}^t \qquad (24)$$

then the smallest allowed value of θ here is

$$\theta = \min \left\{ \begin{array}{ll} \dfrac{a_i{}^t - x_i(t)}{q_{si}{}^t} & , q_{si}{}^t < 0 \end{array} \right. \qquad (25)$$

$$\left\lfloor \quad \frac{b_i{}^t - x_i(t)}{q_{si}{}^t} \quad , q_{si}{}^t > 0. \right.$$

Consider now an increasing $x_s(t)$ from its lower bound, $a_s{}^t$. The basic and nonbasic variables are related by

$$x_i(t)^* = x_i(t) - \theta \cdot q_{si}{}^t, \text{ for } x_i(t) \in X_B \tag{26}$$
$$t = t_c, t_c+1, \cdots, K ; \quad i = 1, 2, \cdots, m_t$$

Since we want (24). Then

$$\theta = \min \left\{ \begin{array}{ll} \dfrac{x_i(t) - b_i{}^t}{q_{si}{}^t} & , q_{si}{}^t < 0 \\[4mm] \dfrac{x_i(t) - a_i{}^t}{q_{si}{}^t} & , q_{si}{}^t > 0. \end{array} \right. \tag{27}$$

In each case, new values for the basic variables are computed from (23) and (26), respectively.

5. Updating Factorization

In this section we will explain how to update correction factor F in the calculation. In a real computation, correction factor F can be stored in computer memory by recording the elements of the nonunit vector column and its position in the matrix (in practice, only the nonzero elements and their row positions need to be stored). These column vectors are obtained from (9) and maintained in the above stated sparse form. In the following discussion, P_s and P_r are columns of A corresponding to $x_s(t)$ and $x_r(t_r)$, respectively.

When updating F, the following four cases will occur.

a) $x_s(t) \in X_S$ and $x_r(t_r) \in X_{\overline{B}-S}$

b) $x_s(t) \in X_S$ and $x_r(t_r) \in X_H$

c) $x_s(t) \in X_{\overline{N}-H}$ and $x_r(t_r) \in X_{\overline{B}-S}$

d) $x_s(t) \in X_{\overline{N}-H}$ and $x_r(t_r) \in X_H$.

In all of these cases, nonunit vector F_j takes the form (7). Therefore, in cases b) and d), $\bar{q}_r = \bar{B}^{-1} P_r$ is a nonunit column vector of F. In addition, for cases c) and d), P_s enters the base $\bar{B}$, then $\bar{q}_s$ becomes a new nonunit column vector of F. The above-mentioned facts demonstrate the following rule.

a) Row replacement: $x_s(t)$ is removed from X_S and enters into $X_{\bar{B}-S}$. On the contrary, $x_r(t_r)$ is removed from $X_{\bar{B}-S}$ and enters into X_S. A row vector of G corresponding to $x_s(t)$ is exchanged with a row vector of H corresponding to $x_r(t_r)$.

b) Column deletion: $x_s(t)$ is removed from X_S and enters into $X_{\bar{B}-S}$, and $x_r(t_r)$ is removed from X_H and enters into $X_{\bar{N}-H}$. Related to this, a nonunit column vector of F corresponding to $x_s(t)$ is exchanged with a unit column vector, and k decreases by 1.

c) Column addition: $x_s(t)$ is removed from $X_{\bar{N}-H}$ and enters into X_H. On the contrary, $x_r(t_r)$ is removed from $X_{\bar{B}-S}$ and enters into X_S. Related to this, a nonunit column vector, $\bar{q}_s$ enters into F, and k increases by 1.

d) Column replacement: $x_s(t)$ is removed from $X_{\bar{N}-H}$ and enters into X_H. On the contrary, $x_r(t_r)$ is removed from X_H and enters into $X_{\bar{N}-H}$. Related to this, $\bar{q}_r$ is eliminated from F and $\bar{q}_s$ enters into F.

This procedure is shown in Fig. 2. PIVOT1 and PIVOT2 in Fig. 2 are the following described processes.

PIVOT1: In time stage t, if there are both of the variables in X_S and X_H, then find the biggest pivot and execute pivoting among these variables. In this case, k decreases by 1.

PIVOT2: The same procedure as PIVOT1 is executed at time stage t_r. In these steps, the inverse of the square blocks, $\bar{B}_{tt}^{-1}$ and $\bar{B}_{tr,tr}^{-1}$, and the correction factor F stored in the computer storage need to be maintained.

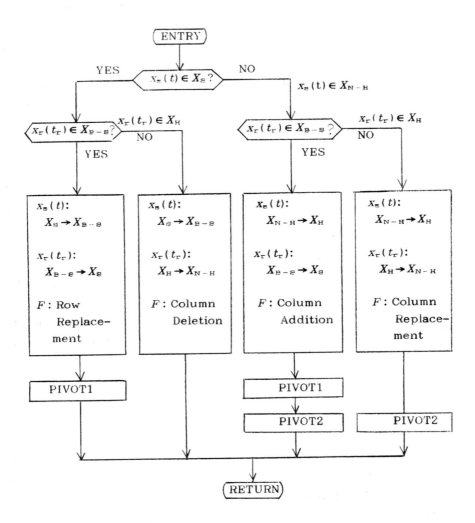

Fig. 2. Flow diagram of the updating factorization.

D. Initialization

When solving the optimization problem for dynamic (time-staged) models, solving the whole problem is costly. This is because from the beginning many artificial variables are introduced and eliminating these variables results in a large overhead due to the computational time. To overcome this shortcoming, we recommend first solving only the first stage problem ($t=1$), then set the initial values of the second stage the same as that of the first stage. Recursively apply this step until the solutions for all of the stages ($t=1,2,\cdots,K$) are obtained.

In the step to obtain the initial values of the ($t+1$)th stage with t-th stage values, the sensitivity analysis method can be used. Hence the dynamic models often have the persistence property [3], i.e., the basic solution will tend to have the same basic activities over several consecutive time stages. Consequently, the basic variables of the ($t+1$)th stage can be initially set the same as that of the t-th stage. The inverse of the basis matrix, $\overline{B}_{t+1,t+1}^{-1}$, remains the same as $\overline{B}_{tt}^{-1}$. The temporal values of the basic variables can be obtained as follows:

$$x_B(t+1) = \overline{B}_{t+1,t+1}^{-1} \{ r^{t+1} - Tx(t) - U_N x_N(t+1) \}. \tag{28}$$

If any basic variables, $x_B(t+1)$, violates its upper or lower bound, then the initial values are obtained. Otherwise artifitial variables are introduced to those variables with very large associated cost coefficients (Charne's 'big M' method [5]) and the simplex procedure starts.

Another merit of this incremental initializing is that the feasibility of the problem can be checked in the earlier stages. On the other hand, in the case of a conventional simplex method, the feasibility is not checked until the solution of the whole problem is obtained. This feature of the proposed algorithm enables plant operators to obtain optimal operational guidance quickly. For example, when the problem is infeasible, the operators can easily detect and change the constraints which cause infeasible solutions and can obtain another operational scheduling quickly.

IV. APPLICATION TO OPTIMAL ENERGY PLANT OPERATION

Preliminary computational experiments are performed to show the efficiency of the proposed algorithm. The results indicate that this method makes a great improvement, i.e., much less computing time and smaller computational storage.

The problem here is to find the optimal economical operating policy of an integrated energy plant containing energy storage units for steel manufacturing.

A schematic diagram of the energy plant is shown in Fig. 3. In this plant three steam tubine generators are used to supply part of the electric power and to furnish steam to the plant. Four boilers are used; two boilers burn blast furnace gas (BFG), one boiler burns coke oven gas (COG), and the other boiler uses oil. The purchased energies are coal, electric power, oil, and liquid propane gas (LPG). Coal is baked in the coke oven and forms COG. In the blast furnace, coke emits BFG. These gases are supplied to the process and the remainder are burned in the boilers. Furthermore, these gases can be stored to the maximum capacity of the gas holders.

The operation problem is to find the most economical amount of power generation and holder level which are subject to the given parameters of cost, constraints of operation, and estimated load duration curves.

Variables are defined as in Fig. 3, and the flow balance of energy should be held. The constraints of operation are as follows.

1) The steam turbine genarators are specified as

$$w_1(t) = 0.05 \ x_1(t) + 0.12 \ y_1(t) - 2.3$$
$$w_2(t) = 0.08 \ x_2(t) + 0.14 \ y_2(t) - 3.2$$
$$w_3(t) = 0.10 \ x_3(t) + 0.18 \ y_3(t) + 0.26 \ z_3(t) - 4.2$$
$$y_3(t) + z_3(t) \leqq 115.0.$$

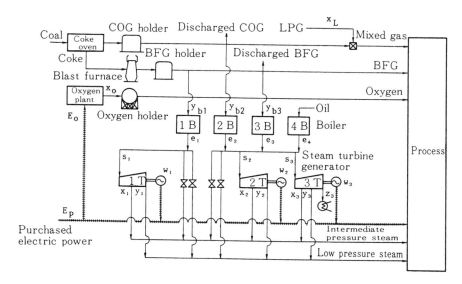

Fig. 3. Skeleton diagram of the in-plant power plant.

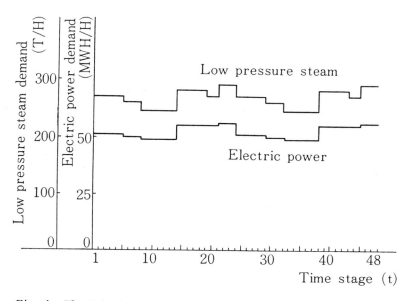

Fig. 4. Electrical power and low-pressure steam demands.

Here, the last equation is introduced as the constraint for the throttle
flow in the 3T turbine.

2) Characteristics of the boilers are assumed as

$$e_1(t) = 1.3 \; y_{b1}(t)$$
$$e_2(t) = 7.9 \; y_{b2}(t)$$
$$e_3(t) = 1.3 \; y_{b3}(t)$$

3) Output rate of the oxygen plant is assumed as

$$x_O(t) = 20.0 \; E_O(t).$$

These equations and the flow balances of each item construct a
problem with 21 constraints and 38 variables for each time stage.

Upper and lower bounds for the holder capacity, boiler output, and
the oxygen plant output are shown in Table I. Similarly, the
operational range for the turbine generators is shown in Table II.
Electric power and low pressure steam demand are shown in Fig. 4.
Mixed gas and oxygen demand are shown in Fig. 5. Intermediate steam
and BFG demand have constant values, 500(T/H) and 150.0(10^3Nm³/H),
respectively, and the charge rates of COG and BFG are 70.0(10^3Nm³/H)
and 260.0(10^3Nm³/H), respectively. The upper limit of purchased
electric power $E_P(t)$ and its cost $C^t{}_E$ are shown in Fig. 6. LPG cost is
constantly 38.0(k¥/10^3Nm³), and the cost of the steam made by 3B is
4.39(k¥/T). Hence, the objective function of this problem is

$$\text{min.} \quad z = \sum_{t=1}^{48} \{ \; C^t{}_E \; E_P(t) + 4.39 \; e_4(t) + 38.0 \; x_L(t) \; \}.$$

The optimal operation of holders at any time stage is given in Fig. 7.
The outputs of the boilers are shown in Fig. 8. The optimal power
generation policy is shown in Fig. 9 and the amount of purchased
electric power and LPG are plotted in Fig. 10.

TABLE I

UPPER AND LOWER BOUNDS FOR THE HOLDER LEVELS,
BOILER OUTPUTS, AND OXYGEN PLANT OUTPUT

| | COG holder | BFG holder | Oxygen holder | Boiler output | | | | Oxygen plant output |
				B1	B2	B3	B4	
Upper bound	30.0	80.0	150.0	100.0	157.0	60.0	120.0	170.0
Lower bound	5.0	10.0	30.0	70.0	118.0	15.0	30.0	140.0

TABLE II

OPERATIONAL RANGE FOR THE TURBINE GENERATORS

| | Output $w_3(t)$ | | Total steam $s_3(t)$ | | Extraction rate $x_3(t)$ | | Extraction rate $y_3(t)$ | | Condensing rate $z_3(t)$ | |
	max.	min.	max.	min.	max.	min.	max.	min.	max.	min.
1T	12.0	3.0	125.0	—	25.0	—	100.0	10.0	—	—
2T	16.5	5.0	126.0	—	42.0	10.0	96.0	10.0	—	—
3T	22.5	5.0	170.0	—	55.0	—	—	—	40.0	8.0

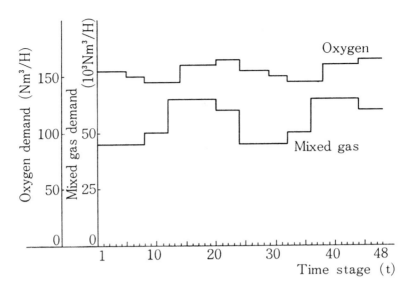

Fig. 5. Mixed gas and oxygen demand.

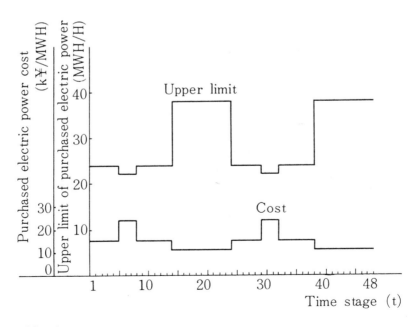

Fig. 6. Upper limit and cost of purchased electric power.

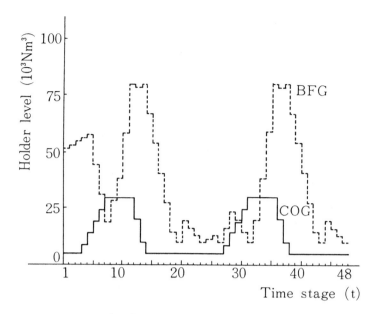

Fig. 7. Optimal holder level.

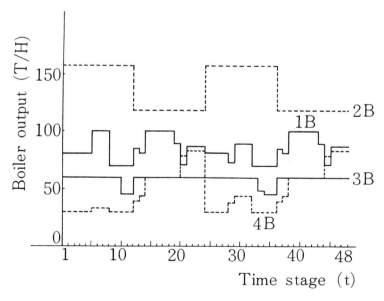

Fig. 8. Optimal boiler output.

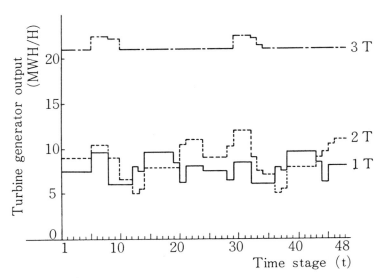

Fig. 9. Optimal power generation policy.

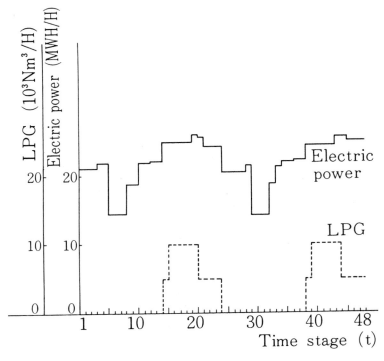

Fig. 10. Purchased amount of electric power and LPG.

COG is discharged when the mixed gas demand increases and as a result the amount of purchased LPG diminishes are shown by the levels in Fig. 7. On the other hand, BFG is discharged when the steam demand increases and when the cost of purchased electric power is high.

Boiler 2B reduces output when the mixed gas demand is at a high level, as the output in Fig. 8 indicates. Boiler 1B supplies steam for turbine 1T only, and 1T's efficiency is worse than that of the other turbines. So that it works as a supplement, Boiler 3B works at its full output except during the interval when steam demand is low.

As turbine 3T has the highest efficiency, it is operated to supply base power, and the other turbines compensate for shortages caused by load demand. In the interval when the cost of the purchased electric power is high, turbine 3T produces high output by steam condensing. As a result of this high output, the amount of purchased electric power is less during that interval as shown in Fig. 10.

We compared the computational efficiency of the proposed method to a general commercially used LP code (MPS- II) which is an implementation of the revised simplex algorithm and used LU factorization method to maintain the inversion of the basis matrix. The results are shown in Table III. The computation was performed using a general purpose computer Hitachi M-180 (4.5 MIPS). The storage requirements and computing time are approximately 65 and 6 percent, respectively, of those of MPS- II.

In solving the problem, the maximum number of structural columns in F is 8. In addition the factorization is almost always minimal. This demonstrates in practical terms that updating correction factor F as shown in Section III can maintain an almost minimal factorization level.

TABLE III

COMPUTATIONAL EFFICIENCY

	Required storage (thousand bytes)	CPU computing time (sec)
MPS-II	551	112.1
Proposed Method	360	7.0

V. SUMMARY AND CONCLUSION

A variant of the basis factorization method is developed for solving multistage linear programming in which some outputs of one stage are inputs to the next. This method is devised to enable the special staircase structure of the original basis to be used. It essentially reduces this basis into a square block triangular working basis using a set of elementary column operations. The use of the working basis in place of the original basis now involves the inversion of the diagonal blocks which are considerably smaller in size (of the order of $1/K$ th of the original in a K-stage problem).

Storing the structural columns corresponding to the variables involved in the original basis but not in the working basis, the relative cost factor of the original nonbasis using that of the working nonbasis can be easily obtained. This can be easily derived using a set of elementary calculations.

The algorithm was coded in FORTRAN and applied to an in-plant energy control system. The results show that this method offers great advantages over the conventional simplex method (MPS- II). MPS- II is an implementation of the revised simplex algorithm which uses LU factorization for basis matrix inversion. Both the original matrix and the inverse of the basis are stored in sparse form.

In practical terms, this computational improvement of factor 1 brings a significant benefit to computer control. Operation scheduling tasks discussed in this paper are one of the biggest loads in computer control.

Conventionally, since minicomputers usually installed for plant operation cannot accept these tasks, they are executed by large-scale general purpose computers. These improvements make it possible for minicomputers to execute these tasks. That means flexible and economical task allocations to computer resources, and an increase in the anatomy of computer control systems.

References

[1] C. B. Glassey,"Nested decomposition and multi-stage linear programs," *Management Sci.*, vol. 20, no. 3, pp. 282-292, 1973.

[2] J. K. Ho and A. S. Manne, "Nested decomposition for dynamic models," *Math. Programming*, vol. 6, pp. 121-140, 1974.

[3] A. F. Perold and G. B. Dantzig,"A basis factorization method for block triangular linear programs," in *Sparse Matrix Proc.*, 1978, SIAM, pp.283-312.

[4] J. K. Ho and E. Loute,"A comparative study of two methods for staircase linear programs," *ACM Trans. Math. Software*, vol. 5, pp.17-30, Dec. 1979.

[5] A. Charnes, W. W. Cooper, and A. Henderson, *An Introduction to Linear Programming.* New York: Wiley, 1953.

[6] L. S. Lasdon, *Optimization Theory for Large Systems.* New York: Macmillan, 1970.

[7] W. W. Garvin, *Introduction to Linear Programming*, New York: McGraw-Hill, 1960.

[8] T. Nishiya and M. Funabashi, "Basis Factorization Method for Multistage Linear Programming Problems with an Application to Optimal Energy Plant Operation," *IEEE Trans. on Automatic Control*, Vol. AC-32, No. 10, pp. 851-857, Oct. 1987.

[9] T. Nishiya, et. al., "Basis Factorization Method for Multi-Stage Linear Problem with an Application to Optimal Paper Manufacturing Plant Operation," *Proc. of IFAC/IFORS Symposium on Large Scale Systems: Theory and Application*, 1986

OPTIMIZATION TECHNIQUES IN HYDROELECTRIC SYSTEMS

G.S. CHRISTENSEN

Elect.Eng.Dept.

The University of Alberta

Edmonton, Alberta, Canada

S.A. SOLIMAN

Elect.Power & Machines Dept.

Ain Shams University

Abbassia, Cairo, Egypt

I.   INTRODUCTION

The hydro optimization problem involves planning the use of a limited resource over a period of time.   The resource is   the water available for hydro generation.   Most hydroelectric plants are multipurpose.   In such cases, it is necessary to meet certain obligations other than power generation.   These may include a maximum forebay elevation, not to be exceeded because of the danger of flooding, and a minimum plant discharge and spillage to meet irrigational and navigational commitments.   Thus, the optimum operation of the hydro system depends upon the conditions that exist over the entire optimization interval [1].

Other distinctions among power systems are the number of hydro stations, their location and special operating characteristics. The problem is quite different if the hydro stations are   located on the same stream or on different ones.   An upstream station will highly influence the operation of the next downstream station.   The latter, however, also influences the upstream plant by its effect on the tail water elevation and effective   head. Close coupling of stations by such a phenomenon is a complicating factor [2].

CONTROL AND DYNAMIC SYSTEMS, VOL. 42

The problem of determining the optimal long-term operation of a multireservoir power system has been the subject of numerous publications over the past forty years, and yet no completely satisfactory solution has been obtained, since in every publication the problem has been simplified in order to be solved.

Aggregation of the multireservoir into a single complex equivalent reservoir and solution by stochastic dynamic programming (SDP) was one of the earlier approaches used [3],[4]. Obviously, such a representation of the reservoir cannot take into account all local constraints on the contents of the reservoir, water flows, and hydroplant generation. This method can perform satisfactorily for systems where reservoirs and inflow characteristics are sufficiently "similar" to justify aggregation into a single reservoir and hydroplant model.

Turgeon has proposed two methods for the solution of the problem [5], [6]. The first is really an extension to the aggregation method, and it breaks the problem down into two levels. At the second level the problem is to determine the monthly generation of the valley. This problem is solved by dynamic programming. The problem at the first level is to allocate that generation to the installation; this is done by finding functions that relate the water level of each reservoir to the total amount of potential energy stored in the valley. The second method is the decomposition method by combining many reservoirs into one reservoir for the purpose of optimization of multireservoir power system connected in series on a river, and using dynamic programming for solving n-1 problems of two state variables each. The solution obtained by this method is a function of the water content of that reservoir and the total energy content of the downstream reservoirs. The main drawback is that the approach avoids answering basic questions as to how the individual reservoirs in the system are to be operated in an optimal fashion. Also the inflows to some reservoirs may be periodic in phase with the annual demand cycle, while other reservoirs have an inflow cycle that lags by a certain time.

Stochastic dynamic programming with successive approximation (DPSA) has been proposed to solve the problem of a parallel multireservoir hydroelectric power system. The successive approximation involves a "one-at-a-time" stochastic optimization of each reservoir. The major drawback of this approach is that it ignores the dependence of the operation of one reservoir on the actual energy content of other reservoirs [6].

In the aggregation-decomposition (AD) approach, the optimation of a system of n reservoirs is broken down into n subproblems in which one reservoir is optimized knowing the total energy content of the rest of the reservoirs. For each subproblem one of the reservoir-hydroplant models is retained and the remaining n-1 are aggregated into an equivalent reservoir hydroplant model. A comparison of the last two approaches, DPSA and AD, has been made on a simulation basis for a six-complex-reservoir system. The results indicate that the AD approach gives better results and the computational effort increases only linearly with the number of reservoirs. More precisely, for each new reservoir added to the system, only one additional dynamic programming problem of two-state variables has to be solved. The computing time for each of the last two approaches was 150 min in CPU units [5], [6].

Linear programming and dynamic programming have been applied to the optimization of the production of hydroelectric power. The solution was obtained in two steps with linear and dynamic programming methods. The models that have been used are deterministic. The first step in the solution was the long-term optimization problem; this problem was solved by a linear programming (LP) method. The variation in the efficiency of the turbines, the variation of the water heads, and the time delays are neglected. The second step is the optimal short-term run of the turbine-generator units; this is determined by dynamic programming (DP). The total computation time on a typical minicomputer was 1-3 min per power station [7].

Successive linear programming, an optimal control algorithm, and a combination of linear programming and dynamic programming (LP-DP) are employed to optimize the operation of mutireservoir hydro systems given a deterministic inflow forecast. The algorithm maximizes the total benefits from the system (maximization of energy produced, plus the estimated value of water remaining in storage at the end of the planning period). The LP-DP algorithm is the least satisfactory: it takes longer to find a solution and produces significantly less hydro power than the other two procedures. Successive linear programming (SLP) appears to find the global maximum and is easily implemented. The optimal control algorithm can find the optimum in about one fifth the time required by SLP for small systems but is more difficult to implement. The computing costs were reasonable with SLP and the optimal control algorithm, and increase only as the square of the number of reservoirs in contrast to the exponential growth of dynamic programming [7].

Marino and Loaiciga applied a quadratic optimization model to a large-scale reservoir system to obtain operation schedules. The model they used has the minimum possible dimensionality, treats spillage and penstock releases as decision variables, and takes advantage of system-dependent features to reduce the size of the decision space. They compared the quadratic model with a simplified linear model [8].

II. HYDRO PLANT MODELING FOR LONG-TERM OPTERATION [1]

Hydro power plants are classified into pumped storage plants and conventional hydro plants.

A. PUMPED STORAGE PLANTS

A pumped storage plant is associated with upper and lower reservoirs. During light load periods water is pumped from the lower to the upper reservoirs using the available energy from other sources as surplus energy. During peak load the water stored in the upper reservoir is released to generate power to save fuel costs of the thermal plants. The pumped storage plant is operated until the added pumping cost exceeds the saving in thermal costs due to the peak shaving operation.

The conventional hydroplants are classifsied into run-of-river plants and storage plants.

B. RUN-OF-RIVER PLANTS

The run-of-river plants have little storage capacity, and use water as it becomes available; water not utilized is spilled. The MWh generated from a run-of-river plant is equal to a constant times the discharge through the turbines.

$$G_k = hu_k \, \text{MWh}$$

where h is a constant measured in MWh/m^3 and referred to as the water conversion factor; u_k is the discharge through the turbine during a period k in m^3; and k is an index used for the period- this period may be a week or a month.

C. STORAGE PLANTS

Storage plants are associated with reservoirs with significant storage capacity. In periods with low power requirements water can be stored and then released when the demand is high.

Modeling of storage plants, for long-term study, depends on the water head variation. For hydro plants in which the water head variation is small, the MWh generated from the plants can be considered as a constant times the discharge, as given in the above equation, and this constant is equal to the average number of MWh generated during a period k by an outflow of 1 m^3. But for the power systems in which the water heads vary by a considerable amount, this assumption is not true, and the water conversion factor, MWh/m^3, varies with the head, which itself is a function of the storage. For a long-term study, the MWh generated can be written as

$$G_k = \alpha u_k + \frac{1}{2}\beta u_k(x_k+x_{k-1}) + \frac{1}{4}\gamma u_k(x_k+x_{k-1})^2 \text{ MWh}$$

where α, β, and γ are constants — these can be obtained by least-squares curve fitting to typical plant data available; and x_k is the storage at the end of period k in m^3.

The above equation is a function of the discharge through the turbines and the average storage between two successive months, k-1 and k, to avoid underestimation for rising water levels and overestimation for falling water levels in the MWh generatd.

1. Reservoir Models

Modeling of reservoirs is of great importance for long-term operation of hydroelectric power systems. Reservoir models contain the storage and the release in a well-known equation, called the continuity equation, or the water conservation equation. For long-term study, the reservoir dynamics may be adequately described by the discrete difference equation [1].

$$x_k = x_{k-1} + I_k - u_k - s_k + z_k$$

where I_k is the natural inflow adjusted for evaporation and seepage losses during a period k in m^3, s_k is the spillage during a period k in m^3. Water is spilt when the discharge u_k is greater than the maximum discharge and the reservoir is filled to capacity, and z_k is the outflow from upstream reservoirs.

The variation of storage of a reservoir of regular shape with the elevation can be computed using the formulas for the volumes of solids. In practice, natural factors will change the reservoir configuration with time. An example is sediment accumulation. It is important to update the reservoir model periodically. A mathematical model for the storage-elevation curve may be obtained by using Taylor's expansion [2]

$$h_k = \sum_{p=0}^{N} \alpha_p (x_k)^p$$

where h_k is the net head at the end of period k in meters and N is the highest order of the approximation. We assume that the tailwater elevation is constant, independent of the discharge. α_p's can be obtained by using curve fitting to data available from typical reservoirs.

2. Operational Constraints [1]

Many dams and associated reservoirs are multipurpose developments. Irrigation, flood control, water supply, stream flow augmentation, navigation, and recreation use are among the possible purposes of water resource development. For these purposes the reservoir is regulated so that full requirements for each element are available under the design drought conditions.

The operational reservoir constraints are [1]

$$x^m \leq x_k \leq x^M \tag{5}$$

$$u_k^m \leq u_k \leq u_k^M \tag{6}$$

The first set of inequality constraints simply states that the reservoir storage (or elevation) may not exceed a minimum level, nor be lower than a minimum level. For the maximum level, this is determined by the elevation of the spillway crest or the top of the spillway gates. The minimum level may be fixed by the elevation of the lowest outlet in the dam or by conditions of operating efficiency for the turbines. The second set of the inequality constraint is determined by the discharge capacity of the power plant as well as its efficiency. [1], [2]

III. OPTIMAL LONG-TERM SCHEDULING OF SERIES RESERVOIRS [9]

The purpose of this section is to present a method for solving the long-term optimal operating problem for n reservoirs in series on a river. The optimal control problem is formulated by constructing a cost function, in which we maximize the total benefits from the system (benefits from the energy generated by the hydropower plants over the planning period plus the expected future returns from water left in storage at the end of that period). The problem is formulated as a minimum norm problem in the framework of functional analytic optimization technique. The optimization is done on a monthly time basis for a period of a year and takes into account the stochasticity of the river flows, we assume that their probability properties are pre-estimated from the past history. The times of water travel between upstream and downstream reservoirs are assumed to be shorter than a month and, for this reason, are not taken into account. Furthermore, the transmission line losses are also neglected.

A.   PROBLEM FORMULATION

1. The System Under Study [5]

The system under study consists of n reservoirs in series on a river, Figure 1, with interconnection lines to the neighbouring system for energy exchange. We will number the installation from upstream to downstream.

2. The Objective Function and the Constraints

The long-term optimal operating problem for the system of Figure 1 is to determine the discha rge $u_{i,k}$ of each reservoir, $i=1,...,n$, $k=1,...,K$ which maximizes the total benefits from the system (benefits from energy generated by a hydropower system over the planning period plus the expected future benefit, from the water left in storage at the end of that period), and at the same time, satisfying the following constraints.

a. The mega-watt hours, MWh, generated per one cubic meter (m^3), the water conversion factor, as a function of the storage is adequately described by

$$H_{i,k} = \alpha_i + \beta_i x_{i,k} + \gamma_i (x_{i,k})^2, \quad i=1,...,n, k=1,...,K \tag{1}$$

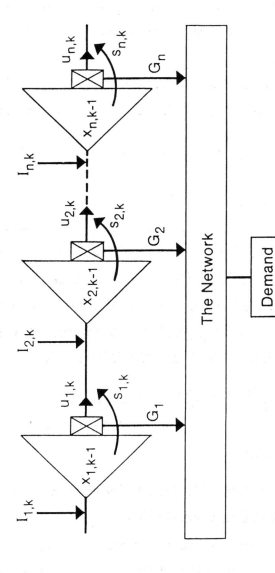

Figure 1. A multireservoir power system

where $H_{i,k}$ is the water conversion factor of reservoir i during a period k measured in MWh/m^3, $x_{i,k}$ is the storage of reservoir i at the end of period k in m^3 and α_i, β_i and γ_i are constants. These were obtained by least error squares curve fitting to typical plant data available.

b. The water conservation equation for each reservoir may be adequately described by the continuity discrete equation

$$x_{i,k} = x_{i,k-1} + I_{i,k} + u_{i-1,k} + s_{i-1,k} - u_{i,k} - s_{i,k}, i=1,\ldots,n; k=1,\ldots,K$$

(2)

where $I_{i,k}$ a random variable representing the natural inflows to the reservoir i during period k. These are statistically independent random variables, we assume that their probability properties are pre-estimated from the past history, and $s_{i,k}$ is the spill from reservoir i during period k. Water is spilt when the reservoir is full and $u_{i,k}$ exceeds the maximum discharge from the turbine. $s_{i,k}$ can be expressed mathematically as

$$s_{i,k} = \begin{cases} (u_{i,k} - u_{i,k}{}^M), \text{if } u_{i,k} > u_{i,k}{}^M \text{ and } x_{i,k} \geq x_i{}^M \\ \\ 0, \text{ otherwise} \end{cases}, i=1,\ldots,n, k=1,\ldots,k$$

(3)

where $u_{i,k}{}^M$ and $x_i{}^M$ are the maximum discharge and storage of reservoir i during period k respectively.

c. In order to be realizable and also to satisfy multipurpose stream use requirements, such as flood control, irrigation, fishing and other purposes, if any, the following upper and lower limits on the variables should be satisfied

(i)   upper and lower bounds on the storage

$$x_i^m \leq x_{i,k} \leq x_i^M, \quad i=1,\ldots,n; k=1,\ldots,K \tag{4}$$

(ii)   upper and lower bounds on the discharge

$$u_{i,k}^m \leq u_{i,k} \leq u_{i,k}^M, \quad i=1,\ldots,n; k=1,\ldots,K \tag{5}$$

where $u_{i,k}^m$ and x_i^m are the minimum discharge and storage of reservoir i during a period k respectively.

The long-term optimal operating scheduling problem can now be written in mathematical terms as: Find the discharge $u_{i,k}$, $i=1,\ldots,n$; $k=1,\ldots,K$ that maximizes

$$J=E\left[\sum_{i=1}^{n} V_i(x_{i,K}) + \sum_{k=1}^{K} \sum_{i=1}^{n} c_k G_{i,k}(u_{i,k}, (x_{i,k}+x_{i,k-1}))\right] \tag{6}$$

while satisfying the equality constraints of Eq. (2) and the inequality constraints of Eqs. (4) and (5). In the above equation $V_i(x_{i,K})$ is the value of water left in storage for reservoir i at the end of last period studied, $G_{i,k}(.)$ is the generation of plant i during period k in MWh and c_k is the value, in dollars, of one MWh produced anywhere on the river during period k. E stands for the expected value, the expectation in Eq. (6) is taken over the random variables $I_{i,k}$.

3. Modelling of the System

The conventional approach for obtaining the equivalent

reservoir and hydroplant is based on the potential energy
concept. Each reservoir on a river is mathematically represented
by an equivalent potential energy balance equation. The potential
energy balance equation is obtained by multiplying both sides of
the reservoir balance-of-water equation by the water conversion
factors of at-site and downstream hydroplants. We may choose the
function $V_i(x_{i,K})$ as [9]:

$$V_i(x_{i,K}) = \sum_{j=1}^{n} x_{i,K}(\alpha_j + \beta_j x_{j,K} + \gamma_j x_{j,K}^2), i=1,\ldots n \qquad (7)$$

In the above equation, we consider c_K at last period is equal to
one, the average cost during the year.

The generation of a hydroelectric plant is a nonlinear
function of the water discharge $u_{i,k}$ and the net head, which
itself is a function of the storage. This can be expressed as
[9]:

$$G_{i,k}(u_{i,k}, \tfrac{1}{2}(x_{i,k} + x_{i,k-1})) = \alpha_i u_{i,k} + \tfrac{1}{2}\beta_i u_{i,k}(x_{i,k} + x_{i,k-1})$$

$$+ \tfrac{1}{4}\gamma_i u_{i,k}(x_{i,k} + x_{i,k-1})^2, i=1,\ldots,n, k=1,\ldots,K \qquad (8)$$

In the above equation an average of begin and end-of-time is used
to avoid under estimation in the MWh generated for rising water
levels and over estimation for falling water levels. Substituting
for $x_{i,k}$ from Eq. (2) into Eq. (8), we obtain

$$G_{i,k} = b_{i,k} u_{i,k} + d_{i,k} u_{i,k} x_{i,k-1} + f_{i,k} u_{i,k}(u_{i-1,k} - u_{i,k})$$

$$+\gamma_i u_{i,k}(x_{i,k-1})^2$$

$$+\frac{1}{4}\gamma_i u_{i,k}(u_{i,k}^2+u_{i-1,k}^2)+\gamma_i u_{i,k}x_{i,k-1}(u_{i-1,k}-u_{i,k})$$

$$-\frac{1}{2}\gamma_i u_{(i-1),k}u_{i,k}^2, \quad i=1,\ldots,n; \ k=1,\ldots,K \qquad (9)$$

where

$$q_{i,k}=I_{i,k}+s_{i-1,k}-s_{i,k} \ ; \ i=1,\ldots,n, \ k=1,\ldots,K \qquad (10)$$

$$b_{i,k}=\alpha_i+\frac{1}{2}\beta_i q_{i,k}+\frac{1}{4}\gamma_i(q_{i,k})^2; \ i=1,\ldots,n, \ k=1,\ldots,K$$
$$\qquad (11)$$

$$d_{i,k}=\beta_i+\gamma_i q_{i,k}, \ i=1,\ldots,n; \ k=1,\ldots,K \qquad (12)$$

$$f_{i,k}=\frac{1}{2}d_{i,k}, \ i=1,\ldots,n; \ k=1,\ldots,K \qquad (13)$$

Now, the cost functional of Eq. (6) can be written as

$$J=E[\sum_{j=i}^{n}\sum_{i=1}^{n} x_{i,K}(\alpha_j+\beta_j x_{j,K}+\gamma_j(x_{j,K})^2+\sum_{k=1}^{K}\sum_{i=1}^{n} c_k\{b_{i,k}u_{i,k}+d_{i,k}u_{i,k}x_{i,k-1}$$

$$+f_{i,k}u_{i,k}(u_{i-1,k}-u_{i,k})+\gamma_i u_{i,k}(x_{i,k-1})^2+\frac{1}{4}\gamma_i u_{i,k}(u_{i,k}^2+u_{i-1,k}^2)$$

$$+\gamma_i u_{i,k}x_{i,k-1}(u_{i-1,k}-u_{i,k})-\frac{1}{2}\gamma_i u_{(i-1),k}u_{i,k}^2\}] \qquad (14)$$

The cost functional of Eq. (14) is a highly nonlinear function, if one defines

$$y_{i,k}=x_{i,k}^2 \ , \ i=1,\ldots,n; \ k=1,\ldots,K \qquad (15)$$

$$z_{i,k}=u_{i,k}^2 \ , \ i=1,\ldots,n; \ k=1,\ldots,K \qquad (16)$$

$$r_{i,k-1}=u_{i,k}x_{i,k-1}, \ i=1,\ldots,n; \ k=1,\ldots,K \qquad (17)$$

Then Eq. (14) can be written as

$$J = E\{ \sum_{i=1}^{n} \sum_{j=i}^{n} x_{i,K}(\alpha_j + \beta_j x_{j,K} + \gamma_j y_{j,K}) + \sum_{k=1}^{K} \sum_{i=1}^{n} [c_k b_{i,k} u_{i,k} + c_k d_{i,k} u_{i,k} x_{i,k-1}$$

$$+ c_k f_{i,k} u_{i,k}(u_{i-1,k} - u_{i,k}) + c_k \gamma_i u_{i,k} y_{i,k-1} + \tfrac{1}{4} c_k \gamma_i u_{i,k}(z_{i,k} + z_{i,k-1})$$

$$+ c_k \gamma_i r_{i,k-1}(u_{i-1,k} - u_{i,k}) - \tfrac{1}{2} c_k \gamma_i z_{i,k} u_{(i-1)k}\}] \tag{18}$$

Subject to satisfying the equality constraints given by Eqs. (2) and (15)-(17) and the inequality constraints given by Eqs. (4) and (5).

B. MINIMUM NORM FORMULATION [12]

To transfer the problem to an unconstrained problem, we adjoin to the cost functional of Eq. (18), the equality constraints via Lagranges' multipliers $\lambda_{i,k}$, $\phi_{i,k}$, $\mu_{i,k}$ and $\psi_{i,k}$, these are obtained so that the corresponding equality constraints are satisfied, and the inequality constraints via Kuhn-Tucker multipliers to obtain [2]

$$J = E[\sum_{j=i}^{n} \sum_{i=1}^{n} x_{i,K}(\alpha_j + \beta_j x_{j,K} + \gamma_j y_{j,K})$$

$$+ \sum_{k=1}^{K} \sum_{i=1}^{n} [c_k b_{i,k} u_{i,k} + c_k d_{i,k} u_{i,k} x_{i,k-1}$$

$$+ c_k f_{i,k} u_{i,k}(u_{i-1,k} - u_{i,k}) + c_k \gamma_i u_{i,k} y_{i,k-1} + \tfrac{1}{4} c_k \gamma_i u_{i,k}(z_{i,k} + z_{i,k-1})$$

$$+ c_k \gamma_i r_{i,k-1}(u_{i-1,k} - u_{i,k})$$

$$- \tfrac{1}{2} c_k \gamma_i z_{i,k} u_{(i-1),k}$$

$$+ \lambda_{i,k}(-x_{i,k} + x_{i,k} + q_{i,k} + u_{(i-1),k} - u_{i,k})$$

$$+\phi_{i,k}(-y_{i,k}+x_{i,k}^2)+\mu_{i,k}(-z_{i,k}+u_{i,k}^2)+$$

$$\psi_{i,k}(-r_{i,k-1}+u_{i,k}x_{i,k-1})+e_{i,k}^m(x_i^m-x_{i,k})+e_{i,k}^M(x_{i,k}-x_i^M)$$

$$+g_{i,k}^m(u_{i,k}^m-u_{i,k})+g_{i,k}^M(u_{i,k}-u_{i,k}^M)\}] \tag{19}$$

where $e_{i,k}^m$, $e_{i,k}^M$, $g_{i,k}^m$ and $g_{i,k}^M$ are Kuhn-Tucker multipliers. These are equal to zero if the constraints are not violated and greater than zero if the constraints are violated [2].

Employing the discrete version of integration by parts [13], [14] and dropping constant terms, Eq. (19) can be written in a vector form as:

$$\tilde{J}=E[x^T(K)(\beta+\phi^T(K)\vec{H})x(K)+(A-\lambda(K))^Tx(K)-\phi^T(K)y(K)$$

$$+x^T(K)\gamma y(K)+x^T(0)\phi^T(0)\vec{H}x(0)+\lambda^T(0)x(0)+\phi^T(0)y(0)]+$$

$$E[\sum_{k=1}^{K}\{x^T(k-1)\phi^T(k-1)\vec{H}x(k-1)+u^T(k)d(k)x(k-1)+\frac{1}{2}u^T(k)f(k)Mu(k)$$

$$+\frac{1}{2}u^T(k)M^Tf(k)u(k)+u^T(k)\mu^T(k)\vec{H}u(k)+u^T(k)\gamma(k)y(k-1)$$

$$+\frac{1}{4}u^T(k)\gamma(k)Nz(k)+r^T(k-1)\gamma(k)Mu(k)-\frac{1}{2}z^T(k)\gamma(k)Lu(k)$$

$$+u^T(k)\psi^T(k)\vec{H}x(k-1)+(\lambda(k)-\lambda(k-1)+\nu(k))^Tx(k-1)+(M^T\nu(k)+B(k)$$

$$+M^T\lambda(k)+\sigma(k))^Tu(k)-\phi^T(k-1)y(k-1)-\mu^T(k)z(k)-\psi^T(k)r(k-1)\}] \tag{20}$$

where we defined

$$\nu(k)=(e^M(k)-e^m(k)) \quad \text{or} \quad \nu_{i,k}=(e_{i,k}^M-e_{i,k}^m), \quad i=1,\ldots,n; k=1,\ldots,K$$

$$\sigma(k)=(g^M(k)-g^m(k)) \quad \text{or} \quad \sigma_{i,k}=(g_{i,k}^{M}-g_{i,k}^{m}), \quad i=1,\ldots,n; k=1,\ldots,K$$

In Eq. (20), $x(k)$, $u(k)$, $y(k)$, $z(k)$, $r(k-1)$, $\lambda(k)$, $\phi(k)$, $\mu(k)$ $\nu(k)$, and $\sigma(k)$ are n-dimensional vectors at the end of month k; their components are $x_{i,k}$, $u_{i,k}$, $y_{i,k}$, $z_{i,k}$, $r_{i,k-1}$, $\lambda_{i,k}$, $\phi_{i,k}$, $\mu_{i,k}$, $\nu_{i,k}$, and $\sigma_{i,k}$ respectively. Also, A, B(k) are n-dimensional vectors; their components are given by $A_i = \sum_{j=i}^{n} \alpha_j$, $B_{i,k}=c_k b_{i,k}$.

$d(k)$, $f(k)$, $\gamma(k)$ are nxn diagonal matrices, their elements are $d_{ii,k}=c_k d_{i,k}$, $f_{ii,k}=c_k f_{i,k}$ and $\gamma_{ii,k}=c_k \gamma_i$, $i=1,\ldots,\ldots,n$.

M is a lower triangular matrix whose elements are given by: $m_{ii}=-1$, $i=1,\ldots,n$; $m_{(j+1)j}=1$; $j=1,\ldots,n-1$.

N is a lower triangular matrix whose elements are given b;y: $n_{ii}=1$, $i=1,\ldots,n$; $n_{(j+1)j}=1$, $j=1,\ldots,n-1$.

L is a lower triangular matrix whose elements are given by: $\ell_{ii}=0$, $i=1,\ldots,n$; $\ell_{(j+1)j}=1$, $jh=1,\ldots,n-1$.

γ is an nxn upper triangular matrix whose elements are $\gamma_{ii}=\gamma_i$, $\gamma_{i(j+1)}=\gamma_{j+1}$, $i=1,\ldots,n$, $j=1,\ldots,n-1$

β is nxn matrix whossse elements are given by: $\beta_{ii}=\beta_i$, $i=1,\ldots,n$, $\beta_{(j+1)i}=\beta_{i(j+1)}=\frac{1}{2}\beta_{(j+1)}$, $i=1,\ldots,n-1$, $j=1,\ldots,n-1$.

$\vec{H}$ is a vector matrix in which the vector index varies from 1 to n while the matrix dimension of H is nxn [15].

If one defines the following vectors, T stands for the transpose

$$Z^T(K)=[x^T(K),y^T(K)]; \quad 1\times 2n \text{ vector} \tag{21}$$

$$
N(K) = \begin{vmatrix} \beta+\phi^T(K)H & \frac{1}{2}\gamma \\[2em] \frac{1}{2}\gamma^T & 0 \end{vmatrix} \quad \text{2nxn matrix} \tag{22}
$$

$$
W^T(K)=[(A-\lambda(K))^T, \ -\phi^T(K)] \ \text{1x2n vector} \tag{23}
$$

and

$$
X^T(k)=[x^T(k-1),y^T(k-1),u^T(k),z^T(k),r^T(k-1)] \ \text{1x5n vector} \tag{24}
$$

$$
R^T(k)=[(\lambda(k)-\lambda(k-1)+\nu(k))^T,-\phi^T(k-1),(B(k)+M^T\lambda(k)+M^T\lambda(k)
$$

$$
+\sigma(k))^T,-\mu^T(k),-\psi^T(k)] \ \text{1x5n vector} \tag{25}
$$

and furthermore define the 5nx5n matrix L(k) as

$$
L(k)= \begin{vmatrix} L_{11}(k) & L_{12}(k) & L_{13}(k) \\[2em] L_{21}(k) & L_{22}(k) & L_{23}(k) \end{vmatrix} \tag{26}
$$

where

$$
L_{11}(k)= \begin{vmatrix} \phi^T(k-1)\vec{H} & 0 \\[1em] 0 & 0 \\[1em] \frac{1}{2}(d(k)+\psi^T(k)H) & \frac{1}{2}\gamma(k) \end{vmatrix} \tag{26a}
$$

$$
L_{12}(k)= \begin{vmatrix} \frac{1}{2}(d(k)+\psi^T(k)\vec{H} & 0 \\[1em] \frac{1}{2}\gamma(k) & 0 \\[1em] (\frac{1}{2}f(k)M+\frac{1}{2}M^Tf(k)+\mu^T(k)\vec{H}) & (\frac{1}{2}\gamma(k)N-\frac{1}{4}L^T\gamma(k) \end{vmatrix} \tag{26b}
$$

$$L_{13}(k) = \begin{vmatrix} 0 \\ 0 \\ \frac{1}{2}M^T\gamma(k) \end{vmatrix} \tag{26c}$$

$$L_{21}(k) = \begin{vmatrix} 0 & 0 \\ \\ 0 & 0 \end{vmatrix} \tag{26d}$$

$$L_{22}(k) = \begin{vmatrix} (\frac{1}{8}N^T\gamma(k) - \frac{1}{4}\gamma(k)L) & 0 \\ \\ \frac{1}{2}\gamma(k)M & 0 \end{vmatrix} \tag{26e}$$

$$L_{23}(k) = \begin{vmatrix} 0 \\ \\ 0 \end{vmatrix} \tag{26f}$$

Then, the augmented cost functional in Eq. (20) can be written as

$$\tilde{J} = E[Z^T(K)N(K)Z(K) + W^T(K)Z(K) + x^T(0)\phi^T(0)\vec{H}x(0) + \lambda^T(0)x(0)$$

$$+ \phi^T(0)y(0)] + E[\sum_{k=1}^{K}[X^T(k)L(k)X(k) + R^T(k)X(k)]] \tag{27}$$

If one defines the vectors

$$Q(K) = N^{-1}(K)W(K) \tag{28}$$

$$V(k) = L^{-1}(k)R(k) \tag{29}$$

Then, the cost function in Eq. (27) can be written in the

following form by a process similar to completing the squares as

$$\tilde{J}=E[(Z(K)+\tfrac{1}{2}Q(K))^T N(K)(Z(K)+\tfrac{1}{2}Q(K))-\tfrac{1}{4}Q^T(K)N(K)Q(K)$$

$$+x^T(0)\phi^T(0)\vec{H}x(0)+\lambda^T(0)x(0)+\phi^T(0)y(0)]$$

$$+E[\sum_{k=1}^{K}\{(X(k)+\tfrac{1}{2}V(k))^T L(k)(X(k)+\tfrac{1}{2}V(k))-\tfrac{1}{4}V^T(k)L(k)V(k)\}]$$

$$(30)$$

Since, it is desired to maximize $\tilde{J}$ with respect to $Z(K)$ and $X(k)$, the problem is equivalent to [1], [2]

$$\underset{[Z(K),X(k)]}{\text{Max.}\tilde{J}}(Z(K).X(K))=\underset{Z(K)}{\text{Max.}}E[(Z(K)+\tfrac{1}{2}Q(K))^T N(K)(Z(K)+\tfrac{1}{2}Q(K))]$$

$$+\underset{X(k)}{\text{Max.}}E[\sum_{k=1}^{K}\{(X(k)+\tfrac{1}{2}V(k))^T L(k)(X(k)+\tfrac{1}{2}V(k))\}]\qquad(31)$$

because $Q(K)$ and $V(K)$ are independent of $Z(K)$ and $X(K)$ and $x(0)$, $y(0)$ are constants [14].

It will be noticed that $\tilde{J}$ in Eq. (31) is composed of a boundary part and a discrete integral part. To maximize $\tilde{J}$ in Eq. (31), one maximizes each term separately [13], [14]. The boundary part in Eq. (31) defines a norm. Hence, we can write this part as

$$\underset{Z(K)}{\text{Max.}J_1}=\underset{Z(K)}{\text{Max.}}E||Z(K)+\tfrac{1}{2}Q(K)||_{N(K)}\qquad(32)$$

Also, the discrete integral part of Eq. (31) can be written as:

$$\underset{X(k)}{\text{Max.}J_2}=\underset{X(k)}{\text{Max.}}E||X(k)+\tfrac{1}{2}V(k))||_{L(k)}\qquad(33)$$

C.   OPTIMAL EQUATIONS

There is only one optimal solution to the problems just formulated in Eqs. (32) and (33). The maximum of J_1 in Eq. (32) is achieved when the norm of this equation is equal to zero, which can be written as

$$E[Z(K)+ Q(K)]=\underline{0} \tag{34}$$

Substituting from Eq. (28) into Eq. (34) for $Q(K)$, we obtain

$$E[W(K)+2N(K)Z(K)]=\underline{0} \tag{35}$$

Writing Eq. (35) explicitly, one obtains

$$E[A-\lambda(K)+2\beta x(K)+2\phi^T(K)\vec{H}x(K)+\gamma y(K)]=\underline{0} \tag{36}$$

$$E[-\phi(K)+\gamma^T x(K)]=\underline{0} \tag{37}$$

The above two equations can be written in component form as

$$E[\sum_{j=i}^{n} \alpha_j - \lambda_{i,K} + 2(\sum_{j=1}^{n} \beta_j)(x_{i,K}) + 2\phi_{i,K} x_{i,K} + (\sum_{j=i}^{n} \gamma_j)y] = 0$$

$$; \quad i=1,\ldots,n \tag{38}$$

$$E[-\phi_{i,K} + \gamma_i \sum_{j=1}^{i} x_{j,K}] = 0, \quad i=1,\ldots,n \tag{39}$$

The maximum of J_2 in Eq. (33) is achieved when the norm of this equation vanishes as

$$E[X(k)+ V(k)]=0 \tag{40}$$

Substituting from Eq. (29) into Eq. (40), we obtain

$$E[R(k)+2L(k)X(k)]=\underline{0} \tag{41}$$

Writing Eq. (41) explicitly, and adding the equality constraints, we obtain

$$E[-x(k)+x(k-1)+I(k)+Mu(k)+Ms(k)]=\underline{0} \tag{42}$$

$$E[-y(k)+x^T(k)\vec{\bar{H}}x(k)]=\underline{0} \tag{43}$$

$$E[-z(k)+u^T(k)\vec{\bar{H}}u(k)]=\underline{0} \tag{44}$$

$$E[-r(k-1)+u^T(k)\vec{H}x(k-1)]=\underline{0} \tag{45}$$

$$E[\lambda(k)-\lambda(k-1)+\nu(k)+2\phi^T(k-1)\vec{H}x(k-1)+d(k)u(k)+\psi^T(k)\vec{\bar{H}}u(k)]=\underline{0} \tag{46}$$

$$E[-\phi(k-1)+\gamma(k)u(k)]=\underline{0} \tag{47}$$

$$E[B(k)+M^T\lambda(k)+M^T\nu(k)+\gamma(k)+d(k)x(k-1)+\psi^T(k)\vec{H}x(k-1)+\gamma(k)y(k-1)$$

$$+f(k)Mu(k)+M^Tf(k)u(k)+2\mu^T(k)\vec{\bar{H}}u(k)+\tfrac{1}{4}\gamma(k)N-\tfrac{1}{2}L^T\gamma(k))z(k)$$

$$+M^T\gamma(k)r(k-1)]=\underline{0} \tag{48}$$

$$E[-\mu(k)+\tfrac{1}{4}N^T\gamma(k)u(k)-\tfrac{1}{2}\gamma(k)Lu(k)]=\underline{0} \tag{49}$$

$$E[-\psi(k)+\gamma(k)Mu(k)]=\underline{0} \tag{50}$$

Equations (42) to (50) can be written in component form as

$$E[-x_{i,k}+x_{i,k-1}+I_{i,k}+u_{(i-1),k}-u_{i,k}+s_{(i-1),k}-s_{i,k}]=0,$$

$$i=1,\ldots,n,\ k=1,\ldots,K \tag{51}$$

$$E[-y_{i,k}+x_{i,k}^2]=0, \quad i=1,\ldots,n; \; k=1,\ldots,\ldots,K \qquad (52)$$

$$E[-z_{i,k}+u_{i,k}^2]=0, \quad i=1,\ldots,n; \; k=1,\ldots,K \qquad (53)$$

$$E[-r_{i,k-1}+u_{i,k}x_{i,k-1}]=0, \quad i=1,\ldots,n; \; k=1,\ldots,K \qquad (54)$$

$$E[-\phi_{i,k-1}+c_k\gamma_i u_{i,k}]=0, \quad i=1,\ldots,n; \; k=1,\ldots,k$$

$$E[-\mu_{i,k}+\tfrac{1}{4}(\sum_{j=i}^{n} c_k\gamma_j u_{j,k})-\tfrac{1}{2}c_k\sum_{j=1}^{i}\gamma_j u_{i,k}]=0, \quad i=1,\ldots,n; \; k=1,\ldots,K \qquad (55)$$

$$E[-\psi_{i,k}+c_k\gamma_i(u_{(i-1),k}-u_{i,k})]=0; \quad ;i=1,\ldots,n; \; k=1,\ldots,K \qquad (56)$$

$$E[\lambda_{i,k}-\lambda_{i,k-1}+\nu_{i,k}+\; 2\phi_{i,k-1}x_{i,k-1}+c_k d_{i,k}u_{i,k}+\psi_{i,k}u_{i,k}]=0,$$
$$i=1,\ldots,n; \; k=1,\ldots,K \qquad (57)$$

$$E[c_k b_{i,k}+\lambda_{i+1,k}-\lambda_{i,k}+\nu_{i+1,k}-\nu_{i,k}+\sigma_{i,k}+c_k d_{i,k}x_{i,k-1}+$$

$$+\psi_{i,k}x_{i,k-1}+c_k\gamma_i y_{i,k-1}+c_k f_{i,k}u_{(i-1),k}-c_k f_{i,k}u_{i,k} \;+$$

$$+c_k f_{i,k}u_{i+1,k}-c_k f_{i,k}u_{i,k}+2\mu_{i,k}u_{i,k}$$

$$+\tfrac{1}{4}c_k\gamma_i \sum_{j=1}^{n} z_{j,k}-\tfrac{1}{2}\sum_{j=i+1}^{n} c_k\gamma_j z_{j,k}]=0, \quad i=1,\ldots,n; \; k=1,\ldots,K \qquad (58)$$

Besides the above equations, we have Kuhn-Tucker exclusion equations which must be satisfied at the optimum [11]

$$e_{i,k}^m(x_i^m-x_{i,k})=0; \quad i=1,\ldots,n; \; k=1,\ldots,K \qquad (59)$$

$$e_{i,k}^M(x_{i,k}-x_i^M)=0; \quad i=1,\ldots,n; \; k=1,\ldots,K \qquad (60)$$

$$g_{i,k}^{m}(u_{i,k}^{m}-u_{i,k})=0; \quad i=1,\ldots,n; \ k=1,\ldots,K \qquad (61)$$

$$g_{i,k}^{M}(u_{i,k}-u_{i,k}^{M})=0; \quad i=1,\ldots,n; \ k=1,\ldots,K \qquad (62)$$

One also has the following limits on the variables

If $x_{i,k} < x_i^{m}$, then we put $x_{i,k} = x_i^{m}$ $\qquad$ (63a)

If $x_{i,k} > x_i^{M}$, then we put $x_{i,k} = x_i^{M}$ $\qquad$ (63b)

If $u_{i,k} < u_{i,k}^{m}$, then we put $u_{i,k} = u_{i,k}^{m}$ $\qquad$ (63c)

If $u_{i,k} > u_{i,k}^{M}$, then we put $u_{i,k} = u_{i,k}^{M}$ $\qquad$ (63d)

Equations (51) to (63) together with Eqs. (38) and (39) completely describe the optimal long-term operation of series reservoirs. In the next section, we discuss the proposed algorithm for solving the above equations.

D. ALGORITHM OF SOLUTION

Assume given: the number of reservoirs on a river, the expected natural inflow to each reservoir, the physical characteristics of each reservoir and the cost of one MWh on the valley, c_k,

Step 1.   Assume initial guess for $u_{i,k}(u_{i,k}^{0})$ such that

$$u_{i,k}^{m} \leq u_{i,k}^{0} \leq u_{i,k}^{M} \quad i=1,\ldots,n, \ k=1,\ldots,K$$

Step 2.   Solve Eq. (51) forward in stages and calculate the spillage, if any, using Eq. (3)

Step 3.   Solve Eqs. (52) to (56) forward in stages.

Step 4.   Solve Eq. (57) backward in stages together with Eqs.
          (38) and (39) as boundary equations.

Step 5.   Check Eq. (58). If it is satisfied within a
          prespecified terminating criterion, terminate the
          iteration. Otherwise update $u_{i,k}$ as

$$u_{i,k}^{I+1} = u_{i,k}^{I} + \alpha(\Delta u_{i,k})^{I}; \quad I = \text{iteration counter}$$

          and go to Step 6.

          where $\Delta u_{i,k}$ is given by Eq. (58) and α is a positive
          scalar that is chosen with consideration to such
          factors as convergence.

Step 6.   Repeat the calculation starting from step 2. Continue
          until the state $x(k)$ and the control $u(k)$ do not change
          significantly from iteration to iteration and the cost
          functional in Eq. (14) is a maximum.

E.   PRACTICAL EXAMPLE [6]

The system in this example consists of four reservoirs
connected in series on a river. The characteristics of the
installations are given in Table 1. If d^k denote the number of
days in month k, then

$$u_{i,k}^{m} = 0.0864 d^k \text{ (minimum effective discharge of plant i in}$$
$$\text{m}^3/\text{sec) Mm}^3$$

$$u_{i,k}^{M} = 0.0864 d^k \text{ (maximum effective discharge of plant i in}$$
$$\text{m}^3/\text{sec) Mm}^3$$

where the minimum and maximum discharges are given in Table 1. In this table the minimum discharge is set equal to zero.

Table 1. Characteristics of the Installations

Site	x_i^M (Mm^3)*	Maximum effective discharge (m^3/sec)	Reservoir's constants		
			α_i	β_i	γ_i
1	9628	400	11.41	0.15226×10^{-2}	-0.19131×10^{-7}
2	570	547	231.53	0.10282×10^{-1}	-0.13212×10^{-5}
3	50	594	215.82	0.12586×10^{-1}	0.2979×10^{-5}
4	3420	1180	432.20	-0.12972×10^{-1}	-0.12972×10^{-5}

* 1 Mm^3=10^6 m^3

Table 2. Monthly Inflows and the Associated Probabilities for One

of the Reservoirs in the System

Month	k	$I_{1,k}$ (Mm^3)	P	Month	k	$I_{1,k}$ (Mm^3)	P
October	1	437.2	0.0668	April	7	46.8	0.0668
		667.8	0.2417			118.6	0.2417
		886.2	0.3830			186.6	0.3830
		1104.6	0.2417			254.6	0.2417
		1335.3	0.0668			326.4	0.0668
November	2	376.5	0.0668	May	8	120.2	0.0668
		550.3	0.2417			425.8	0.2417
		714.9	0.3830			715.2	0.3830
		879.4	0.2417			1004.6	0.2417
		1053.2	0.0668			1310.2	0.3830
December	3	323.9	0.0668	June	9	491.5	0.0668
		414.1	0.2417			857.8	0.2417
		499.6	0.3830			1204.6	0.3830
		585.0	0.2417			1551.5	0.2417
		675.2	0.0668			1917.8	0.0668
January	4	212.4	0.0668	July	10	448.1	0.0668
		269.3	0.2417			739.1	0.2417
		323.1	0.3830			1014.7	0.3830
		377.0	0.2417			1290.2	0.2417
		757.0	0.0668			1581.3	0.0668
February	5	130.3	0.0668	August	11	255.1	0.0668
		168.8	0.2417			547.7	0.2417
		205.2	0.3830			824.7	0.3830
		241.6	0.2417			1101.8	0.2417
		280.0	0.0668			1394.4	0.0668
March	6	109.0	0.0668	September	12	263.4	0.0668
		147.5	0.2417			551.5	0.2417
		183.9	0.3830			824.8	0.3830
		220.3	0.2417			1097.2	0.2417
		258.7	0.0668			1385.4	0.0668

Table 3. Optimal Monthly Releases from the Four Reservoirs
and the Profits realized

Month k	$u_{1,k}$ (Mm^3)	$u_{2,k}$ (Mm^3)	$u_{3,k}$ (Mm^3)	$u_{4,k}$ (Mm^3)	Profits ($)
1	0	315	474	1867	817,745
2	1025	1289	1378	2270	1,566,862
3	1071	1256	1305	2029	1,596,784
4	1071	1367	1420	1894	1,997,250
5	953	1035	1058	2459	2,151,664
6	1071	1035	1058	2459	2,151,664
7	1024	999	1022	1039	1,137,618
8	0	234	273	950	533,420
9	0	483	660	2194	1,042,055
10	337	675	773	1370	703,741
11	0	305	407	1060	457,890
12	0	305	407	1695	723,348

Value of water remaining in the reservoirs
at the end of the year   9,380,718
Total profits   24,537,888

Table 4. Optimal Monthly Reservoir Storage

Month	$x_{1,k}$ (Mm^3)	$x_{2,k}$ (Mm^3)	$x_{3,k}$ (Mm^3)	$x_{4,k}$ (Mm^3)
1	7575	570	0	3254
2	7265	570	0	3409
3	6693	570	13	3417
4	5945	393	0	3417
5	5196	342	0	2235
6	4308	446	0	1104
7	3470	540	0	1360
8	4185	570	50	1731
9	5390	533	22	1962
10	6067	570	50	2853
11	6892	570	50	3408
12	7717	570	50	3328

The natural inflows in Mm^3 $(10^6 m^3)$ to the first reservoir and the associated probabilities during each month are given in Table 2. In Table 3, we give the monthly release from each reservoir and the profits realized during each month, also we give the total profit at the end of the year. In Table 4, we give the corresponding optimal monthly storage during the same year. The computing time to obtain the optimal solution for a period of a year was 0.9 sec. in CPU units on the Amdahl 470V/6 computer.

F.   CONCLUSION

In this section, the problem of the long-term optimal operation of a series of reservoirs with a nonlinear model is discussed. The nonlinear model used was a quadratic function of the average storage to avoid underestimation in the hydroelectric production for rising water levels and overestimation for falling water levels.

The   section deals with a nonlinear cost functional and explained how to deal with such cost functionals; we introduced a set of pseudostate variables to cast the problem into a quadratic form that can be solved by the minimum norm formulation.

IV. OPTIMAL OPERATION OF MULTICHAIN POWER SYSTEMS FOR CRITICAL
      WATER CONDITIONS

The period in which reservoirs are drawn down from full to empty is referred to as the "critical period", and the stream flows that occur during the critical period are called "critical period stream flow" because they are the lowest on record. The duration of the critical period is determined by the amount of reservoir storage in the hydroelectric system and on the amount of energy support available from thermal, gas turbine plants, and possible purchase, and it depends on how these resources are committed to support the hydroelectric system.

Critical period (CP) regulation requirements are to determine the Firm Energy Load Carrying Capability (FELCC) of the system and to define energy content curves (ECC) and critical rule curves (CRC) for guiding actual reservoir operation. FELCC is defined as the minimum generation, uniformly shaped during each period similar to the hydropower system flow periods on record, while optimally drawing the available reservoir storage from full to empty. The critical period is the specific period during the historical stream-flow record that the hydro-power system produces the FELCC.

Over the past several years a number of methods have been developed to solve the problem. These methods are limited to either dynamic programming, linear programming, or a combination of them, but they suffer from major problems when they are applied to multidimensional systems, including excessive demands on computing time and storage requirements [16]-[20].

In this section the optimal long-term operation of multichain power systems is discussed, where the minimum norm formulation of functional analysis is used. We apply this optimization technique to solve the optimal long-term operation of the B.P.A.

(Bonneville Power Administration) hydro-electric power system for critical water conditions. The B.P.A. system is considered one of the largest hydro power systems in the world. The system consists of 51 run-of-river hydro plants and 37 storage plants, and is characterized by having a variable tail-water elevation.

A. PROBLEM FORMULATION

1. The System Under Study

The system under consideration consists of m rivers, with one or several reservoirs in series on each. These rivers, may or may not be dependent on each other, and interconnection lines to the neighboring system through which energy may be exchanged, Figure 2.

2. The Optimization Objective and Constraints

The optimization objective for the hydro-electric system given in Figure (2), is to calculate the discharge $u_{ij,k}$, $i=1,\ldots,n_j$; $j=1,\ldots,m$, $k=1,\ldots,K$ from reservoir i located on river j during the optimization interval k, that maximizes the expected total system energy during the critical period. In mathematical terms, the optimal long-term optimization problem is to find the discharge $u_{ij,k}$ that maximizes [19], [20]

$$J=E\left[\sum_{k=1}^{K} \sum_{i=1}^{n_j} \sum_{j=1}^{m} G_{ij,k}(u_{ij,k}, (x_{ij,k}+x_{ij,k-1}))\right] \text{ MWh} \qquad (64)$$

where E stands for the expected value, $G_{ij,k}$ is the generation of reservoir i located on river j at the end of period k. It is a function of the discharge $u_{ij,k}$ and the average storage between

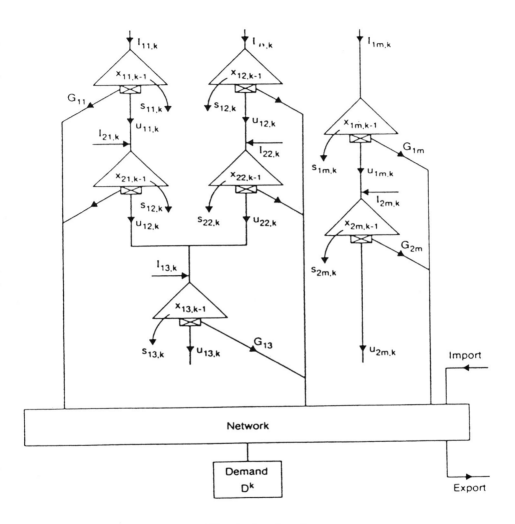

Figure 2. The hydroelectric system.

two successive periods to avoid underestimation in the production of energy for rising water levels and overestimation for falling water levels [20], and $x_{ij,k}$ is the storage of reservoir i located on river j at the end of period k. Subject to satisfying the following constraints [21]

(1) The water conservation equation for each reservoir during the optimization interval k is adequately described by the continuity-type difference equation

$$x_{ij,k} = x_{ij,k-1} + I_{ij,k} - u_{ij,k} - s_{ij,k} + \sum_{\substack{\ell \in Ru \\ j \in Rr}} (u_{\ell j,k} + s_{\ell j,k}),$$

$$,i=1,\ldots,n_j, \ j=1,\ldots,m, \ k=1,\ldots,K \qquad (65)$$

where

$I_{ij,k}$ is the natural inflow to the reservoir i located on river j during the period k. These are statistically independent random variables. Their statistical parameters can be determined from historical records. In reality, there is a statistical correlation between inflows of successive months. However, it is assumed here that each inflow $I_{ij,k}$ is characterized by a discrete distribution given by

$$\text{Prob}[I_{ij} = I_x] = p_x > 0, \text{ where } \sum_{x=1}^{N} p_x = 1.$$

$s_{ij,k}$ is the spillage from reservoir i on river j at the end of period k. Water is spilt when the reservoir is full and the discharge through the turbine exceeds the maximum discharge. This can be expressed mathematically as

$$s_{ij,k} = \begin{cases} (u_{ij,k} - u_{ij,k}^{M}), \text{if } u_{ij,k} > u_{ij,k}^{M} \text{ and } x_{ij,k} \geq x_{ij,k}^{M} \\ \qquad\qquad\qquad i=1,\ldots,n, j=1,\ldots,m, k=1,\ldots,K \\ 0, \text{ otherwise} \end{cases}$$

(66)

R_u, R_r are the sets of immediately upstream reservoirs and rivers to the reservoir ℓ on river j

K   is the last period studied

n_j is the number of reservoirs on river j

(2) To satisfy the constraint, that reflect physical limits, bank erosion, coordiation agreement among various ownerships, and at the same time satisfying the multipurposes stream use requirements such as irrigation, navigation, fishing and flood control, the reservoir variables are bounded as

(a) upper and lower limits on the reservoir storages

$$x_{ij,k}^{m} \leq x_{ij,k} \leq x_{ij,k}^{M}, \ k=1,\ldots,n_j, \ j=1,\ldots,m, k=1,\ldots,K$$

(67)

(b) upper and lower limits on the reservoir outflow

$$u_{ij,k}^{m} \leq u_{ij,k} \leq u_{ij,k}^{M}, \ i=1,\ldots,n_j, j=1,\ldots,m, k=1,\ldots,K$$

(68)

where $x_{ij,k}{}^m$ and $x_{ij,k}{}^M$ are the minimum and maximum storages of reservoir i located on river j at the end of period k respectively. $u_{ij,k}{}^m$, $u_{ij,k}{}^M$ are the minimum and maximum discharges from reservoir i located on river j during the period k respectively. These are given by

$$u_{ij,k}{}^m = 0.0864d^k \text{ (minimum effective discharge in } m^3/sec)$$

$$u_{ij,k}{}^M = 0.0864d^k \text{ (maximum effective discharge in } m^3/sec)$$

and d^k is the number of days in period k.

(3) To prevent excessive soil erosion around the reservoir the forebay elevation is bounded as

$$f_{ij,k-1} - f_{ij,k} \le d_{ij,k}; \; i=1,\ldots,n_j, \; j=1,\ldots,m,k=1,\ldots,K$$

(69)

where $f_{ij,k}$ is the forebay elevation of reservoir i located on river j at the end of period k and $d_{ij,k}$ is the maximum draft from reservoir i on river j at the end of period k.

(4) The forebay elevation as a function of the storage can be written as

$$f_{ij,k} = a_{ij} + b_{ij}x_{ij,k} + c_{ij}x_{ij,k}{}^2, \; i=1,\ldots,n_j, j=1,\ldots,m,k=1,\ldots,K$$

(70)

where a_{ij}, b_{ij} and c_{ij} are constants, these can be obtained by least error squares fitting to plant data available.

Now, the inequality constraints of Eq. (69) can be written as

$$b_{ij}(x_{ij,k-1}-x_{ij,k})+c_{ij}(x_{ij,k-1}^2-x_{ij,k}^2)\leq d_{ij,k}, k=1,\ldots,n_j,$$

$$j=1,\ldots,m,k=1,\ldots,K \qquad (71)$$

Equation (71) stands for the draft constraints and it is a nonlinear state-dependent inequality constraint.

(5) The water conversion factor, MWh/m^3 as a function of the storage can be written as

$$H_{ij,k}(x_{ij,k})=\alpha_{ij}+\beta_{ij}x_{ij,k}+\gamma_{ij}x_{ij,k}^2, i=1,\ldots,n_j, j=1,\ldots,m,$$

$$k=1,\ldots,K \qquad (72)$$

where $H_{ij,k}(x_{ij,k})$ is the water conversion factor measured in MWh/Mm^3, and α_{ij}, β_{ij}, γ_{ij} are constants. These are obtained from the least error squares curve fitting to typical data available. For a typical hydro-electric power system, the following tables should be available to model the water conversion factor given by Eq. (72)

(i) forebay elevation as a function of the reservoir contents. This is called (f_i, x_i) table.

(ii) tailwater elevation as a function of the total discharge, This is called $(t_i, \sum (u_i+s_i))$ table.

(iii) water-to-energy conversion factor as a nonlinear function of reservoir contents and turbine discharge. This is called (H_i, u_i, x_i) table.

(iv) maximum generation at each plant as a function of full-gate flow.

The steps to model $H_{i,k}$ for a reservoir i are

(a) Assume values for the storage x_i, solve Eq. (1) to find
     the corresponding total discharge $\sum (u_i + s_i)$.

(b) From the table of forebay elevation and the storage
     find the corresponding forebay elevation, also from the
     table of the tailwater elevation and the total
     discharge find the corresponding tailwater elevation
     using the cubic spline subroutine (Appendix A).

(c) Construct a new table between the storage and the head,
     where the head $h_i = f_i - t_i$, t_i is the tailwater elevation.

(d) Using this table with the table of the water conversion
     factor and the head, calculate a new table of storage
     x_i and the water conversion factor H_i using the cubic
     spline subroutine, (H_i, x_i) table.

(e) Calculate α_i, β_i and γ_i from the available (H_i, x_i)
     table by using the least error squares fitting
     technique.

During this process, more weight has been given to the points
in each table, that satisfy the boundary constraints.

Now, in mathematical terms, the cost functional in Eq. (64) can
be explicitly written as

$$J = E\left[\sum_{k=1}^{K} \sum_{i=1}^{n_j} \sum_{j=1}^{m} \{ \alpha_{ij} u_{ij,k} + \frac{1}{2}\beta_{ij} u_{ij,k}(x_{ij,k} + x_{ij,k-1}) \right.$$

$$\left. + \frac{1}{4}\gamma_{ij} u_{ij,k}(x_{ij,k} + x_{ij,k-1})^2 \} \right] \qquad (73)$$

Subject to satisfying the equality constraints given by Eq. (65)
and inequality constraints given by Eqs. (67), (68) and (71).

Substituting for $x_{ij,k}$ from Eq. (65) into EEq. (73), one obtains

$$J = E[\sum_{k=1}^{K} \sum_{i=1}^{n_j} \sum_{j=1}^{m} \{b_{ij,k} u_{ij,k} + u_{ij,k} d_{ij,k} x_{ij,k-1}$$

$$+ u_{ij,k} f_{ij,k} (\sum_{\substack{\ell \in Ru \\ j \in Rr}} u_{\ell j,k} - u_{ij,k})$$

$$+ \gamma_{ij} u_{ij,k} x_{ij,k-1} + \frac{1}{4} u_{ij,k} \gamma_{ij} [u_{ij,k}^2 + (\sum_{\substack{\ell \in Ru \\ j \in Rr}} u_{\ell j,k})^2] +$$

$$+ \frac{1}{2} \gamma_{ij} u_{ij,k} x_{ij,k-1} [\sum_{\substack{\ell \in Ru \\ j \in Rr}} u_{\ell j,k} - u_{ij,k}]$$

$$- \frac{1}{2} \gamma_{ij} u_{ij,k}^2 (\sum_{\substack{\ell \in Ru \\ j \in Rr}} u_{\ell j,k})\}], i=1,\ldots,n_j; j=1,\ldots,m; k=1,\ldots,K$$

$$\tag{74}$$

where we defined

$$b_{ij,k} = \alpha_{ij} + \frac{1}{2}\beta_{ij} q_{ij,k} + \frac{1}{4}\gamma_{ij} q_{ij,k}^2, \quad i=1,\ldots,n_j; j=1,\ldots,m, k=1,\ldots,K$$

$$\tag{75}$$

$$q_{ij,k} = I_{ij,k} + \sum_{\substack{\ell \in Ru \\ j \in Rr}} s_{\ell j,k} - s_{ij,k}, \quad k=1,\ldots,n; j=1,\ldots,m; k=1,\ldots,K$$

$$\tag{76}$$

$$d_{ij,k} = \beta_{ij} + \alpha_{ij} q_{ij,k} = i=1,\ldots,n; j=1,\ldots,m; k=1,\ldots,K \tag{77}$$

$$f_{ij,k} = \frac{1}{2} d_{ij,k}, \quad i=1,\ldots,n; j=1,\ldots,m; k=1,\ldots,K \tag{78}$$

The cost functional in Eq. (74) is a highly nonlinear function. To cast the problem into a quadratic one, we may define pseudo-variables as [15]

$$y_{ij,k} = x_{ij,k}^2; \quad i=1,\ldots,n_j; j=1,\ldots,m; k=1,\ldots,K \tag{79}$$

$$z_{ij,k} = u_{ij,k}^2; \quad i=1,\ldots,n_j; j=1,\ldots,m; k=1,\ldots,K \tag{80}$$

$$r_{ij,k-1} = u_{ij,k} x_{ij,k-1}; \quad i=1,\ldots,n_j; j=1,\ldots,m; k=1,\ldots,K \tag{81}$$

Now, the cost functional of Eq. (74) becomes

$$J=E[\sum_{k=1}^{K}\sum_{i=1}^{n_j}\sum_{j=1}^{m}\{b_{ij,k}u_{ij,k}+u_{ij,k}d_{ij,k}x_{ij,k-1}+$$

$$+u_{ij,k}f_{ij,k}(\sum_{\substack{\ell\in Ru\\j\in Rr}}u_{\ell j,k}-u_{ij,k})$$

$$+\gamma_{ij}u_{ij,k}y_{ij,k-1}+\frac{1}{2}u_{ij,k}\gamma_{ij}(\sum_{\substack{\ell\in Ru\\j\in Rr}}z_{\ell j,k}+z_{ij,k})$$

$$+\gamma_{ij}r_{ij,k-1}[\sum_{\substack{\ell\in Ru\\j\in Rr}}u_{\ell j,k}-u_{ij,k}]$$

$$-\frac{1}{2}\gamma_{ij}z_{ij,k}(\sum_{\substack{\ell\in Ru\\j\in Rr}}u_{\ell j,k})\}] \tag{82}$$

Subject to satisfying the equality constraints given by Eqs. (65), (79), (80) and (81), and the inequality constraints given by Eqs. (67), (68) and (70).

B. MINIMUM NORM FORMULATION [12]

We can now form an augmented cost functional by adjoining the equality constraints via Lagrange multipliers, $\lambda_{ij,k}$, $u_{ij,k}$, $\phi_{ij,k}$, and $\psi_{ij,k}$ and the inequality constraints via Kuhn-Tucker multipliers to obtain

$$\tilde{J}=E[\sum_{k=1}^{K}\sum_{i=1}^{n_j}\sum_{j=1}^{m}\{b_{ij,k}u_{ij,k}+u_{ij,k}d_{ij,k}x_{ij,k-1}$$

$$+u_{ij,k}f_{ij,k}(\sum_{\substack{\ell\in Ru\\j\in Rr}}u_{\ell j,k}-u_{ij,k})$$

$$+\gamma_{ij}u_{ij,k}y_{ij,k-1}+\frac{1}{2}u_{ij,k}\gamma_{ij}(\sum_{\substack{\ell\in Ru\\j\in Rr}}z_{\ell j,k}+z_{ij,k})$$

$$+\gamma_{ij}r_{ij,k-1}[\sum_{\substack{\ell\in Ru\\j\in Rr}}u_{\ell j,k}-u_{ij,k}]$$

$$-\frac{1}{2}\gamma_{ij}z_{ij,k}(\sum_{\substack{\ell\in Ru\\j\in Rr}}u_{\ell j,k})+\lambda_{ij,k}(-x_{ij,k}+x_{ij,k-1}+q_{ij,k}$$

$$+(\sum_{\substack{\ell \in Ru \\ j \in Rr}} u_{\ell j,k} - u_{ij,k}))$$

$$+\mu_{ij,k}(y_{ij,k} - x_{ij,k}^2)k + \phi_{ij,k}(z_{ij,k} - u_{ij,k}^2)$$

$$+\mu_{ij,k}(r_{ij,k-1} - u_{ij,k}x_{ij,k-1})$$

$$+e_{ij,k}^1(x_{ij,k}^m - x_{ij,k}) + e_{ij,k}^2(x_{ij,k} - x_{ij,k}^M)$$

$$+g_{ij,k}^1(u_{ij,k}^m - u_{ij,k})$$

$$+g_{ij,k}^2(u_{ij,k} - u_{ij,k}^M) - \ell_{ij,k}[b_{ij}(\sum_{\substack{\ell \in Ru \\ j \in Rr}} u_{\ell j,k} - u_{ij,k})$$

$$+b_{ij}q_{ij,k} + c_{ij}(y_{ij,k} - y_{ij,k-1}) + d_{ij,k}]]] \tag{83}$$

where $e_{ij,k}^1$, $e_{ij,k}^2$, $g_{ij,k}^1$, $g_{ij,k}^2$ and $\ell_{ij,k}$ are Kuhn-Tucker multipliers. These are equal to zero, if the constraints are not violated and greater than zero, if the constraints are violated [2].

Applying the discrete version of integration by parts, and dropping constant terms, one obtains [21]

$$\tilde{J} = E[\sum_{i=1}^{nj} \sum_{j=1}^{m} \{-\mu_{ij,K}x_{ij,K}^2 - \lambda_{ij,K}x_{ij,K} + (\mu_{ij,K} - c_{ij,K})y_{ij,K}$$

$$+\mu_{ij,0}x_{ij,0}^2$$

$$+\lambda_{ij,0}x_{ij,0} - (\mu_{ij,0} - c_{ij,0})y_{ij,0}\}$$

$$+\sum_{k=1}^{K} \sum_{i=1}^{nj} \sum_{j=1}^{m} \{-\mu_{ij,k-1}x_{ij,k-1}^2$$

$$+u_{ij,k}d_{ij,k}x_{ij,k-1} + u_{ij,k}f_{ij,k}(\sum_{\substack{\ell \in Ru \\ j \in Rr}} u_{\ell j,k} - u_{ij,k})$$

$$+\gamma_{ij}u_{ij,k}y_{ij,k-1}$$

$$+\frac{1}{4}u_{ij,k}\gamma_{ij}(z_{ij,k}+\sum_{\substack{\ell\varepsilon Ru\\j\varepsilon Rr}}z_{\ell j,k})+\gamma_{ij}r_{ij,k-1}(\sum_{\substack{\ell\varepsilon Ru\\j\varepsilon Rr}}u_{\ell j,k}-u_{ij,k})$$

$$-\frac{1}{2}\gamma_{ij}z_{ij,k}(\sum_{\substack{\ell\varepsilon Ru\\j\varepsilon Rr}}u_{\ell j,k})-\phi_{ij,k}u_{ij,k}^2-\psi_{ij,k}u_{ij,k}x_{ij,k-1}$$

$$+(\lambda_{ij,k}-\lambda_{ij,k-1}+\theta_{ij,k})x_{ij,k-1}+(b_{ij,k}+\nu_{ij,k})u_{ij,k}$$

$$+(\lambda_{ij,k}+\theta_{ij,k}-B_{ij,k})(\sum_{\substack{\ell\varepsilon Ru\\j\varepsilon Rr}}u_{\ell j,k}-u_{ij,k})+\mu_{ij,k}y_{ij,k-1}$$

$$+\phi_{ij,k}z_{ij,k}+\psi_{ij,k}r_{ij,k-1}\}] \qquad (84)$$

where we defined

$$B_{ij,k}=b_{ij}\ell_{ij,k}, \quad i=1,\ldots,n_j; \quad j=1,\ldots,m;k=1,\ldots,K \qquad (85)$$

$$C_{ij,k}=c_{ij}\ell_{ij,K}; \quad i=1,\ldots,n_j;j=1,\ldots,m;k=1,\ldots,K \qquad (86)$$

$$\theta_{ij,k}=e_{ij,k}^2-e_{ij,k}^1; \quad i=1,\ldots,n_j; \quad j=1,\ldots,m;k=1,\ldots,K \qquad (87)$$

$$\nu_{ij,k}=g_{ij,k}^2-g_{ij,k}^1; \quad i=1,\ldots,n_j; \quad j=1,\ldots,m,k=1,\ldots,K \qquad (88)$$

Equation (84) can be written in vector form as

$$\tilde{J}=E[\{-x^T(K)u^T(K)\vec{H}x(K)-\lambda^T(K)x(K)+(\mu(K)-C(K))^Ty(K)$$

$$+x^T(0)u^T(0)\vec{H}x(0)+\lambda^T(0)x(0)-(\mu(0)-C(0))^Ty(0)\}$$

$$+\sum_{k=1}^{K}\{-x^T(k-1)u^T(k-1)\vec{H}x(k-1)+u^T(k)d(k)x(k-1)$$

$$+u^T(k)f(k)Mu(k)+u^T(k)\gamma y(k-1)+\frac{1}{4}u^T(k)\gamma Nz(k)$$

$$+r^T(k-1)\gamma Mu(k)-\frac{1}{2}z(k)Qu(k)-u^T(k)\phi^T(k)\vec{H}u(k)$$

$$-u^T(k)\psi^T(k)\vec{H}x(k-1)+(\lambda(k)-\lambda(k-1)+\theta(k))^Tx(k-1)$$

$$+(b(k)+\nu(k)+M^T\lambda(k)+M^T\theta(k)-M^TB(k))u(k)$$

$$+\mu^T(k-1)y(k-1)+\phi^T(k)z(k)+\psi^T(k)r(k-1))\}] \tag{89}$$

In the above equation, Eq. (89), $x(k)$, $u(k)$, $y(k)$, $z(k)$, $r(k)$, $\phi(k)$, $\psi(k)$, $\lambda(k)$, $\mu(k)$, $\nu(k)$, $b(k)$, $\theta(k)$, $C(k)$ and $B(k)$ are n-dimensional vectors, $n=\sum_{j=1}^{m} n_j$, at the end of period k; their components are $x_{ij,k}$, $u_{ij,k}$, $y_{ij,k}$, $z_{ij,k}$ $r_{ij,k}$, $\phi_{ij,k}$, $\psi_{ij,k}$, $\lambda_{ij,k}$, $\mu_{ij,k}$, $\nu_{ij,k}$, $b_{ij,k}$, $\theta_{ij,k}$, $C_{ij,k}$ and $B_{ij,k}$, $i=1,\ldots,n_j$; $j=1,\ldots,m$; $k=1,\ldots,K$ respectively, and $d(k)$, $f(k)$, γ are nxn diagonal matrices whose elements are $d_{\sigma\sigma,k}=d_{ij,k}$, $f_{\sigma\sigma,k}=f_{ij,k}$, $\gamma_{\sigma\sigma}=\sigma_{ij}$, $i=1,\ldots,n_j$, $j=1,\ldots,m$, $k=1,\ldots,K$. $\vec{H}$ is a vector matrix in which the vector index varies from 1 to n, while the matrix dimension of $\vec{H}$ is nxn. Furthermore, M, N and Q are nxn matrices, their elements depend on the topological arrangement of the reservoirs and rivers, and they vary between -1 to 1.

Define the nx1, $n=\sum_{j=1}^{m} n_j$, vectors $X(k)$, $R(k)$ and $L(k)$ as

$$X^T(k)=[x^T(k-1),y^T(k-1), r^T(k-1), z^T(k), u^T(k)] \tag{90}$$

$$R^T(k)=[(\lambda(k)-\lambda(k-1)+\theta(k))^T, \mu^T(k-1),\psi^T(k),\phi^T(k),(b(k)+\nu(k)$$

$$+M^T\lambda(k)+M^T\theta(k)-M^TB(k))^T] \tag{91}$$

$$L(k)=\begin{vmatrix} L_{11}(k) & L_{12}(k) \\ \\ L_{21}(k) & L_{22}(k) \end{vmatrix} \tag{92}$$

where

$$L_{11}(k)= \begin{vmatrix} -\mu^T(k-1)\vec{H} & 0 & 0 \\ 0 & 0 & 0 \\ 0 & 0 & 0 \end{vmatrix} \qquad (92a)$$

$$L_{12}(k)= \begin{vmatrix} 0 & (\tfrac{1}{2}d(k)-\tfrac{1}{2}\psi^T(k)\vec{H} \\ 0 & \tfrac{1}{2}\gamma \\ 0 & \tfrac{1}{2}\gamma M \end{vmatrix} \qquad (92b)$$

$$L_{21}(k)= \begin{vmatrix} 0 & 0 & 0 \\ \\ (\tfrac{1}{2}d(k)-\tfrac{1}{2}\psi^T(k)\vec{M} & \tfrac{1}{2}\gamma & \tfrac{1}{2}M^T\gamma \end{vmatrix} \qquad (92c)$$

$$L_{22}(k)= \begin{vmatrix} 0 & (\tfrac{1}{8}N^T\gamma - \tfrac{1}{4}\gamma Q) \\ \\ (\tfrac{1}{8}\gamma N-\tfrac{1}{4}Q^T\gamma) & (\tfrac{1}{2}M^T f(k)+\tfrac{1}{2}f(k)M-Q^T(k)\vec{H} \end{vmatrix} \qquad (92d)$$

Then, the cost functional in Eq. (89) becomes

$$\tilde{J}=E[-\mathbf{x}^T(K)\mu^T(K)\vec{H}\mathbf{x}(K)-\lambda^T(K)x(K)+(\mu(K)-C(K))^T y(K)]$$

$$+E[\sum_{k=1}^{K} \{X^T(k)L(k)X(k)+R^T(k)X(k)\}] \qquad (93)$$

Since the initial storage is assumed to be given, $x(0)$ is given, in advance, constant terms are dropped in Eq. (93). If one

defines the nx1 vector $V(k)$ as

$$V(k)=L^{-1}(k)R(k) \qquad (94)$$

Then, Eq. (93) can be written by the process of completing the squares as

$$\tilde{J}=E[-\mathbf{x}^{T}(K)\mu^{T}(K)\overrightarrow{H\mathbf{x}}(K)-\lambda^{T}(K)\mathbf{x}(K)+(\mu(K)-C(K))^{T}y(K)]$$

$$+E[\sum_{k=1}^{K} \{(X(k)+\tfrac{1}{2}V(k))^{T}L(k)(X(k)+\tfrac{1}{2}V(k))-\tfrac{1}{4}V^{T}(k)L(k)V(k)\}]$$

$$(95)$$

The last term in Eq. (95) is constant independent of X(k). If we drop this term, then Eq. (95) becomes

$$\tilde{J}=E[-\mathbf{x}^{T}(K)\mu^{T}(K)\overrightarrow{H}\mathbf{x}(K)-\lambda^{T}(K)\mathbf{x}(K)+(\mu(K)-C(K))^{T}y(K)]$$

$$+E[\sum_{k=1}^{K} \{(X(k)+\tfrac{1}{2}V(k))^{T}L(k)(X(k)+\tfrac{1}{2}V(k))\}] \qquad (96)$$

Equation (96) is composed of two parts, the boundary part and the discrete integral part, which are independent of each other. To maximize $\tilde{J}$ in Eq. (96), one may maximize each part separately

$$\text{Max.}\tilde{J}(x(K),y(K),X(k))=\underset{x(K),y(K)}{\text{Max.}}E[-\mathbf{x}^{T}(K)\mu^{T}(K)\overrightarrow{H}\mathbf{x}(K)-$$

$$-\lambda^{T}(K)\mathbf{x}(K)+(\mu(K)-C(K))^{T}y(K)]+\underset{X(k)}{\text{Max.}}E[(X(k)+\tfrac{1}{2}V(k))^{T}L(k)(X(k)$$

$$+\tfrac{1}{2}V(k))] \qquad (97)$$

C. OPTIMAL EQUATIONS

There is only one solution to the problem just formulated in Eq. (97) namely the optimal solution. To maximize $\tilde{J}$ in Eq. (97), one maximizes each part separately. The maximum of the boundary part is clearly achieved when

$$E[\lambda(K)+2\mu^T(K)Hx(K)]=\underline{0} \tag{98}$$

$$E[\mu(K)-C(K)]=\underline{0} \tag{99}$$

because $\delta x(K)$ and $\delta y(K)$ are arbitrary. Equations (98) and (99) give the optimal value of μ and λ at last period studied.

The discrete integral part of Eq. (97) can be written as a norm

$$J_2 = \underset{X(k)}{\text{Max.E}} \left| \left| X(k)+\frac{1}{2}V(k) \right| \right|_{L(k)} \tag{100}$$

Maximization of Eq. (100), is equivalent to minimization of the norm of this equation. According to the norm axioms, the norm in Eq. (100) is a minimum if and only if

$$E[X(k)+\frac{1}{2}V(k)]=\underline{0} \tag{101}$$

Substituting for $V(k)$ from Eq. (94) into Eq. (101), we obtain

$$E[R(k)+2L(k)X(k)]=\underline{0} \tag{102}$$

Equation (102) is the optimality condition for long-term optimal operation of multichain power systems. Writing Eq. (102) explicitly, and adding the equality constraints, one obtains

$$E[\lambda(k)-\lambda(k-1)+\theta(k)-2\mu^T(k-1)\vec{H}x(k-1)+d(k)u(k)-\psi^T(k)\vec{H}u(k)]=0 \tag{103}$$

$$E[\mu(k-1)+\gamma u(k)]=0 \tag{104}$$

$$E[\psi(k)+\gamma Mu(k)]=\underline{0} \tag{105}$$

$$E[\,\phi(k)+\tfrac{1}{4}N^T\gamma u(k)-\tfrac{1}{2}\gamma Qu(k)]=\underline{0} \qquad (106)$$

$$E[\,b(k)+\nu(k)+M^T\lambda(k)+M^T\theta(k)-M^TB(k)$$

$$+d(k)x(k-1)-\psi^T(k)\vec{H}x(k-1)+\gamma y(k-1)+M^T\gamma r(k-1)$$

$$+\tfrac{1}{4}\gamma Nz(k)-\tfrac{1}{2}Q^T\gamma z(k)+M^Tf(k)u(k)+f(k)Mu(k)$$

$$-2\phi^T(k)\vec{H}u(k)]=0 \qquad (107)$$

$$E[-x(k)+x(k-1)+q(k)+Mu(k)]=\underline{0} \qquad (108)$$

$$E[-y(k)+x^T(k)\vec{H}x(k)]=\underline{0} \qquad (109)$$

$$E]-z(k)+u^T(k)\vec{H}u(k)]=\underline{0} \qquad (110)$$

$$E[r(k-1)+u^T(k)\vec{H}x(k-1)]=\underline{0} \qquad (111)$$

Besides the above equations, we have the Kuhn-Tucker exclusion equations, which must be satisfied at the optimum [3]

$$e_{ij,k}^{\ 1}(x_{ij,k}^{\ m}-x_{ij,k})=0; \quad i=1,\ldots,n_j; \quad j=1,\ldots,m; k=1,\ldots,K \qquad (112)$$

$$e_{ij,k}^{\ 2}(x_{ij,k}-x_{ij,k}^{\ M})=0; \quad i=1,\ldots,n_j; \quad j=1,\ldots,m; k=1,\ldots,K \qquad (113)$$

$$g_{ij,k}^{\ 1}(u_{ij,k}^{\ m}-u_{ij,k})=0; \quad i=1,\ldots,n_j; \quad j=1,\ldots,m; k=1,\ldots,K \qquad (114)$$

$$g_{ij,k}^{\ 2}(u_{ij,k}-u_{ij,k}^{\ M})=0; \quad i=1,\ldots,n_j; \quad j=1,\ldots,m; k=1,\ldots,K \qquad (115)$$

Equations (103)-(115) together with eqs. (98) and (99) completely specify the optimal long-term operation of multichain power systems. In the next section, we discuss the algorithm used to solve these equations.

D.   ALGORITHM OF SOLUTION

Assume given the system structure as Figure 2, the initial storage for each reservoir on each river $x_{ij,0}$ and the physical constraints on the system.

1.   Assume an initial guess for $u_{ij,k}{}^{0}$, $i=1,..,n_j$, $j=1,...,m$, $k=1,..,K$ such that

$$u_{ij,k}{}^{m} \leq u_{ij,k} \leq u_{ij,k}{}^{M}$$

Also assume that no initial spillage is present.

2.   Solve Eq. (108) forward in stages starting with $k=1$ to $k=K$, and calculate $s_{ij,k}$, if any, using Eq. (66) taking into account the Kuhn-Tucker exclusion equations.

3.   Solve Eqs. (104)-(106) and Eqs. (109)-(111) forward in stages using the values of $x_{ij,k}$ and $u_{ij,k}$ obtained from Step (2).

4.   Solve Eq. (104) backward in stages with Eqs. (98) and (99) as boundary conditions.

5.   Check Eq. (107); if it is satisfied with a prescribed terminating criterion, terminate the iteration. Otherwise update $u_{ij,k}$ as

$$(u_{ij,k})_{new} = (u_{ij,k})_{old} + \alpha (\Delta u_{ij,k})$$

and go to Step 2.

Where $(\Delta u_{ij,k})$ is given by Eq. (107) and α is a positive number with consideration given to this factor for convergence.

6. Continue the iteration until the state $x_{ij,k}$ and the control $u_{ij,k}$ do not change significantly from iteration to iteration and J in equation (73) is a maximum.

E. PRACTICAL EXAMPLE

The proposed algorithm is used to solve the B.P.A.[*] hydropower system. This system consists of 38 storage reservoirs and 50 run-of-river plants. The optimization was done in half-month periods for four years (96 optimization intervals) including 42 months critical period.

Because of the nature of some projects, six projects have been grouped so that each two are considered as if they are one project. This decreases the computational effort; and will not affect the distinctive features for any of the projects.

Also, a special subroutine has been written to suite the individual topological configuration and characteristics of the Canal Plant and the Corra Linn projects.

The cubic spline technique is a good fit for all of the field measured tables in all the 88 projects except for two tables of the tailwater elevation versus the total discharge (discharge through the turbine plus the spillage) for the Yale and The Upper Baker projects.

---

[*] B.P.A. = Bonneville Power Administration

The program takes 9.7 seconds of central processing unit time on the MTS-AMDAHL-5870 computer for each trial (3 iterations). We have performed 14 trials until we reached the global maximum for the proposed technique. After each trial we modify $\hat{u}_{ij,k}$ and $\hat{x}_{ij,k}$ (See Appendix A) especially those which have been violated by the program, to satisfy the system hard constraints. The results show:

(1) no violation of any of the hard constraints

(2) very small (unavoidable we believe) amount of spill compared to the results from the B.P.A. rule curves, and more importantly,

(3) an increase in the system total energy capability of 4.7% over that given when applying the B.P.A. rule curves.

The discretized system state and consequently, the computational effort, increases linearly with the number of state variables. This is a very important feature of this technique that enables it to deal with highly complicated problems. For example, in [17], the number of the system states increases exponentially with the number of state variables.

The hard constraints have covered two essential types:

(1) constraints that enforce feasibility due to physical and/or technical features in the system, and

(2) constraints that guarantee a complete satisfaction of the contractual agreements and regulations related to flood control, wildlife, fisheries requirements, water quality, recreational

use, etc. Table 5 gives a comparison between our results and the results from applying the B.P.A. rule-curve during the critical period.

Table 5. Comparison between the energy capability measured when applying the results of the proposed technique to those obtained from the B.P.A. rule curves during the critical period

Period	Average Power in GW(10^9watts)	
	B.P.A. (Refs. 16,17)	Calculated
1st 10 months critical period	12.373	12.5941
next 12 months critical period	11.801	12.511
next 12 months critical period	12.321	12.211
next 8 months critical period	12.163	13.974
Total energy in GWh	372.512	398.891
Average power in GW	12.155	12.725
The % increase	4.7%	

D. CONCLUSIONS

In this section, a new modelling and formulation for long-term optimal operation of hydro-electric power systems is offered. The new formulation is based on the minimum norm formulation of functional analysis. This formulation takes into account the tail-water variation and the soil erosion around the reservoir. We discuss, in this section, how to model the water

conversion   factor   from the field measurements and   by   using   a
cubic   spline   function.   The proposed technique is applied to   a
very   large   scale   power system consisting   of   51   run-of-river
plants and 37-reservoir plants (B.P.A. system)

V.   OPTIMIZATION OF POWER SYSTEM OPERATION WITH A SPECIFIED
      MONTHLY GENERATION [22]

In the previous sections, we maximized the generation from a
hydro-electric power system during the critical water conditions.
In this maximization, the required load on the system was not
accounted for; that load may be higher in winter than in summer.
This section is devoted to the solution of the long-term optimal
operating problem of the multireservoir power system for the
critical period with a monthly variable load; this load is equal
to a certain percentage of the total generation at the end of the
year, and at the same time the total generation during each year
of the critical period should be equal and maximum. To meet all
these requirements, we maximize the generation from the system
during each year of the critical period taking into account this
specified load on the system. In other words, we minimize the
difference between the monthly generation and the monthly load on
the system, which is equal to the monthly percentage load
required multiplied by the total generation at the end of the
year.

A. PROBLEM FORMULATION

1. The System Under Study

The system under consideration consists of m independent rivers, with one or several reservoirs and power plants in series on each, and interconnection lines to the neighboring system through which energy may be exchanged, Figure 2.

2. The Objective Function

The long-term optimal operating problem, under critical water conditions, aims to find the dischage $u_{ij,k}$, $i\ R_h$, $j\ R_r$ as a function of time that minimizes

$$J(k)=E[\ \sum_{j=1}^{m}\ \sum_{i=1}^{n}\ G_{ij,k}(u_{ij,k},\ (x_{ij,k}+x_{ij,k-1}))$$

$$-a(k)\ \sum_{j\epsilon Rr}\ \sum_{i\epsilon Rh}\ \sum_{k=1}\ G_{ij,k}(u_{ij,k},\ (x_{ij,k}+x_{ij,k-1}))]\qquad (116)$$

In the above equation $a(k)$ is the given percentage load on the system during a period k, subject to satisfying the hydro constraints given by the following:

(1) The water conservation equation, continuity equation, may be adequately described by the difference equation

$$x_{ij,k}=x_{ij,k-1}+I_{ij,k}+u_{i(i-1)j,k}+s_{(i-1)j,k}-u_{ij,k}-s_{ij,k}\qquad (117)$$

where $s_{ij,k}$ is given by

$$
s_{ij,k} = \begin{cases} (x_{ij,k-1} + I_{ij,k} + s_{(i-1)j,k} + u_{(i-1)j,k} - x_{ij,k}) - u_{ij,k}^{M} \\[2ex] \text{if } (x_{ij,k-1} + I_{ij,k} + S_{(i-1)j,k} + u_{(i-1)j,k} - x_{ij,k}) > u_{ij,k}^{M} \\ \text{and } x_{ij,k} \geq x_{ij}^{M} \\[2ex] 0 \text{ otherwise} \end{cases} \tag{118}
$$

Equation (118) states that water is spilt only when the reservoir is filled to capacity and the discharge exceeds $u_{ij,k}^{M}$. This spillage has nothing to do with the generation of that reservoir, and it is considered as a constant inflow to the downstream reservoir. The monthly inflows into the reservoirs form a random sequence that depends on natural phenomena such as rainfall and snowfall. The statistical parameters of the inflow can be determined from historical records. In reality, there is a statistical correlation between inflows of successive months. However, it is assumed here that the inflow in each month is statistically independent of previous inflows. In particular, it is assumed that each inflow $I_{ij,k}$ is characterized by a discrete distribution given by

$$
\text{Prob}[I_{ij} = I_x] = p_x > 0
$$

where

$$
\sum_{x=1}^{N} p_x = 1
$$

(2)   To satisfy the multipurpose stream use requirements, such as flood control, irrigation, fishing, and other purposes if

any, the following upper and lower limits on the variables should
be satisfied at the optimum:

(a) Upper and lower bounds on the storages:

$$u_{ij}^{m} \leq u_{ij,k} \leq x_{ij}^{M}, \ i \ R_h, \ j \ R_r \qquad (119)$$

(b) Upper and lower bounds on the discharge:

$$u_{ij,k}^{m} \leq u_{ij,k} \leq u_{ij,k}^{M} \qquad (120)$$

In the cost functional given by Eq. (116) we simply minimize
the difference between the monthly generation and the monthly
load on the system, in such a way that the total generation at
the end of the year is as large as possible since we can store
water for the second year of the critical period. The expectation
in Eq. (116) is taken with respect to the random inflow $I_{ij,k}$.

In the above equation, R_h and R_r stand for the set of the
hydro reservoirs connected in series on the same river and the
set of the rivers respectively.

3. Modelling of the System

The generation of a hydroplant is a nonlinear function of
the discharge $u_{ij,k}$ and the storage. To avoid overestimation of
production for falling water levels and underestimation for
rising water levels, an average of begin and end-of-time step
storage is used. We may choose the generating function $G_{ij,k}$ as

$$G_{ij,k} = \alpha_{ij} u_{ij,k} + \tfrac{1}{2} \beta_{ij} u_{ij,k} (x_{ij,k} + x_{ij,k-1})$$

$$+\frac{1}{4}\gamma_{ij}u_{ij,k}(x_{ij,k}+x_{ij,k-1})^2 \tag{121}$$

where α_{ij}, β_{ij}, and γ_{ij} are constants. These were obtained by least-squares curve fitting to typical plant data available. Substituting for $x_{ij,k}$ from Eq. (117) into Eq. (121), one obtains

$$G_{ij,k}=b_{ij,k}u_{ij,k}+u_{ij,k}d_{ij,k}x_{ij,k-1}+u_{ij,k}f_{ij,k}(u_{(i-1)j,k}-u_{ij,k})$$

$$+\gamma_{ij}u_{ij,k}y_{ij,k}+\frac{1}{4}u_{ij,k}\gamma_{ij}(z_{(i-1)j,k}+z_{ij,k})$$

$$+\gamma_{ij}r_{ij,k-1}(u_{(i-1)j,k}-u_{ij,k})-\frac{1}{2}\gamma_{ij}z_{ij,k}u_{(i-1)j,k} \tag{122}$$

where

$$q_{ij,k}=I_{ij,k}+s_{(i-1)j,k}-s_{ij,k} \tag{123}$$

$$b_{ij,k}=\alpha_{ij}+\frac{1}{2}\beta_{ij}q_{ij,k}+\frac{1}{4}\gamma_{ij}(q_{ij,k})^2 \tag{124}$$

$$d_{ij,k}=\beta_{ij}+\gamma_{ij}q_{ij,k} \tag{125}$$

$$f_{ij,k}=\frac{1}{2}d_{ij,k} \tag{126}$$

and the following are pseudo-state variables

$$y_{ij,k}=(x_{ij,k})^2 \tag{127}$$

$$z_{ij,k}=(u_{ij,k})^2 \tag{128}$$

$$r_{ij,k-1}=u_{ij,k}x_{ij,k-1} \tag{129}$$

B.   THE OPTIMAL SOLUTION

1.   A Minimum Norm Formulation

The augmented cost functional is obtained by adjoining the equality constraints via Lagrange multipliers and the inequality constraints via Kuhn-Tucker multipliers. One thus obtains

$$
\begin{aligned}
J(k)=E[&\sum_{j\in Rr} \sum_{i\in Rn} \{b_{ij,k}u_{ij,k}+u_{ij,k}d_{ij,k}x_{ij,k-1} \\
&+u_{ij,k}f_{ij,k}(u_{(i-1)j,k}-u_{ij,k})+\gamma_{ij}u_{ij,k}y_{ij,k-1} \\
&+\tfrac{1}{4}u_{ij,k}\gamma_{ij}(z_{(i-1)j,k}+z_{ij,k})+\gamma_{ij}r_{ij,k-1}(u_{(i-1)j,k}-u_{ij,k}) \\
&-\tfrac{1}{2}\gamma_{ij}z_{ij,k}u_{(i-1)j,k}\} \\
-a(k)&\sum_{h\in Rr} \sum_{i\in Rh} \sum_{k=1}^{K} \{b_{ij,k}u_{ij,k}+u_{ij,k}d_{ij,k}x_{ij,k-1} \\
&+u_{ij,k}f_{ij,k}(u_{(i-1)j,k}-u_{ij,k})+\gamma_{ij}yu_{ij,k}y_{ij,k-1} \\
&+\tfrac{1}{4}\gamma_{ij}u_{ij,k}(z_{(i-1)j,k}+z_{ij,k})+\gamma_{ij}r_{ij,k-1}(u_{(i-1)j,k}-u_{ij,k}) \\
&-\tfrac{1}{2}\gamma_{ij}z_{ij,k}u_{(i-1)j,k}+\mu_{ij,k}(-y_{ij,k}+(x_{ij,k})^2) \\
&+\phi_{ij,k}(-z_{ij,k}+(u_{ij,k})^2)+\psi_{ij,k}(-r_{ij,k-1}+u_{ij,k}x_{ij,k-1}) \\
&+\lambda_{ij,k}(-x_{ij,k}+x_{ij,k-1}+q_{ij,k}+u_{(i-1)j,k}-u_{ij,k}) \\
&+e_{ij,k}(x_{ij}^{m}-x_{ij,k}) \\
&+e_{ij,k}^{1}(x_{ij,k}-x_{ij}^{M})+g_{ij,k}(u_{ij,k}^{m}-u_{ij,k}) \\
&+g_{ij,k}^{1}(u_{ij,k}-u_{ij,k}^{M})\}]
\end{aligned}
\tag{130}
$$

where $u_{ij,k}$, $\phi_{ij,k}$, $\psi_{ij,k}$, and $\lambda_{ij,k}$ are Lagrange multipliers. They are determined such that the corresponding equality constraints are satisfied and $e_{ij,k}$, $e_{ij,k}^{1}$, $g_{ij,k}$, and $g_{ij,k}^{1}$ are Kuhn-Tucker multipliers. These are equal to zero if the constraints are not violated and greater than zero if the constraints are violated.

The cost functional in Eq. (130) can be written in vector form as

$$J(k)=E[\{B^{T}(k)u(k)+u^{T}(k)D(k)x(k-1)+u^{T}(k)F(k)MU(k)$$

$$+u^{T}(k)C(k)y(k-1)+\frac{1}{4}u^{T}(k)C(k)Nz(k)$$

$$+r^{T}(k-1)C(k)Mu(k)-\frac{1}{2}z^{T}(k-1)C(k)Lu(k)\}$$

$$-\sum_{k=1}^{K}\{b^{T}(k)u(k)+u^{T}(k)d(k)x(k-1)$$

$$+u^{T}(k)f(k)Mu(k)+u^{T}(k)Cy(k-1)+\frac{1}{4}u^{T}(k)CNz(k)$$

$$+r^{T}(k-1)CMu(k)-\frac{1}{2}z^{T}(k)CLu(k)$$

$$+u^{T}(k)(-y(k)+x^{T}(k)\vec{H}x(k))$$

$$+\phi^{T}(k)(-z(k)+u^{T}(k)\vec{H}u(k))+\psi^{T}(k)(-r(k-1)$$

$$+x^{T}(k-1)\vec{H}u(k))+\lambda^{T}(k)(-x(k)$$

$$+x(k-1)+q(k)+Mu(k))+\nu^{T}(k)(x(k-1)$$

$$+q(k)+Mu(k))+\sigma^{T}(k)u(k)\}] \tag{131}$$

In the above equation $\vec{H}$ is a vector matrix in which the vector index varies from 1 to n while the matrix dimension of $\vec{H}$ is nxn, and $B(k)$, $b(k)$, $u(k)$, $x(k)$, $y(k)$, $z(k)$, $r(k)$, $\mu(k)$, $\phi(k)$, $\lambda(k)$, $\nu(k)$ and $\sigma(k)$ are nx1, $n = \sum_{j=1}^{m} n_j$, column vectors. Their components are

$(b_{ij,k}/a(k))$, $b_{ij,k}$, $u_{ij,k}$, $x_{ij,k}$, $y_{ij,k}$, $z_{ij,k}$, $r_{ij,k}$, $\mu_{ij,k}$, $\phi_{ij,k}$, $\lambda_{ij,k}$, $\nu_{ij,k}$ and $\sigma_{ij,k}$; $i\varepsilon R_h$, $j\varepsilon R_r$ respectively

where

$$\nu_{ij,k} = e^1_{ij,k} - e_{ij,k} \tag{132}$$

$$\sigma_{ij,k} = g^1_{ij,k} - g_{ij,k} \tag{133}$$

Furthermore, define the nxn diagonal matrices as

$$d(k) = \mathrm{diag}(d_{ij,k}; \; i\varepsilon R_h, j\varepsilon R_r) \tag{134}$$

$$D(k) = \mathrm{diag}(d_{ij,k}/a(k); i\varepsilon R_h, j\varepsilon R_r \tag{135}$$

$$f(k) = \mathrm{diag}(f_{ij,k}; i\varepsilon R_h, j\varepsilon R_r) \tag{136}$$

$$F(k) = \mathrm{diag}(f_{ij,k}/a(k); i\varepsilon R_h, j\varepsilon R_r) \tag{137}$$

$$C(k) = \mathrm{diag}(\gamma_{ij}/a(k); i\varepsilon R_h, j\varepsilon R_r) \tag{138}$$

$$C = \mathrm{diag}(\gamma_{ij}; i\varepsilon R_h, j\varepsilon R_r) \tag{139}$$

$$M = \mathrm{diag}(M_j; j\varepsilon R_r) \tag{140}$$

where $M_1, \ldots, M_m$ are lower triangular matrices, whose elements are given by

1. $m_{ii} = -1, \ i\varepsilon R_h;$

2. $m_{(v+1)v} = 1, \ v=1,\ldots,n_j-1.$

$$N = diag(N_j, j \ R_r) \tag{141}$$

where $N_1,\ldots,N_m$ are lower triangular matrices whose elements are given by

1. $n_{ij} = 1, \ i\varepsilon R_r, \quad j\varepsilon R_m;$

2. $n_{(v+1)v} = 1, \quad v=1,\ldots,n_j-1.$

$$L = diag(L_j, j \ R_r) \tag{142}$$

where the elements of any matrix L_j, $j\varepsilon R_r$ are given by

1. $\ell_{(v+1)v} = 1, \quad v=1,\ldots,n_j-1, \ j\varepsilon R_r;$

2. The rest of the elements are equal to zero.

Employing the discrete version of integration by parts and dropping the constant terms, Eq. (131) becomes

$$J^k = E[\mathbf{x}^T(K)_\mu{}^T(K)\vec{H}\mathbf{x}(K) - \lambda^T(K)\mathbf{x}(K) - \mu^T(K)y(K)$$

$$-\mathbf{x}^T(0)_\mu{}^T(0)\vec{H}\mathbf{x}(0) + \lambda^T(0)\mathbf{x}(0) + \mu^T(0)y(0)$$

$$+\{B^T(k)u(k) + u^T(k)D(k)\mathbf{x}(k-1)$$

$$+u^T(k)F(k)Mu(k) + u^T(k)C(k)y(k-1) + \frac{1}{4}u^T(k)C(k)Nz(k)$$

$$+r^T(k-1)C(k)Mu(k) - \frac{1}{2}z^T(k)C(k)Lu(k)\}$$

$$- \sum_{k=1}^{K} \{b^T(k)u(k) + u^T(k)d(k)\mathbf{x}(k-1) + u^T(k)f(k)Mu(k)$$

$$+u^T(k)Cy(k-1) + \frac{1}{4}u^T(k)CNz(k)$$

$$+(\lambda(k)-\lambda(k-1))^{T}x(k-1)+\lambda^{T}(k)Mu(k)+r^{T}(k-1)CMu(k)$$

$$-\frac{1}{2}z^{T}(k)CLu(k)-\mu^{T}(k-1)y(k-1)$$

$$+x^{T}(k-1)\mu^{T}(k-1)\vec{H}x(k-1)+\nu^{T}(k)x(k-1)-\phi^{T}(k)z(k)$$

$$+u^{T}(k)\phi^{T}(k)\vec{H}u(k)-\psi^{T}(k)r(k-1)$$

$$+x^{T}(k-1)\phi^{T}(k)\vec{H}u(k)+\psi^{T}(k)u(k)+\nu^{T}(k)Mu(k)\}] \tag{143}$$

We define the 5nx1 vectors as

$$X^{T}(k)=[x^{T}(k-1),y^{T}(k-1),u^{T}(k),z^{T}(k),r^{T}(k-1)] \tag{144}$$

$$R^{T}(k)=[(\lambda(k)-\lambda(k-1)+\nu(k))^{T},-\mu^{T}(k-1), (b(k)+M^{T}\lambda(k)+\sigma(k)$$

$$+M^{T}\nu(k))^{T},-\phi^{T}(k),-\psi^{T}(k)] \tag{145}$$

and

$$Q^{T}(k)=[0,0,B^{T}(k),0,0] \tag{146}$$

Furthermore, define the 5nx5n rectangular matrices as

$$L(k)=\left|\begin{array}{cc} L_{11}(k) & L_{12}(k) \\ \\ \\ L_{21}(k) & L_{22}(k \end{array}\right| \tag{147}$$

where we define

$$L_{11}(k)=\left|\begin{array}{ccc} \mu^{T}(k-1)\vec{H} & 0 & (\frac{1}{2}d(k)+\frac{1}{2}\psi^{T}(k)\vec{H}) \\ 0 & 0 & \frac{1}{2}c(k) \\ \frac{1}{2}(d(k)+\psi^{T}(k)\vec{H}) & \frac{1}{2}c & (\frac{1}{2}f(k)M+\frac{1}{2}M^{T}f(k)+\phi^{T}(k)\vec{H} \end{array}\right|$$

$$L_{12}(k)= \begin{vmatrix} 0 & 0 \\ & 0 \\ \frac{1}{8}CN-L^TC & \frac{1}{2}M^TC \end{vmatrix} \qquad (147b)$$

$$L_{21}(k) \begin{vmatrix} 0 & 0 & (\frac{1}{8}N^TC-\frac{1}{4}CL) \\ \\ 0 & 0 & \frac{1}{2}CM \end{vmatrix} \qquad (147c)$$

$$L_{22}(k)= \begin{vmatrix} 0 & 0 \\ \\ 0 & 0 \end{vmatrix}$$

and

$$W(k)= \begin{vmatrix} 0 & 0 & \frac{1}{2}D(k) & 0 & 0 & 0 \\ 0 & 0 & \frac{1}{2}C(k) & 0 & 0 & 0 \\ \frac{1}{2}D(k) & \frac{1}{2}C(k) & (\frac{1}{2}F(k)M+\frac{1}{2}M^TF(k) & (\frac{1}{8}C(k)N-\frac{1}{4}L^TC(k)) & \frac{1}{2}M^TC(k) \\ 0 & 0 & (\frac{1}{8}N^TC(k)-\frac{1}{4}C(k)L & 0 & 0 & 0 \\ 0 & 0 & \frac{1}{2}C(k)M & 0 & 0 & 0 \end{vmatrix}$$

$$(148)$$

Then, the cost functional in Eq. (143) becomes

$$J(k)=E[\{x^T(K) \mu^T(K)\vec{H}x(K)-\lambda^T(K)x(K)-\mu^T(K)y(K)$$

$$-x^T(0) \mu^T(0)Hx(0)+\lambda^T(0)+\mu^T(0)y(0)]$$

$$+[X^T(k)W(k)X(k)+Q^T(k)X(k)$$

$$-\sum_{k=1}^{K} (X^T(k)L(k)X(k)+R^T(k)X(k))\}] \qquad (149)$$

The cost functional in Eq. (149) can be written as

$$J(k)=J_1(K)+J_2(k) \tag{150}$$

where

$$J_1(K)=E[x^T(K)\mu^T(K)\vec{H}x(K)-\lambda^T(K)x(K)-\mu^T(K)y(k)$$

$$-x^T(0)\mu^T(0)\vec{H}x(0)+\lambda^T(0)x(0)+\mu^T(0)y(0)] \tag{151}$$

and

$$J_2(k)=E[X^T(k)W(k)X(k)+Q^T(k)X(k)$$

$$-\sum_{k=1}^{K} \{X^T(k)L(k)X(k)+R^T(k)X(k)\}] \tag{152}$$

2. Optimal Equations

To minimize $J(k)$ in Eq. (150), one minimizes $J_1(K)$ and $J_2(k)$ separately, because the variables $x(K)$ and $y(K)$ are independent of the variables $X(k)$ in $J_2(k)$. The minimum of $J_1(K)$ is clearly achieved when

$$E[\lambda(K)]=[0] \tag{153}$$

and

$$E[\mu(K)]=[0] \tag{154}$$

Since $x(K)$ and $y(K)$ are free, $\delta x(K) \neq 0$, $\delta y(K) \neq 0$ and $x(0)$, $y(0)$ are fixed (we assume that the storages of the reservoirs are known at the beginning of the optimization interval as in the real case). Eqs. (153) and (154) give the values of Lagrange multipliers at the last period studied.

If one defines

$$A(k)=W(k)-KL(k) \tag{155}$$

$$P(k)=Q(k)-KR(k) \tag{156}$$

then the discrete integral part in (152) can be written as

$$J_2 = \sum_{k=1}^{K} J_2(k) = E[\sum_{k=1}^{K} \{X^T(k)A(k)X(k)+P^T(k)X(k)\}] \tag{157}$$

In the above equaiton, if $J_2(k)$ is a minimum (equal to zero), which is our objective, then the sum of $J_2(k)$, J_2, over the optimization interval is also a minimum.

Now define

$$V(k)=A^{-1}(k)P(k) \tag{158}$$

Then, Eq. (157) can be written in the following form by a process similar to completing the squares:

$$J_2 = E[\sum_{k=1}^{K} \{(X(k)+\frac{1}{2}V(k))^T A(k)(X(k)+\frac{1}{2}V(k))-\frac{1}{4}V^T(k)A(k)V(k)\}] \tag{159}$$

The last term of the above equation does not depend explicitly on $X(k)$. Thus one needs only to consider minimizing

$$J_2 = E[\sum_{k=1}^{K} \{(X(k)+\frac{1}{2}V(k))^T A(k)(X(k)+\frac{1}{2}V(k))\}] \tag{160}$$

Equation (160) defines a norm in Hilbert space. Hence, one can write Eq. (160) as

$$\min_{X(k)} J_2 = \min E||X(k)+V(k)||_{A(k)} \tag{161}$$

The minimum of J_2 is clearly achieved when the norm in Eq. (161) is equal to zero:

$$E[X(k)+\frac{1}{2}V(k)]=[0] \tag{162}$$

Substituting from Eqs. (155), (156) and (158) into Eq. (162) one obtains the optimal solution as

$$E[2(L(k)-\frac{1}{K}W(k))X(k)+(R(k)-\frac{1}{K}Q(k))]=[0] \tag{163}$$

Writing Eq. (163) explicitly and adding Eqs. (127)-(129), one obtains

$$E[-x(k)+x(k-1)+I(k)+Mu(k)+Ms(k)]=[0] \tag{164}$$

$$E[x^T(k)\vec{H}x(k)-y(k)=[0] \tag{165}$$

$$E[u^T(k)\vec{H}u(k)-z(k)]=[0] \tag{166}$$

$$E[u^T(k)\vec{H}x(k-1)-r(k-1)]=[0] \tag{167}$$

$$E[\lambda(k)-\lambda(k-1)+2\mu^T(k-1)\vec{H}x(k-1)+(\Delta(k)$$

$$+\psi^T(k)\vec{H})u(k)+\upsilon(k)]=[0] \tag{168}$$

$$E[\Gamma(k)u(k)-\mu(k-1)=[0] \tag{169}$$

$$E[\beta(k)+M^T\lambda(k)+M^T\upsilon(k)+\sigma(k)+(\Delta(k)+\psi^T(k)\vec{H})x(k-1)$$

$$+\Gamma(k)y(k-1)+M^T\Gamma(k)r(k-1)+(\theta(k)M+M^T\theta(k)+2\phi^T(k)H)u(k)$$

$$+(\frac{1}{4}\Gamma(k)N-\frac{1}{2}L^T\Gamma(k))z(k)]=[0] \tag{170}$$

$$E[-\phi(k)+\frac{1}{4}N^T\Gamma(k)u(k)-\frac{1}{2}\Gamma(k)Lu(k)]=[0_1 \tag{171}$$

$$E[-\psi(k)+\Gamma(k)Mu(k)]=[0] \tag{172}$$

where

$$\beta(k)=b(k)-(1/K)B(k)$$

$$\Delta(k)=d(k)-(1/K)D(k)$$

$$\Gamma(k)=C-(1/K)C(k) \tag{173}$$

$$\theta(k)=f(k)-(1/K)F(k)$$

Besides the above equations, one has the following Kuhn-Tucker exclusion equations that must be satisfied at the optimum:

$$e_{ij,k}(x_{ij}^{m}-x_{ij,k})=0 \tag{174}$$

$$e_{ij,k}^{1}(x_{ij,k}-x_{ij}^{M})=0 \tag{175}$$

$$g_{ij,k}(u_{ij,k}^{m}-u_{ij,k})=0 \tag{176}$$

$$g_{ij,k}^{1}(u_{ij,k}-u_{ij,k}^{M})=0 \tag{177}$$

One also has the following limits on the variables

$$\left.\begin{array}{ll} \text{If } x(k)>x^{M}, & \text{then we put } x(k)=x^{M} \\ \text{If } x(k)<x^{m}, & \text{then we put } x(k)=x^{m} \\ \text{If } u(k)>u^{M}(k), & \text{then we put } u(k)=u^{M}(k) \\ \text{If } u(k)<u^{m}(k), & \text{then we put } u(k)=u^{m}(k) \end{array}\right\} \tag{178}$$

Equations (164)-(178) with Eqs. (153) and (154) completely specify the optimal long-term scheduling of the system during the critical period. The algorithm of section IV, D may be used to solve the above equations. In the next section we offer an example of a system in operation. The system in this example is characterized by having a monthly variable load.

C. PRACTICAL EXAMPLE

The algorithm of the last section has been used to determine the optimal monthly operation of a real system in operation consisting of two rivers (m=2); each river has two series reservoirs (n_j=2, j=1,2). The characteristics of the installations are given in Table 6.

Table 6. Characteristics of the Installations

Site name	Capacity of the reservoirs (Mm^3)	Minimum storage (x) (Mm^3)	Maximum effective discharge (m^3/sec)	Minimum effective discharge (m^3/sec)
R_{11}	24763	9949	1119	85.0
R_{21}	5304	3734	1583	85.0
R_{12}	74255	33195	1877	283.2
R_{22}	0	0	1930.3	283.2

Reservoir constants			
Site name	α_{ij} (MWh/Mm^3)	β_{ij} [$MWh/(Mm^3)^2$]	γ_{ij} [$MWh/(Mm^3)^3$]
R_{11}	212.11	146.956×10^{-4}	$-20503142.65 \times 10^{-14}$
R_{21}	117.20	569.71×10^{-4}	$-368119890.482 \times 10^{-14}$
R_{12}	232.46	359.449×10^{-4}	$-1603544.3196 \times 10^{-14}$
R_{22}	100.74	0	0

Table 7. Expected Monthly Inflows to the Reservoirs in the
Critical Period[*]

Month k	$I_{11,k}$ (Mm^3)	$I_{21,k}$ (Mm^3)	$(_{12,k}$ (Mm^3)	$I_{22,k}$ (Mm^3)
1	796	373	1805	23
2	369	184	910	7
3	288	140	645	8
4	207	74	781	8
5	190	81	452	6
6	313	70	485	6
7	947	521	866	7
8	1456	849	3898	53
9	2833	1307	9175	73
10	4611	1714	4877	61
11	3148	895	1798	23
12	1285	426	1585	22
13	811	387	1320	15
14	363	183	1299	15
15	219	97	872	8
16	188	67	880	8
17	84	102	493	6
18	213	76	473	6
19	411	288	1740	22
20	1798	1024	4103	53
21	3428	1666	7120	88
22	2950	1168	3989	53
23	2700	834	2184	23
24	1798	683	1549	22
25	918	493	2124	23
26	545	321	1431	22
27	293	225	1024	15
28	227	193	727	9
29	190	185	647	8
30	236	146	497	6
31	241	157	559	7
32	1623	956	4194	53
33	3346	1637	7560	73
34	3572	1229	3534	46
35	2526	789	1737	23
36	1270	433	1262	15
37	658	275	1380	15
38	340	162	837	7
39	224	161	689	8

40	149	105	465	6
41	149	84	351	5
42	234	80	411	5
43	465	291	837	7

If we let $d(k)$ denote the number of days in month k, then

$u_{ij,k}^m = 0.0864d(k)$ (minimum effective discharge in $m^3/sec)Mm^3$

$u_{ij,k}^M = 0.0864d(k)$ (maximum effective discharge in $m^3/sec)Mm^3$

where the minimum and maximum discharges are given in Table 6.

The expected natural inflows to the sites during the critical period are given in Table 7.

In Tables 8-11 we give the optimal releases from the turbines, the profits realized, and the percentage required load. From these tables one can observe that the calculated percentage load is equal to the given percentage load. Also, one can observe that the benefits during each year of the critcal period are also equal. In Tables 12-15 give the optimal storage.

The dimension of this example during the critical period is as follows:

Number of states = 860

Number of dual variables = 1032

---

*These values are obtained at any month k from the following equation:

$$I^k = \sum_{x=1}^{N} (i_x \cdot p_x)$$

where i_x is the value of the random variable, p_x is the probability that this random variable takes a value of i_x, and N is the number of measurements during a month k.

The computing time to get the optimal solution during the 43-month critical period was 6.85 sec in CPU on the Amdahl 470V/6 computer, which is very small compared to what has been done so far using other approaches.

Table 8.   Optimal Releases from the Turbines, the Profits Realized, and the Percentage Load Required During the First Year of the Critical Period

Month k	$u_{11,k}$ (Mm^3)	$u_{21,k}$ (Mm^3)	$u_{12,k}$ (Mm^3)	$u_{22,k}$ (Mm^3)	Profits (MWh)	Percent load Calculated	Percent load Given
1	1470	2119	2494	2517	2,619,563	8.340	8.34
2	2038	2048	2404	2411	2,786,490	8.871	8.87
3	1454	2552	2992	3001	2,949,777	9.391	9.39
4	1842	2424	3107	3114	3,068,649	9.769	9.77
5	1764	1837	2957	2963	2,748,784	8.751	8.75
6	1849	1906	3091	3097	2,839,532	9.040	9.04
7	1573	2064	2614	2622	2,510,435	7.992	7.99
8	1498	2079	2488	2541	2,430,745	7.738	7.74
9	1420	1966	2355	2428	2,352,717	7.490	7.49
10	1232	2455	2042	2103	2,333,564	7.429	7.43
11	1314	2209	2192	2215	2,399,492	7.639	7.64
12	1364	1906	2275	2297	2,371,382	7.549	7.55

Total benefits from the generation during the first year of the critical period: 32,411,168

Table 9. Optimal Releases from the Turbines, Profits Realized,

and the Percentage Load Required during the Second Year

of the Critical Period

Month k	$u_{11,k}$ (Mm^3)	$u_{21,k}$ (Mm^3)	$u_{12,k}$ (Mm^3)	$u_{22,k}$ (Mm^3)	Profits (MWh)	Percent load Calculated	Given
1	1491	2150	2529	2544	2,619,398	8.339	8.34
2	2077	2078	2438	2453	2,785,707	8.869	8.87
3	1475	2594	3041	3049	2,949,907	9.391	9.39
4	1910	2318	3221	3229	3,069,785	9.773	9.77
5	1793	1888	3007	3013	2,748,036	8.749	8.75
6	1872	1935	3179	3185	2,839,552	9.040	9.04
7	1653	1913	2751	2773	2,509,918	7.991	7.99
8	1529	2126	2542	2595	2,431,018	7.739	7.74
9	1435	2006	2383	2471	2,352,861	7.491	7.49
10	1267	2435	2101	2154	2,333,571	7.429	7.43
11	1367	2201	2268	2291	2,399,486	7.639	7.64
12	1370	2053	2288	2310	2,371,875	7.551	7.55

Total benefits from the generation
during the second year of the
critical period   31,411,056

Table 10. Optimal Releases from the Turbines, Profits Realized,

and the Percentage Load Required during the Third

Year of the Critical Period

Month k	$u_{11,k}$ (Mm^3)	$u_{21,k}$ (Mm^3)	$u_{12,k}$ (Mm^3)	$u_{22,k}$ (Mm^3)	Profits (MWh)	Percent load Calculated	Given
1	1529	2184	2579	2602	2,619,493	8.339	8.34
2	2049	2251	2487	2509	2,786,245	8.870	8.87
3	1513	2630	3091	3106	2,949,134	9.389	9.39
4	1879	2698	3173	3182	3,068,526	9.769	9.77
5	1830	2023	3077	3085	2,748,968	8.752	8.75
6	1771	1909	3461	3467	2,840,465	9,143	9.04
7	255	390	4796	4803	2,509,759	7,990	7.99
8	1583	2203	2633	2686	2,432,202	7.743	7.74
9	1488	2066	2473	2546	2,353,636	7.493	7.49
10	1324	2407	2199	2245	2,332,929	7.427	7.43
11	1410	2199	2355	2378	2,399,289	7.638	7.64
12	1436	2012	2399	2414	2,370,568	7.547	7.55

Total benefits from the
generation during the
third year of the
critical period   3,141,152

Table 11. Optimal Releases from the Turbines, Profits Realized,

and the Percentage Load Required during the Rest of

the Critical Period

Month k	$u_{11,k}$ (Mm^3)	$u_{21,k}$ (Mm^3)	$u_{12,k}$ (Mm^3)	$u_{22,k}$ (Mm^3)	Profits (MWh)	Percent load	
						Calculated	Given
1	779	1441	3519	3534	2,466,724	8.393	8.34
2	1180	2203	3099	3106	2,602,812	8.856	8.87
3	1444	1785	3579	3587	2,756,584	9.379	9.39
4	1207	1313	4373	4379	2,868,556	9.769	9.77
5	1624	1608	3314	3319	2,559,364	8.710	8.75
6	683	862	4826	4831	2,675,327	9.102	9.04
7	465	755	4410	4417	2,337,184	7.956	7.99

Total benefits from the
generation during the
rest of the critical period: 18,266,544

Table 12. Optimal Reservoir Storage during the First Year of the

Critical Period

Month k	$x_{11,k}$ (Mm^3)	$x_{21,k}$ (Mm^3)	$x_{12,k}$ (Mm^3)
1	24088	5027	73565
2	22419	5120	72071
3	21253	4242	69724
4	19618	3734	67398
5	18044	3742	64892
6	16507	3755	62285
7	15881	3784	60537
8	15838	4053	61947
9	17252	4813	68766
10	10630	5304	71600
11	22463	5304	71205
12	22384	5187	70515

Table 13. Optimal Reservoir Storage during the Second Year of
the Critical Period

month k	$x_{11,k}$ (Mm^3)	$x_{21,k}$ (Mm^3)	$x_{12,k}$ (Mm^3)
1	21704	4915	69305
2	19989	5097	68165
3	18733	4075	65996
4	17011	3734	63654
5	15401	3742	61139
6	13742	3754	58433
7	12499	3782	57421
8	12766	4210	58982
9	14759	5304	63718
10	15442	5304	65521
11	17775	5304	65521
12	18203	5304	64810

Table 14. Optimal Reservoir Storage during the Third Year of the
Critical Period

Month k	$x_{11,k}$ (Mm^3)	$x_{21,k}$ (Mm^3)	$x_{12,k}$ (Mm^3)
1	17591	5142	64325
2	16085	5262	63269
3	14865	4369	610202
4	13213	3743	58755
5	11572	3734	56324
6	10037	3741	53360
7	10022	3764	49123
8	10061	4100	50683
9	11919	5158	55770
10	14167	5304	57105
11	15282	5304	65486
12	15115	5161	55348

Table 15. Optimal Reservoir Storage during the Rest of the
          Critical Period

Month k	$x_{11,k}$ (Mm^3)	$x_{21,k}$ (Mm^3)	$x_{12,k}$ (Mm^3)
1	14994	4774	53210
2	14153	3914	50947
3	12932	3734	48057
4	11874	3734	44148
5	10399	3833	41184
6	9949	3734	36769
7	9949	3734	33196

D.   CONCLUSIONS

In the solution presented in this section we have presented
an efficient approach for solving the long-term optimal operating
problem of series-parallel reservoirs for the critical period
with specified monthly load; this load is equal to a certain
percentage of the total generation at each year of the critical
period. The generating function used is a nonlinear function of
the discharge and the average storage between two successive
months k, k-1. The resulting problem is a highly nonlinear
problem; we defined a set of pseudostate variables to overcome
these nonlinearities.

The proposed approach takes into account the stochasticity of
the river flows; we assume that their probability properties were
preestimated from past history. We use the expected values for

the random inflow. The proposed approach also has the ability to deal with large-scale coupled power systems. Most of the current approaches either use composite reservoirs or do not consider the coupling of reservoirs.

The dimension of the given example consisting of two rivers, each river having two series reservoirs and a 43-month critical period, is as follows:

Number of states = 860

Number of dual variables = 1032

The computing time to get the optimal operation for this example was 6.85 sec in CPU units on the Amdahl 470V/6 computer, which is very small compared to what has been done so far using other approaches.

IIV. EFFICIENT LOAD FOLLOWING SCHEDULING [23]

This section presents an efficient technique for solving the load following scheduling problem for large scale hydro-electric power systems, to obtain a maximum and most uniform surplus power, while satisfying the various environmental, physical, legal and contractual constraints. The technique used to solve this problem is based on the minimum norm formulation of functional analysis reported in refs. [9], [10], [14] and [20] and explained earlier in the previous sections. The proposed algorithm takes into account the variations of tail water elevation and forebay elevation (Section IV, A,2). The MWh generated from each reservoir is considered as a quadratic function of the average storage times the discharge through the turbine.

A. PROBLEM FORMULATION

1. The System Under Study

The system under study consists of m rivers, with one or several reservoirs in series on each river. These rivers may or may not be dependent on each other, and have interconnection lines to the neighboring system through which energy may be exchanged, see for instance Figure 4.

2. The Objective Function and the Hydro-Constraints

The object for the hydro-electric system of Figure 4, is to determine the optimal discharge $u_{ij,k}$, $i=1,\ldots,n_j$, $j=1,\ldots,m$; $k=1,\ldots,K$ that maximizes the average period generation shaped uniformly to the load. This can be expressed mathematically as

Maximize

$$J = E\left[\sum_{k=1}^{K} \sum_{j=1}^{m} \sum_{i=1}^{n_j} G_{ij,k}(u_{ij,k} + (x_{ij,k} + x_{ij,k-1}))\right] \qquad (179)$$

In the above equation E stands for the expected value, expectation is taken with respect to the random inflow $I_{ij,k}$. Subject to satisfying the following constraints.

(1) The generation should match the load and the losses of the system, for long-term study we assume that the losses are equal to a certain percentage of the load and hence added to it.

(2) Surplus power is as uniform as possible. This can be expressed by the inequality constraints

$$E[\sum_{j=1}^{m} \sum_{i=1}^{n_j} G_{ij,k}(.) - \ell_k - C] \geq 0, \; k=1,\ldots,K \qquad (180)$$

where ℓ_k is the dependable load during a period k and C is a constant which represents the expected surplus energy calculated at each iteration so as to assure, if satisfied, the most uniform surplus power during the optimization interval k. In Eq. (180) inequality is preferred to equality because the total surplus energy varies from iteration to iteration, and hence it is difficult to predict the value of C exactly. Also, due to the characteristics of the problem, as we will see later, when the technique is applied to a system in operation, it is almost impossible to obtain a uniform surplus power due to non-power constraints which have to be satisfied first.

(3) The water conservation equation for each reservoir may be adequately described by the continuity-type difference equation as

$$x_{ij,k} = x_{ij,k-1} + I_{ij,k} - u_{ij,k} - s_{ij,k} + \sum_{\substack{\ell \in Ru \\ j \in Rr}} (u_{\ell j,k} + s_{j,k})$$

$$;i=1,\ldots,n_j, \; j=1,\ldots,m, k=1,\ldots,K \qquad (181)$$

where R_u and R_r is the set of upstream reservoirs on river set R_r and hydraulicaly coupled to the reservoir i on river j, $s_{ij,k}$ is given by

$$s_{ij,k} = \begin{cases} (u_{ij,k} - u_{ij,k}^M) \text{ if } u_{ij,k} > u_{ij,k}^M \text{ and } x_{ij,k} \geq x_{ij,k}^M \\ \\ 0, \text{ otherwise} \end{cases} \qquad (182)$$

(4) To satisfy multi-purpose stream use requirements, such as flood control, navigation, irrigation, fishing, water quality, recreational activities and other purposes, if any, the plant variables must satisfy the following inequality constraints.

(i) upper and lower bounds on reservoir contents

$$x_{ij,k}^{m} \leq x_{ij,k} \leq x_{ij,k}^{M}, \; i=1,\ldots,n_j; j=1,\ldots,m$$
$$;k=1,\ldots,K \qquad (183)$$

(ii) upper and lower bounds on the reservoirs discharge

$$u_{ij,k}^{m} \leq u_{ij,k} \leq u_{ij,k}^{M}, i=1,\ldots,n_j; j=1,\ldots,m, k=1,\ldots,K$$
$$(184)$$

(5) To prevent excessive soil erosion around the reservoirs the forebay elevation is bounded by

$$f_{ij,k-1} - f_{ij,k} \leq d_{ij,k}^{M}, \; i=1,\ldots,n_j; j=1,\ldots,m; k=1,\ldots,K$$
$$(185)$$

where the forebay elevation as a function of the storage is given by

$$f_{ij,k} = a_{ij} + b_{ij} x_{ij,k} + c_{ij} (x_{ij,k})^2,$$
$$i=1,\ldots,n_j; j=1,\ldots,m, k=1,\ldots,K \qquad (186)$$

where a_{ij}, b_{ij} and c_{ij} are constants. They can be obtained by using the least error squares curve fitting to typical data plant available.

Equation (185) can be rewritten as a functional of the storage as a nonlinear state dependent inequality constraint

$$b_{ij}(x_{ij,k-1}-x_{ij,k})+c_{ij}(x_{ij,k-1}{}^2-x_{ij,k}{}^2)\leq d_{ij,k}{}^M \qquad (187)$$

(6) the MWh generated from each reservoir during a period k is given by [1].

$$G_{ij,k}[u_{ij,k},\tfrac{1}{2}(x_{ij,k}+x_{ij,k-1})]=\alpha_{ij}u_{ij,k}+\tfrac{1}{2}\beta_{ij}u_{ij,k}(x_{ij,k}+x_{ij,k-1})$$

$$+\tfrac{1}{4}\gamma_{ij}u_{ij,k}(x_{ij,k}+x_{ij,k-1})^2, i=1,\dots,n_j, j=1,\dots,m,k=1,\dots,K$$

$$(188)$$

where α_{ij}, β_{ij} and γ_{ij} are constants, they were obtained from the least error squares curve fitting to typical plant data available (Section IV, A,2). If one substitutes for $x_{ij,k}$ from Eq. (181) into Eq. (188), we obtain

$$G_{ij,k}(.)=b_{ij,k}u_{ij,k}+u_{ij,k}d_{ij,k}x_{ij,k-1}+u_{ij,k}f_{ij,k}(\sum_{\substack{\ell\varepsilon Ru\\ j\varepsilon Rr}}u_{\ell j,k}-u_{ij,k})$$

$$+\gamma_{ij}u_{ij,k}x_{ij,k-1}{}^2+\tfrac{1}{4}u_{ij,k}\gamma_{ij}[(\sum_{\substack{\ell\varepsilon Ru\\ j\varepsilon Rr}}u_{\ell j,k})^2+u_{ij,k}{}^2]$$

$$+\gamma_{ij}u_{ij,k}x_{ij,k-1}[\sum_{\substack{\ell\varepsilon Ru\\ j\varepsilon Rr}}u_{\ell j,k}-u_{ij,k}]-\tfrac{1}{2}\gamma_{ij}u_{ij,k}{}^2(\sum_{\substack{\ell\varepsilon Ru\\ j\varepsilon Rr}}u_{\ell j,k})$$

$$,i=1,\dots,n_j,j=1,\dots,m,k=1,\dots,K \qquad (189)$$

where

$$b_{ij,k}=\alpha_{ij}+\tfrac{1}{2}\beta_{ij}q_{ij,k}+\tfrac{1}{4}\gamma_{ij}q_{ij,k}{}^2, i=1,\dots,n_j,j=1,\dots,m,k=1,\dots,K$$

$$(190)$$

$$q_{ij,k}=I_{ij,k}+\sum_{\substack{\ell\varepsilon Ru\\ j\varepsilon Rr}}s_{\ell j,k}-s_{ij,k}; i=1,\dots,n_j;j=1,\dots,m;k=1,\dots,K$$

$$(191)$$

$$d_{ij,k}=\beta_{ij}+\gamma_{ij}q_{ij,k}; i=1,\dots,n_j;j=1,\dots,m;k=1,\dots,K \qquad (192)$$

$$f_{ij,k} = \frac{1}{2} d_{ij,k}; \quad i=1,\ldots,n_j; j=1,\ldots,m; k=1,\ldots,K \tag{193}$$

Now, the cost functional in Eq. (179) becomes. Maximize

$$J = E\left[\sum_{i=1}^{n_j} \sum_{j=1}^{m} \sum_{k=1}^{K} \{ b_{ij,k} u_{ij,k} + u_{ij,k} d_{ij,k} x_{ij,k-1} \right.$$

$$+ u_{ij,k} f_{ij,k} \left(\sum_{\substack{\ell \in Ru \\ j \in Rr}} u_{\ell j,k} - u_{ij,k} \right)$$

$$+ \gamma_{ij} u_{ij,k} x_{ij,k-1}^2 + \frac{1}{4} u_{ij,k} \gamma_{ij} \left[\left(\sum_{\substack{\ell \in Ru \\ j \in Rr}} u_{\ell j,k} \right)^2 + u_{ij,k}^2 \right]$$

$$+ \gamma_{ij} u_{ij,k} x_{ij,k-1} \left[\sum_{\substack{\ell \in Ru \\ r \in Rr}} u_{\ell j,k} - u_{ij,k} \right]$$

$$- \frac{1}{2} \gamma_{ij} u_{ij,k}^2 \left(\sum_{\substack{\ell \in Ru \\ j \in Rr}} u_{\ell j,k} \right) \} \right] \tag{194}$$

The cost functional in Eq. (194) is a highly nonlinear function; to cast the problem into a quaadratic one, we may define pseudo-variables as [20]

$$y_{ij,k} = x_{ij,k}^2; \quad i=1,\ldots,n_j; j=1,\ldots,m; k=1,\ldots,K \tag{195}$$

$$z_{ij,k} = u_{ij,k}^2; \quad i=1,\ldots,n_j; j=1,\ldots,m; k=1,\ldots,K \tag{196}$$

$$r_{ij,k-1} = u_{ij,k} x_{ij,k-1}; \quad i=1,\ldots,n_j; j=1,\ldots,m; k=1,\ldots,K \tag{197}$$

Substituting from Eqs. (195)-(197) into Eq. (194), one obtains

$$J = E\left[\sum_{i=1}^{n_j} \sum_{j=1}^{m} \sum_{k=1}^{K} \{ b_{ij,k} u_{ij,k} + u_{ij,k} d_{ij,k} x_{ij,k-1} \right.$$

$$+ u_{ij,k} f_{ij,k} \left(\sum_{\substack{\ell \in Ru \\ j \in Rr}} u_{\ell j,k} - u_{ij,k} \right)$$

$$+\gamma_{ij}u_{ij,k}y_{ij,k-1}+\frac{1}{4}u_{ij,k\ j}(z_{ij,k}+\sum_{\substack{\ell\in Ru\\j\in Rr}}z_{\ell j,k})$$

$$+\gamma_{ij}r_{ij,k-1}(\sum_{\substack{\ell\in Ru\\j\in Rr}}u_{\ell j,k}-u_{ij,k})-\frac{1}{2}\gamma_{ij}z_{ij,k}(\sum_{\substack{\ell\in Ru\\j\in Rr}}u_{\ell j,k})\}]$$

$$(198)$$

Subject to satisfying the equality constraints of Eqs. (195)-
(197) and Eq. (181), and the inequality constraints of Eqs.
(180), (183), (184) and (187).

B.   MINIMUM NORM FORMULATION [19,20]

The augmented cost functional can be obtained by adjoining
to the cost functional of Eq. (198) the equality constraints of
Eqs. (181) and (195)-(197) via Lagranges' multipliers
$\lambda_{ij,k}$, $\mu_{ij,k}$, $\phi_{ij,k}$, $\psi_{ij,k}$, $i=1,\ldots,n_j, j=1,\ldots,m,k=1,\ldots,K$ and the
inequality constraints via Kuhn-Tucker multipliers as

$$J=E[\sum_{k=1}^{K}(1-h(k))\sum_{j=1}^{m}\sum_{i=1}^{n_j}\{b_{ij,k}u_{ij,k}+u_{ij,k}d_{ij,k}x_{ij,k-1}$$

$$+u_{ij,k}f_{ij,k}(\sum_{\substack{\ell\in Ru\\j\in Rr}}u_{\ell j,k}-u_{ij,k}+\gamma_{ij}u_{ij,k}y_{ij,k-1}$$

$$+\frac{1}{4}u_{ij,k}\gamma_{ij}(z_{ij,k}+\sum_{\substack{\ell\in Ru\\j\in Rr}}z_{\ell j,k})$$

$$+\gamma_{ij}r_{ij,k-1}(\sum_{\substack{\ell\in Ru\\j\in Rr}}u_{\ell j,k}-u_{ij,k})-\frac{1}{2}\gamma_{ij}z_{ij,k}(\sum_{\substack{\ell\in Ru\\j\in Rr}}u_{\ell j,k})$$

$$+\lambda_{ij,k}(-x_{ij,k}+x_{ij,k-1}+q_{ij,k}+(\sum_{\substack{\ell\in Ru\\j\in Rr}}u_{\ell j,k}-u_{ij,k}))$$

$$+\mu_{ij,k}(-y_{ij,k}+x_{ij,k}^{2})$$

$$+\phi_{ij,k}(-z_{ij,k}+u_{ij,k}^{2})+\psi_{ij,k}(-r_{ij,k-1}+u_{ij,k}x_{ij,k-1})$$

$$-g_{ij,k}^{M}(b_{ij}q_{ij,k}+b_{ij}(\sum_{\substack{\ell\varepsilon Ru \\ j\varepsilon Rr}} u_{\ell j,k}-u_{ij,k})-c_{ij}(y_{ij,k-1}-y_{ij,k})$$

$$-d_{ij,k}^{M}+e_{ij,k}^{m}(x_{ij,k}^{m}-x_{ij,k})+e_{ij,k}^{M}(x_{ij,k}-x_{ij,k}^{M})$$

$$+f_{ij,k}^{m}(u_{ij,k}^{m}-u_{ij,k})+f_{ij,k}^{M}(u_{ij,k}-u_{ij,k}^{M})+h(k)(\ell_{f}+C)]]$$

$$(199)$$

In Eq. (194) $e_{ij,k}^{m}$, $e_{iij,k}^{M}$, $f_{ij,k}^{m}$, $f_{ij,k}^{M}$, $g_{ij,k}^{M}$ and $h(k)$ are Kuhn-Tucker multipliers. They are equal to zero, if the constraints (inequality) are not violated and greater than zero if the constraints are violated.

Employing the discrete integration by parts[*], and dropping constant terms, Eq. (200) is obtained

$$\tilde{J}=E[(1-h(k))[\mu_{ij,K}x_{ij,K}^{2}-\mu_{ij,K}y_{ij,K}-\lambda_{ij,K}x_{ij,K}$$

$$-\mu_{ij,0}x_{ij,0}^{2}+\mu_{ij,0}y_{ij,0}$$

$$+C_{ij,K}y_{ij,K}-C_{ij,0}y_{ij,0}+\lambda_{ij,0}x_{ij,0}]+\sum_{k=1}^{K}(1-h(k))\sum_{j=1}^{m}\sum_{i=1}^{n_j}\{b_{ij,k}u_{ij,k}$$

$$+u_{ij,k}d_{ij,k}x_{ij,k-1}$$

$$+u_{ij,k}f_{ij,k}(\sum_{\substack{\ell\varepsilon Ru \\ j\varepsilon Rr}} u_{\ell j,k}-u_{ij,k})+\gamma_{ij}u_{ij,k}y_{ij,k-1}$$

$$+\frac{1}{4}u_{ij,k} ij(z_{ij,k}+\sum_{\substack{\ell\varepsilon Ru \\ j\varepsilon Rr}} z_{\ell j,k})$$

$$+ ij r_{ij,k-1}(\sum_{\substack{\ell\varepsilon Ru \\ j\varepsilon Rr}} u_{\ell j,k}-u_{ij,k})-\frac{1}{2}\gamma_{ij}z_{ij,k}(\sum_{\substack{\ell\varepsilon Ru \\ j\varepsilon Rr}} u_{\ell j,k})$$

---

[*] $\sum_{k=1}^{K}\lambda_{ij,k}x_{ij,k}=\lambda_{ij,K}x_{ij,K}-\lambda_{ij,0}x_{ij,0}+\sum_{k=1}^{K}\lambda_{ij,k-1}x_{ij,k-1}$

$$+(\lambda_{ij,k}-\lambda_{ij,k-1}+\theta_{ij,k})x_{ij,k-1}+\lambda_{ij,k}(\sum_{\substack{\ell\varepsilon Ru\\j\varepsilon Rr}}u_{\ell j,k}-u_{ij,k})$$

$$-\mu_{ij,k-1}y_{ij,k-1}+\mu_{ij,k-1}x_{ij,k-1}^2+\phi_{ij,k}(-z_{ij,k}+u_{ij,k}^2)$$

$$+\psi_{ij,k}(-r_{ij,k-1}+u_{ij,k}x_{ij,k-1})+\theta_{ij,k}(\sum_{\substack{\ell\varepsilon Ru\\j\varepsilon Rr}}u_{\ell j,k}-u_{ij,k})$$

$$+\nu_{ij,k}u_{ij,k}-B_{ij,k}(\sum_{\substack{\ell\varepsilon Ru\\j\varepsilon Rr}}u_{\ell j,k}-u_{ij,k})\} \qquad (200)$$

where we defined

$$\sigma_{ij,k}=e_{ij,k}^M-e_{ij,k}^m; \quad i=1,\ldots,n_j; j=1,\ldots,m; k=1,\ldots,K \qquad (201)$$

$$\nu_{ij,k}=f_{ij,k}^M-f_{ij,k}^m; \quad i=1,\ldots,n_j; j=1,\ldots,m; k=1,\ldots,K \qquad (202)$$

$$C_{ij,k}=c_{ij}g_{ij,k}^M; \quad i=1,\ldots,n_j; j=1,\ldots,m; k=1,\ldots,K \qquad (203)$$

$$B_{ij,k}=b_{ij}g_{ij,k}^M; \quad i=1,\ldots,n_j; j=1,\ldots,m; k=1,\ldots,K \qquad (204)$$

Equation (200) can be written in vector form as

$$\tilde{J}=E[\{x^T(K)\mu^T(K)\vec{H}x(K)-\lambda^T(K)x(K)-\mu^T(K)y(K)+C^T(K)y(K)$$

$$-x^T(0)\mu^T(0)Hx(0)+\lambda^T(0)x(0)+\mu^T(0)y(0)-C^T(0)y(0)\}$$

$$+\sum_{k=1}^{K}\{x^T(k-1)\mu^T(k-1)\vec{H}x(k-1)+u^T(k)\phi^T(k)\vec{H}u(k)+u^T(k)\psi^T(k)\vec{H}x(k-1)$$

$$+u^T(k)d(k)x(k-1)+u^T(k)\gamma y(k-1)+\frac{1}{2}u^T(k)f(k)Mu(k)+\frac{1}{2}u^T(k)M^Tf(k)u(k)$$

$$+\frac{1}{4}u^T(k)\gamma Mz(k)+r^T(k-1)\gamma Mu(k)-\frac{1}{2}z^T(k)\gamma Nu(k)$$

$$(b(k)-M^TB(k)+M^T\lambda(k)+M^T\theta(k)+\nu(k))^Tu(k)+(\lambda(k)-\lambda(k-1)+$$

$$\theta(k))^Tx(k-1)-\mu^T(k-1)y(k-1)-\phi^T(k)z(k)-\psi^T(k)r(k-1))] \qquad (205)$$

In the above equation all the variables are nx1 column vectors, $n=\sum_{j=1}^{m}n_j$ and γ, $d(k)$, $f(k)$ are nxn diagonal matrices. Furthermore, M, N are nxn matrices, their elements depend on the topological arrangement of the reservoirs and rivers, they vary between -1 to 1, and $\vec{H}$ is a vector matrix in which the vector index varies from 1 to n, while the matrix dimension of H is nxn.

Define the nx1, $n=\sum_{j=1}^{m}n_j$, vectors

$$X^T(k)=[x^T(k-1),\ y^T(k-1),\ r^T(k-1),\ z^T(k),\ u^T(k)] \qquad (206)$$

$$R^T(k)=[(\lambda(k)-\lambda(k-1)+\theta(k))^T,-\mu^T(k-1),-\psi^T(k),-\phi^T(k),$$

$$(b(k)+\nu(k)-M^TB(k)+M^T\lambda(k)+M^T\theta(k))^T] \qquad (207)$$

and define the nxn matrix

$$L(k)=\begin{vmatrix} \mu^T(k-1)\vec{H} & 0 & 0 \\ 0 & 0 & 0 \\ 0 & 0 & 0 \\ 0 & 0 & 0 \\ (\frac{1}{2}d(k)+\frac{1}{2}\psi^T(k)H) & \frac{1}{2}\gamma & \frac{1}{2}M^T\gamma \end{vmatrix}$$

$$
\left.
\begin{array}{cc}
0 & (\tfrac{1}{2}d(k)+\tfrac{1}{4}{}^{T}(k)\vec{H}) \\[4pt]
0 & \tfrac{1}{2}\gamma \\[4pt]
0 & \tfrac{1}{2}\gamma M \\[4pt]
0 & (\tfrac{1}{8}M^{T}\gamma-\tfrac{1}{4}\gamma N) \\[4pt]
(\tfrac{1}{8}\gamma M-\tfrac{1}{4}N^{T}\gamma) & (\phi^{T}(k)H+\tfrac{1}{2}f(k)M+\tfrac{1}{2}M^{T}f(k))
\end{array}
\right|
$$

$$(208)$$

Then, the cost functional in eq. (205) can be written as

$$
\begin{aligned}
\tilde{J}=E[\,\{ & x^{T}(K)_{\mu}^{T}(K)\vec{H}x(K)-\lambda^{T}(K)x(K)+(C(K)-\mu(K))^{T}y(K) \\[4pt]
& -x^{T}(0)_{\mu}^{T}(0)\vec{H}x(0)+\lambda^{T}(0)x(0)+(\mu(0)-C(0))^{T}y(0)\} \\[4pt]
& +\sum_{k=1}^{K}(x^{T}(k)L(k)X(k)+R^{T}(k)X(k))\}\,]
\end{aligned}
$$

$$(209)$$

If one obtains

$$
V(k)=L^{-1}(k)R(k) \tag{210}
$$

Then Eq. (209) can be written by a process similar to completing the squares as [20]

$$
\begin{aligned}
\tilde{J}=E[\,\{ & x^{T}(K)\mu^{T}(K)\vec{H}x(K)-\lambda^{T}(K)x(K)+(C(K)-\mu(K))^{T}y(K) \\[4pt]
& -x^{T}(0)\mu^{T}(0)\vec{H}x(0)+\lambda^{T}(0)x(0)+(\mu(0)-C(0))^{T}y(0)\} \\[4pt]
& +\sum_{k=1}^{K}\{(X(k)+\tfrac{1}{2}V(k))^{T}L(k)(X(k)+\tfrac{1}{2}V(k))-\tfrac{1}{4}V^{T}(k)L(k)V(k)\}\,] \quad (211)
\end{aligned}
$$

Since $x(0)$ is a known vector and the term $\frac{1}{4}V^{T}(k)L(k)V(k)$ is constant independent of $X(k)$, then Eq. (211) can be written as

$$\tilde{J}=E[x^T(K)\mu^T(K)\vec{H}x(K)-\lambda^T(K)x(K)+(C(K)-\mu(K))^Ty(K)]$$

$$+E[\sum_{k=1}^{K}\{(X(k)+\frac{1}{2}V(k))^TL(k)(X(k)+\frac{1}{2}V(k))\}] \qquad (212)$$

Equation (212) is composed of two parts, the boundary part and the discrete integral part, which are independent of each other. To maximize $\tilde{J}$ in Eq. (212), one maximizes each part separately [21,22]

$$\underset{[x(K),y(K),X(k)]}{Max.\tilde{J}} = \underset{x(K),y(K)}{Max.E}[x^T(K)\mu^T(K)\vec{H}x(K)-\lambda^T(K)x(K)$$

$$+(C(K)-\mu(K))^Ty(K)]+$$

$$\underset{X(k)}{Max.E}\sum_{k=1}^{K}\{(X(k)+\frac{1}{2}V(k))^TL(k)(X(k)+\frac{1}{2}V(k))\}]$$

$$(213)$$

The discrete integral part of Eq. (213) defines a norm in Hilbert space, this part can be written as

$$\underset{X(k)}{Max.J_2}=\underset{X(k)}{Max.E}\,||X(k)+\frac{1}{2}V(k)\,||\,L(k) \qquad (214)$$

1.   Optimal Equations

The maximum of the boundary part of Eq. (213) is clearly achieved when

$$E[2\mu^T(K)\vec{H}x(K)-\lambda(K)]=0 \qquad (215)$$

$$E[C(K)-\mu(K)]=0 \qquad (216)$$

Equations (215) and (216) give the values of Lagranges' multipliers $\lambda(K)$ and $\mu(K)$ at last period studied.

Maximumization of Eq. (214) is equivalent to minimization of the norm of this equation. The norm of Eq. (214) is minimum if and only if

$$E[X(k)+\tfrac{1}{2}V(K)]=0 \tag{217}$$

or

$$E[R(k)+2L(k)X(k)]=0 \tag{218}$$

Writing Eq. (218) explicitly, arranging and adding the equality constraints, one obtains

$$E[-x(k)+x(k-1)+q(k)+Mu(k)]=\underline{0} \tag{219}$$

$$E[-y(k)+x^{T}(k)\vec{H}x(k)]=\underline{0} \tag{220}$$

$$E[-z(k)+u^{T}(k))\vec{H}u(k)]=\underline{0} \tag{221}$$

$$E[r(k-1)+u^{T}(k)\vec{H}x(k-1)]=0 \tag{222}$$

$$E[-\mu(k-1)+\gamma u(k)]=0 \tag{223}$$

$$E[-\psi(k)+\gamma Mu(k)]=0 \tag{224}$$

$$E[-\phi(k)+(\tfrac{1}{4}M^{T}\gamma-\tfrac{1}{2}\gamma N)u(k)]=0 \tag{225}$$

$$E[\lambda(k)-\lambda(k-1)+\theta(k)+2\mu^{T}(k-1)\vec{H}x(k-1)+(d(k)+\psi^{T}(k)\vec{H})u(k)]=0 \tag{226}$$

$$E[b(k)+\upsilon(k)-M^T B(k)+M^T \lambda(k)+M^T \theta(k)$$

$$+(d(k)+\psi^T(k)\vec{H})x(k-1)+\gamma y(k-1)+M^T \gamma r(k-1)+(\tfrac{1}{4}\gamma M-\tfrac{1}{2}N^T \gamma)z(k)$$

$$+(2\phi^T(k)\vec{H}+f(k)M+M^T f(k))u(k)]=0 \qquad (227)$$

Besides the above equations, one has the Kuhn-Tucker exclusion equations, which must be satisfied at the optimum as, $i=1,..,n_j$; ;$j=1,...,m$, $1=1,...,K$ [20,21,22]

$$e_{ij,k}{}^m(x_{ij,k}{}^m - x_{ij,k})=0 \qquad (228)$$

$$e_{ij,k}{}^M(x_{ij,k} - x_{ij,k}{}^M)=0 \qquad (229)$$

$$f_{ij,k}{}^m(u_{ij,k}{}^m - u_{ij,k})=0 \qquad (230)$$

$$f_{ij,k}{}^M(u_{ij,k} - u_{ij,k}{}^M)=0 \qquad (231)$$

$$g_{ij,k}{}^M(b_{ij}(x_{ij,k-1} - x_{ij,k})+c_{ij}(x_{ij,k-1}{}^2 - x_{ij,k}{}^2)-d_{ij,k}{}^M)=0 \qquad (232)$$

$$h(k)[\sum_{j=1}^{m}\sum_{i=1}^{n_j} G_{ij,k}(u_{ij,k},\tfrac{1}{2}(x_{ij,k}+x_{ij,k-1}))-\ell_k-C]=0 \qquad (233)$$

Equations (219)-(233) together with Eqs. (215) and (216) specify the optimal long-term load following operation of multichain power systems. The above equations are solved using the proposed algorithm mentioned in references [19,20,21,22].

C.   APPLICATION TO THE B.P.A. SYSTEM [Figure 3]

The problem of the load following optimization of the B.P.A. hydro system is carried out considering all the requirements

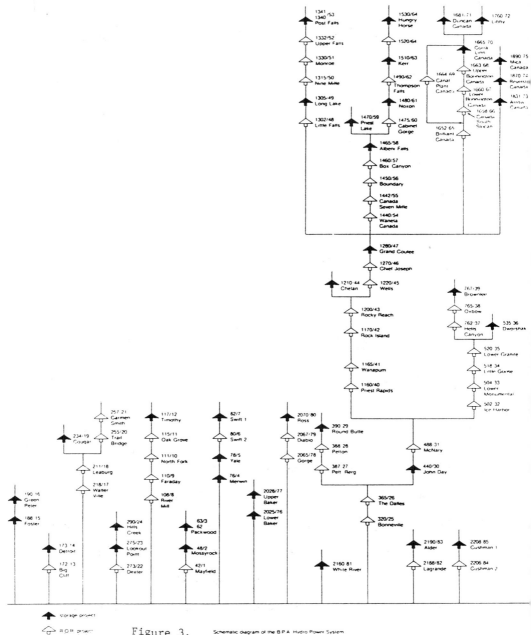

Figure 3.   Schematic diagram of the B.P.A. Hydro Power System

(hard constraints). As we can see in Figs. 4 and 6 there is an essential need for a load following optimization technique. Fig. 4 compares the dependable load and the corresponding hydro generation obtained from the given rule curves used in the B.P.A. hydro corporation. By "the dependable load" we mean the value of load obtained after subtracting from the total firm load an amount representing the required thermal, nuclear, solar, wind, wave and other sources of power generation. The fulfilment of this dependable load means no curtailment of firm load is necessary and there is no need to import expensive energy to meet the deficit. Fig. 5 presents the resulting total generation versus the load using the results obtained from only maximizing the total generation regardless of the shape of the load. Fig. 6 presents the same comparison for the load following algorithm presented in this chapter. A comparison between Figs. 4, 5 and 6 proves the capability of this algorithm to maximize the total energy generated while it guarantees the most uniform surplus power.

In Fig. 6 there is some nonuniformity which cannot be avoided due to the imposed constraints on the scheduling process. One of the major causes of the nonuniformity is the water budget requirements. For example, the required minimum flow at Grand Coulee and Priest Rapid Dams jumps from 50,000 or 60,000 CFS to 134,000 CFS during May to satisfy the Water Budget minimum flow on the Columbia River. This makes the total generation during the

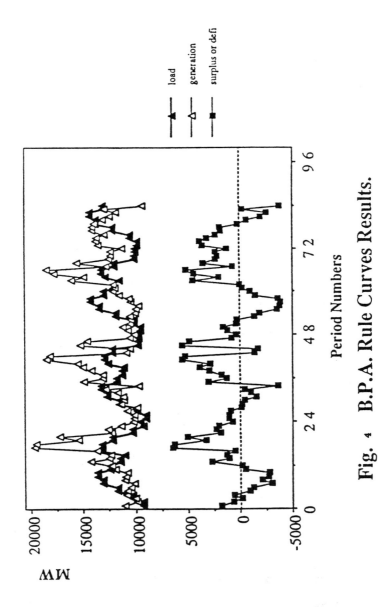

Fig. 4 B.P.A. Rule Curves Results.

load

generation

surplus or defi

MW

Period Numbers

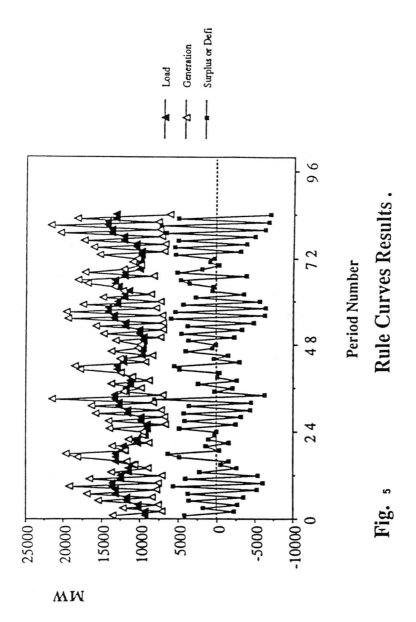

Fig. 5

Rule Curves Results.

463

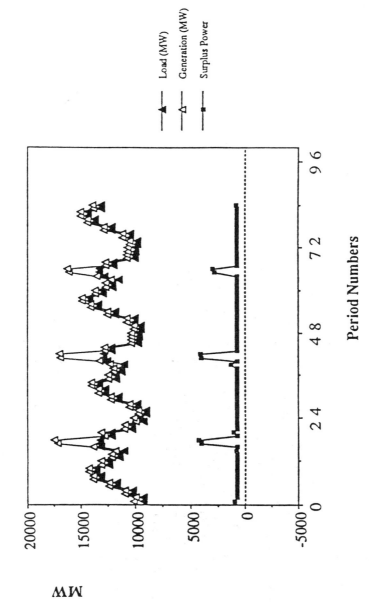

Period Numbers

MW

Load (MW)

Generation (MW)

Surplus Power

Fig 6 Load Following Results .

May peak more than during any other period and cause this unavoidable nonuniformity.

In the programming process, we have compiled six projects in pairs to decreases the computational effort. Also, a special subroutine is made to suite the special configuration of the Canal Plant Dam and the required distribution of the water budget between the Canal Plant and Corra Linn Dams. Furthermore, a special subroutine is made to calculate the corresponding tailwater for a given release to Yale and Upper Baker Dams because of the sharp changes occurring in their field measured tables. In Table 15, we compare the amount of the uniform surplus power obtained by B.P.A. rule curves and that obtained without considering the load following problem to that obtained using the load following algorithm. In Table 17 a comparison of the average power generated by the B.P.A. rule curves and the calculated results obtained without the consideration of the load following problem and that obtained by the load following technique is carried out. In Table 18 a comparison of the total energy capability for the B.P.A. rule curves and the results obtained without considering the load following technique and that by using the load following technique is carried out.

Table 16. A comparison of the violation of the minimum discharge requirement, the spill and the amount the uniform surplus power for the B.P.A. rule curves results, the results obtained regardless of load following and that using the load following technique

	$u^{ij,km} < u^{ij,k}$	Spillage MCF	Uniform Surplus Power
B.P.A. rule curve results	this happens 68 times during the 84 critical period for the 88 project $\sum (u_{ij,k}{}^{m} - u_{ij,k})$ = 48846	4.378E6	−3,828.2MW (deficit)
No load following	0	1.773E6	−6,908.7MW (deficit)
Load following results	0	1.774E6	709.4MW (surplus)

Table 17. A comparison of the average power generated between
the B.P.A. rule curve results, the calculated results
without load following and the load following results.

Period		Average Power Generated by		
from	to	B.P.A.	Ref.[22]	Load Following
Sept. 1, 1928	June 31, 1929	12.373	12.594	13.044
July 1, 1929	June 31, 1930	11.801	12.511	12.377
July 1, 1930	June 31, 1931	12.321	12.211	12.699
July 1, 1931	Feb. 29, 1932	12.163	13.974	12.464

Table 18. A comparison of the total energy capability for the
B.P.A. rule curve results, the results without load
following, and the load following results.

Comparison of	B.P.A.	Ref. [22]	Load Following
Total Energy (GWH)	372,512	389,981	387,258
Average Power (GW)	12.155	12,725	12,636
% Increase over the B.P.A.		4.689%	3.967%

D.   CONCLUSIONS

We have presented a very general framework for the
formulation and solution of the hydro system scheduling problem.
The framework allows for the incorporation of virtually all types
of constraints that are imposed in the planning and the actual

G. S. CHRISTENSEN AND S. A. SOLIMAN

operation of hydro systems. The algorithm described proves its ability to produce the maximum generation with the most uniform surplus power. The algorithm accurately models many of the hydro-plant parameters and constraints and optimizes the schedules by means of functional analysis and the minimum norm formulation technique.

In this Section, the generation from each reservoir is considered as a quadratic function of the storage times the discharge through the turbine. Tailwater elevation is taken as a nonlinear function of the total release [23]. Forebay elevation is considered as a nonlinear function of the storage.

The basic feature of this new procedure is its ability to produce the maximum uniform surplus, which satisfies the system constraints. The procedure can compute the system parameters for hundreds of sets of rule curves of 51 run-of-river plants and 37 reservoir and 96 time periods. Typical numerical results for an actual hydro system illustrate the capabilities of the proposed solution scheme.

References

1. S.A. Soliman, "Optimal Long-term Operation of Hydro-Reservoirs", Ph.D. Thesis, The University of Alberta, Edmonton, Canada (1986).

2. M.E. El-Hawary, "Application of Functional Analysis Optimization Techniques to the Economic Scheduling of Hydro-Thermal Electric Power Systems", Ph.D. Thesis, The University of Alberta, Edmonton, Canada (1972).

3. N.V. Arvanitidies and J. Rosing, "Composite Representation of a Multireservoir Hydroelectric Power System", IEEE Transactions on PAS, Vol. PAS.89 (2), 319-326 (1970).

4. N.V. Arvanitidies and J. Rosing, "Optimal Operation of Multireservoir Systems Using a Composite Representation", "IEEE Transactions on PAS, Vol. PAS-89(2), 327-335 (1970).

5. A. Turgeon, "Optimal Operation of Multireservoir Power Systems with Stochastic Inflow", Water Resources Research Vol.16, No.2 275-283 (1980).

6. A. Turgeon, "A Decomposition Method for Long-Term Scheduling of Reservoirs in Series", Water Resources Research Vol.17, No.6, 1565-1570 (1981).

7. S. Olcer, "et.al", "Application of Linear and Dynamic Programming to the Optimization of the Production of Hydroelectric Power", Optimal Control Application and Methods 6, 43-56 (1985).

8. M.A. Marino and H.A. Ioaiciga, "Dynamic Model for Multireservoir Operation", Water Resources Research 21 (5), 619-630 (1985).

9. S.A. Soliman and G.S. Christensen, "Discrete Stochastic Optimal Long-Term Scheduling of Series Reservoirs", Applied Simulation and Modelling, ASM'86 098-014, 159-163 (1986).

10. S.A. Soliman and G.S. Christensen, "Optimization of Reservoirs in Series on a River with Nonlinear Storage Curve for Long-Term Regulation by Using the Minimum Norm Formulation of Functional Analysis", Journal of Optimization Theory and Applications(JOTA), Vol.58(1), 109-126 (1988).

11. S.A. Soliman, et.al., "Optimal Operation of Multireservoir Power Systems Using Functional Analysis", JOTA, Vol.49(3), 449-461 (1986).

12. W.A. Porter, "Modern Foundations of Systems Engineering", The Macmillan Company, New York, (1966).

13. A.P. Sage and C.C. White, "Optimal System Control", Prentice Hall Inc., Englewood Cliffs, New Jersey, 1977.

14. A.P. Sage, "Optimal Systems Control", Prentice Hall Inc, Englewood Cliffs, New Jersey, 1968.

15. A. Shamaly, "et.al.", "A Transformation for Necessary Optimality Conditions for System with Polynomial Nonlinearities", IEEE Trans.on Automatic Control, Vol.AC-245 (6), 983-985, (1979).

16. R.H. Hicks, C.R. Gagnon, S.L.S. Jacoby and J.S. Kawalik, "Large Scale Nonlinear Optimization of Energy Capability for Pacific Northwest Hydroelectric Systems", IEEE Trans. on PAS Vol. PAS-93 (5), 1604-1612, (1974).

17. C.R. Gagnon and J.F. Bolton, "Optimal Hydro Scheduling at the Bonneville Power Administration", IEEE Trans. on PAS Vol. PAS-97, (3), 772-775, (1978).

18. D. Sjelvgren, "et.al.", "Optimal Operations Planning in a Large Hydro-Thermal Power System", IEEE Trans. on PAS Vol. PAS-102, (11), 3644-3651, (1983).

19. G.S. Christensen and S.A. Soliman, "Modeling and Optimization of Parallel Reservoir Having Nonlinear Storage Curve Under Critical Water Conditions for Long-Term Regulation Using Functional Analysis", JOTA, Vol.55, (3), 359-376 (1987).

20. G.S. Christensen and S.A. Soliman, "On the Application of Functional Analysis to the Optimization of the Production of Hydroelectric Power", IEEE Trans. Vol. PWRS-2, (4), 841-847, (1987).

21. G.S. Christensen, "et.al.", "Optimal Long-Term Operation of the B.P.A. Hydro Electric Power System", Submitted for Publication to the Int. Jour.of Electrical Power & Energy Systems, (1989).

22. G.S. Christensen and S.A. Soliman, "Long-Term Optimal Operation of Series Parallel Reservoirs for Critical Period with Specified Monthly Generation and Average Monthly

Storage", JOTA, Vol. 63(3), (1989).

23. A.M. Atallah and G.S. Christensen, "Efficient Load Following Scheduling for a Large-Scale Hydro-System", Proceeding of Canadian Conference on Electrical and Computer Engineering 480-483 , Vancouver, B.C., Canada, Nov.3-4, (1988).

INDEX

A

Abnormal faults, voltage stability, 154
Accuracy bulk power system reliability
    assessment, Monte Carlo simulations, 224
AC load flow, bulk power system reliability
    modeling
  decoupled flow, 211
  full flow, 210–211
Active constraints, optimal operational
    planning, 312
Adaptive transmission protection, 88–105
  characteristics of, 88–92
  communication and control issues, 93–94
  defined, 88
  distributed computing approach, 97, 99–100
  future directions, 101–105
  hardware components, 92–93
  human factors, 95–96
  overcurrent relay settings, 89–90
  parallelism in relay coordination, 97–98
  software issues, 94–95
  supercomputing approach, 100
Adequacy, reliability of power systems and,
    164
Adequacy indices, generating system reliability
    indices and, 174–175
Aggregation-deposition (AD) approach,
    multireservoir power systems
    optimization, 373–374
Aggregation method, optimization techniques,
    372–373
Algebraic equations
  power flow analysis, 20–25
  voltage stability criteria, 136
Alliant FX/8 computer, parallel algorithm
    implementation problems, 39–43
Amdahl's law, 22

Analytic approximations, distribution
    coordination systems
  loss, 257–258
  voltage at node, 257–258
Antithetic variables, bulk power system
    reliability assessment, 226
Approximate integer voltage-constrained case
    distribution system coordination, 262–263
  solution strategy, 264–265
Approximate integer voltage-unconstrained
    case distribution system coordination,
    265–268
  algorithms, 268–274
  bounding results, 266
  rounding off solution, 266–268
Approximate switched voltage-constrained
    case, 260–263
Approximation formulas
  distribution coordination systems, 249–259
    analytic approximation to loss function,
      257–258
    analytic approximation to voltage at node,
      257–258
    base-case loadflow, 250–253
    computational and communication
      concerns, 253–254
    numerical tests, 257–259
    skeleton distribution system, 255–256
  optimal coordination problems, 248–249
Artificial intelligence (AI)
  adaptive transmission protection, 91
  protection with, 106
Automatic capacitor switching, voltage collapse
    prevention, 159
Automatic generation control (AGC), voltage
    stability, 125
Automatic load shedding, voltage collapse
    prevention, 159

Automatic LTC blocking, voltage collapse
  prevention, 159
Automatic tap changers, voltage stability, 155

B

Back-Tracking (BT) technique, transmission
  protection, 65
Backup constraints, subsystem coordination,
  81–82
Backwards difference method, parallel
  processing for transient stability, 27–28
Balance-of-water equation, optimal long-term
  scheduling, 384
Base-case methods
  distribution coordination systems, 248–249
  approximation formulas, 249–253
Basis factorization, multistage linear
  programming, 344–346
B-coefficients, optimal operational planning,
  303–305
Benchmarks, distribution system coordination
  algorithms, 277–278
Blast furnace gas boilers, optimization of
  operations, 360–368
Block-diagonal structure
  implicit integration, 36–37
  parallel processing, 2–3
  for transient stability, 31
Boundary conditions, subsystem coordination,
  79–80
Boundary expectations, approximate integer
  voltage-unconstrained case, 266
BPA hydro power system, load flow scheduling
  efficiency, 459–468
Branching nodes, distribution coordination
  systems, 255
Branch participations, voltage stability, 133
Break point (BP) set, transmission protection
  loop enumeration, 66
  topological analysis, 64–65
Bulk Power Energy Curtailment Index
  (BPECI), 176–177
Bulk power systems
  reliability assessment methods, 215–229
    comparisons of methods, 227–229
    hybrid approaches, 226–227
    Monte Carlo simulation, 221–226
      random number generation, 221–222
      sample size and accuracy, 224
      sequential sampling, 222–223
      snapshot sampling, 222
      variance reduction, 225–226

  state enumeration, 215–221
    computation reduction, 218–221
    severity ranking of failure states,
      219–220
    state-space truncation, 218–219
    system reliability indices, 215–217
    upper and lower boundaries, 220–221
  reliability modeling, 202–213
    common-mode failures, 203–204
    failure modes of overhead transmission
      lines, 203
    failures in originating stations, 208
    load flow methods in state evaluation,
      210–212
      approximate AC load flows, 212
      DC load flow, 211
      decoupled AC load flow, 211
      full AC load flow, 210–211
      transportation model, 212
    protection system malfunctions, 206–208
    remedial measures in state evaluation,
      212–213
    scheduled outages of transmission
      components, 209–210
    system load models, 204–205
    weather conditions, 205–206
  reliability programs, comparison of,
    227–229
Bus test system, 119–120
Butterfly topology, parallel processing and, 8
BX algorithm, power flow analysis, 16–17

C

Capacitive susceptance, distribution
  coordination systems
  base-case loadflow, 250–251
  control variables, 260
Capacitors
  distribution coordination systems, 246–248
  settings
    approximate integer voltage-constrained
      case, 264
    integer voltage-unconstrained
      distinguished problem, 271–272
Capacity outage probability, reliability
  assessment, 193–194
Chronological load model, system reliability
  modeling, 190–191
Circuit simulators, parallel processing for
  transient stability, 26
Coherency
  fault length and, 46–47

parallel processing for transient stability, 34
Coke over gas (COG) boilers, optimization of operations, multistage linear programming, 360–368
Column addition rule, simplex method, 357
Column deletion rule, simplex method, 357
Common-mode failures, reliability modeling, 203–204
Communication issues
  adaptive transmission protection, 93–94
  algorithm efficiency and, 11
  distribution coordination systems, 253–254
  parallel algorithm implementation and, 38–39
Compensation
  approximation formulas, 253–254
  distribution coordination systems, 246
Complexity issues
  adaptive transmission protection, 101–103
  networks, voltage stability, 123
Component modeling, operational reliability, 233–235
Composite systems, reliability analysis, 170–171
Computational efficiency
  bulk power system reliability assessment, 218–221
  distribution coordination systems, 253–254
  optimization of operations, 367–368
Computer-aided coordination systems, 59
Concurrent processing. See Parallel processing
Constant matrix, parallel processing for transient stability, 31
Constraints
  approximate integer voltage-constraints, 274–275
  distribution system coordination, 246
  optimal operational planning, 307–308
  subsystem coordination, 81
Contingency analysis, parallel processing and, 47
Continuity-type difference equation
  hydro-electric power systems, load flow scheduling efficiency, 448–449
  multichain power system optimality, 404–410
  optimal operation of, with critical water conditions
    forebay elevation, 406–410
    inequality constraints, 406–408
    reservoir storage limits, 405–406
    water conservation equations, 404–405
Continuous optimization algorithms, distribution coordination systems, 255

Continuous relaxation, distribution coordination systems, 248
Control issues, adaptive transmission protection, 93–94
Control variables
  bulk power system reliability assessment, 225–226
  distribution system coordination, 260
  optimal operational planning, 309–311
Convergence
  optimal operational planning
    6–bus system, 319, 321
    14–bus system, 326–327
    30–bus system, 336–337
  parallel processing for transient stability, 26–27
  function space iterates, 34
  parallel-in-time methods, 37–38
Coordination problems
  approximation formulas, 260–263
    approximate switched voltage-constrained problem, 260–262
    approximate voltage-constrained problem, 262–265
    algorithms for, 274–276
    approximate voltage-unconstrained problem, 265–268
    algorithms for, 268–274
    bounding result, 266
    integer voltage-unconstrained distinguished problem, 270–274
  control variable, 260
  overcurrent relay coordination, 74
Coordination Time Interval (CTI), 71–73
Correction factor, multistage linear programming
  simplex method, 350
  pivoting, 353–356
  updating, 356–358
Cost functionals
  hydro-electric power systems
    load flow scheduling efficiency, 451–452
    minimum norm formulation, 452–459
  multichain power system optimality, 408–410
  optimal long-term scheduling, series reservoirs, 384–386
    minimum norm formulation, 390–391
    specified monthly generation, 427–433
Critical period, defined, 400
Critical period stream flow, defined, 400
Critical rule curves (CRC), 401
Critical voltage levels, 123

Critical water conditions
  multichain power system optimality and,
    400–422
    cubic spline technique, 419–422
    minimum norm formulation, 410–415
    optimal equations, 415–417
    optimization objective and constraints,
      402–410
    solution algorithms, 418–419
Cubic spline technique, critical period
  operation, 419–422
Cumulant method, reliability assessment,
  194–195
Customer loads, reliability modeling, 204
Cycling operations, system reliability
  modeling, 186–188

                    D

Damped Newton technique, 27–28
Dantzig-Wolfe decomposition principle, 342
DataBase Management System (DBMS)
  computer-aided coordination systems and,
    59
  "internal model," 61–62
  role of, in RELAY programming, 60–63
  "user model," 61–62
DC load flow, bulk power system reliability
  modeling, 211
Depth-First Search (DFS)
  loop enumeration, 66
  transmission protection, 65
Design criteria, reliability of power systems
  and, 166–167
Design issues, distribution system coordination
  and, 285–286
Deterministic algorithms
  approximate integer voltage-constrained
    case, 275–276
  approximate integer voltage-unconstrained
    case, 267–274
  distribution system coordination, 282–284
  integer voltage-unconstrained distinguished
    case, 272–274
Deterministic methods
  operational reliability, 166–167
  generating system operating reserve,
    235–237
Diagnostician program, adaptive transmission
  protection, 95
Diagonal matrices, specified monthly
  generation, 429–433

Difference equation, specified monthly
  generation, 423–424
Differential algebraic equations (DAE)
  parallel processing, 2
  transient stability, 27–28
Digital relay settings, adaptive transmission
  protection, 102–103
Direct implicit method, for transient stability,
  27–28
Discrete Event simulation, adaptive
  transmission protection, 95
Discretized equations, for transient stability, 32
Dispatcher aids, adaptive transmission
  protection, 95
Distance relay coordination, transmission
  protection, 75–78
  man-machine dialog, 77–78
  models, 76
  successive zone coordination algorithm,
    76–77
  zone 1 setting, 76
DistFlow formulation, distribution coordination
  systems, 251–252
Distinguished nodes, distribution coordination
  systems, 255
Distributed algorithms, distribution
  coordination systems, 254
Distributed processing
  adaptive transmission protection, 97, 99–100
  vs. parallel processing, 8–9
Distribution automation systems
  distribution system coordination, 245
  loss, constraints, 246
Distribution protection, transmission combined
  with, 105–106
Distribution system coordination
  capacitors and regulators, integer quadratic
    optimization, 245–287
  algorithms, 268–276
  approximate integers
    voltage-constrained case, 262–265,
      274–276
    voltage-unconstrained case, 265–274
  approximate switched voltage-constrained
    problem, 260–262
  approximation formulas, 249–259
  bounding results, 266
  control variables, 260
  design problems and, 285–286
  losses, constraints and distribution
    automation, 246
  organization, 248–249
  solution strategies, 264–25

switched capacitors and regulators,
  246–248
    test results, 276–285
  design issues and, 285–286
Domain stability simulations, 118
Dynamic analysis,
  behavioral analysis, 116–117
  contingency analysis, 47
  voltage stability, 135
Dynamic programming
  optimization techniques, multireservoir
    power systems, 372–373
  with successive approximation (DPSA),
    373–374

E

EDF survey of voltage stability, 112
  criteria for, 136–138
Edgeworth expansion, reliability assessment,
  194–195
EHV transmission, voltage stability, 111
Eigenvalues, voltage stability modal analysis,
  128–133
Eigenvectors, voltage stability modal analysis,
  128–133
Electric distribution systems, integer quadratic
  optimization, 245
Energy content curves (ECC), critical period
  regulations, 401
Energy curtailment index (ECI), 175
Energy limited units, reliability assessment, 199
Energy Management System (EMS), adaptive
  transmission protection, 94
EPRI production grade program, 39
Equality constraints
  hydro-electric power systems, 458–459
  optimal long-term scheduling, 383–386
Event trees, reliability modeling, 207–208
Execution times, distribution system
  coordination, 283–285
Expected energy not supplied (EENS)
  generating system reliability, 197–198
    interconnected systems, 199
  reliability indices and, 175
  short- and mid-term time frames, 238–239
  state enumeration, 217
Explicit search procedures, distribution system
    coordination
  algorithms, 277–278
  execution times, 283–285
External system, subsystem coordination, 82

F

Factorization path tree, power flow analysis,
  21–22
Factor path length
  partial factorization, 21–22
  power flow analysis, 20–21
Failure effects analysis (FEA)
  large composite systems, 170–171
  reliability indices and, 168–170
  state enumeration, 215–221
Failure probability distributions, state
  enumeration, 218–219
Failure rates, reliability modeling, protected
  components, 206
Fast auto reclosure, voltage collapse prevention,
  158–159
Fast decoupled power flow analysis
  algebraic equations, 23
  bulk power system reliability modeling, 211
  classic iteration scheme, 18–19
Fault current pairs
  overcurrent relay coordination, 72–73
  subsystem coordination, 82–83
Fault length, parallel algorithm implementation
  and, 46–47
Firm Energy Load Carrying Capability
  (FELCC), 401
Forced outage rate (FOR), system reliability
  modeling, 184–185
Forebay elevation
  hydro-electric power systems, 446, 449–450
  multichain power system optimality,
    406–410
Forward/backward substitution, 20–21
Frequency and duration method, reliability
  assessment, 196–197
Fuel cost minimization, optimal operational
  planning, 294–295
Full-gate flow, critical period operation,
  407–408
Functional analytic optimization, series
  reservoirs, 379–400

G

Gaussian elimination, transient stability, 27–28
Gauss-Jacobi algorithm
  parallel processing for transient stability, 28
  time-point-pipelining (TPP) algorithm,
    35–36
  WR algorithm, 29, 33–34

Gauss-Jacobi algorithm (*continued*)
  power flow analysis, 18–19
Gauss-Seidel algorithm
  bulk power system reliability modeling,
    210–211
  parallel processing for transient stability, 28
    iteration, 29, 31–32
    time-point-pipelining (TPP) algorithm,
      35–36
  power flow analysis, 18–19
Generating system reliability, 184–191
  assessment methods, 192–200
    energy limited units, 199
    frequency and duration method, 196–197
    interconnected systems, 198–199
    loss of load methods, 192–196
      loss of load expectation, 195–196
      loss of load probability, 192–195
    operating considerations, 200
    operating reserve of, 235–237
  distribution system coordination, 245
  probability model, 188–189
  system load models, 189–191
    chronological load model, 190–191
    load duration curve, 189–190
  unit models, 184–188
    multi-state model, 185–186
    peaking and cycling operations, 186–188
    two-state model, 184–185
  voltage stability, 149–152
Generator data
  optimal operational planning
    6-bus system, 313, 316
    14-bus system, 323
    30-bus system, 330
  voltage stability
    modal analysis, 133
    modeling, 124
Global information, distribution system
  coordination, 277–278
Gradient projection method, optimal
  operational planning, 295–296, 311–313
  P-optimization module, 306
  Q-optimization module — linear form,
    307–308
Grain size, algorithm efficiency and, 12
Gram-Charlier expansion
  generating system reliability assessment
    loss of load probability (LOLP), 194–195
Granularity
  algorithm efficiency and, 12
  parallel processing
    algebraic equations, 21–22

algorithm implementation and, 38–39
transient stability, 27–29
Graphics
  distance relay coordination, 77–78
  overcurrent relay coordination, 73–74
  transmission protection, 68
"Guess" waveform, transient stability, 35

                                    H

Hard constraints, critical period operation,
  420–421
Hardware components, adaptive transmission
  protection, 92–93
Hermite polynomials, reliability assessment,
  194–195
High load, distribution system coordination,
  282–284
Human factors, adaptive transmission
  protection, 95–96
HV transmission, voltage stability, 111
Hybrid approaches, reliability assessment,
  226–227
Hydro-constraints, hydro-electric power
  systems, 447–452
Hydroelectric systems
  optimization techniques, 371–468
    load following scheduling efficiency,
      446–468
      B.P.A. system application, 459–468
      minimum norm formulation, 452–459
      objective function and hydro-
        constraints, 447–452
    long-term operation modeling, 375–379
      pumped storage plants, 375–376
      run-of-river plants, 376
      storage plants, 376–379
    multichain power systems with critical
      water conditions, 400–422
      cubic spline technique, 419–421
      minimal norm formulation, 410–415
      optimal equations, 415–417
      optimization objectives and constraints,
        402–410
      solution algorithm, 418–419
    overview, 371–375
    series reservoirs long-term scheduling,
      379–400
      inflows and probabilities, 396–398
      installation characteristics, 397
      mega-watt hours (MWh), 380–382
      minimum norm formulation, 386–391

monthly releases, 396, 399
monthly reservoir storage, 399–400
objective function and constraints,
    380–383
optimal equations, 392–395
solution algorithm, 395–396
system modeling, 383–386
water conservation equation, 382
specified monthly generation, 422–446
    expected monthly inflows, 438
    installation characteristics, 437
    minimum norm formulation, 427–433
    optimal equations, 433–437
    optimal reservoir storage, 439, 444–445
    optimal turbine releases, 439–443
    system modeling, 425–426
schematic of, 402–404
Hypercube technique
    parallel processing and, 8
    transient stability iteration, 29, 31–32
Hypernets, parallel processing and, 8
Hypertree topology, parallel processing and, 8

I

Identification algorithm, subsystem
    coordination, 83–85
IEEE 14–bus system, optimal operational
    planning, 319, 322–326
initial loadflow, 324
line and generator data, 323
on-line diagram, 319, 322
P-optimization, 325–326
P-Q optimization, 325–328
Q-optimization, 319, 325–326
IEEE 30–bus system, optimal operational
    planning, 326, 329–335
generator data, 330
initial loadflow, 331–332
line data, 329
P-optimization, 331, 333–334
P-Q optimization, 331, 333–334
Q-optimization, 331, 333–334
Implicit integration technique, transient
    stability, 27–28
Importance sampling, reliability assessment,
    225
Incremental cost function, optimal operational
    planning, 305–306
Independent Power Producers (IPP), 105–106
Inequality constraints
    hydro-electric power systems, 447–449

long-term operations reservoir modeling, 379
multichain power system optimality,
    406–408
series reservoirs, 383–386, 384–386
Initialization, multistage linear programming,
    359
In-plant power plants, multistage linear
    programming, 360–361
Installation characteristics
    optimal long-term scheduling, series
        reservoirs, 397
    specified monthly generation, 437
Instantaneous tap setting
    adaptive transmission protection, 88–92
    overcurrent relay coordination, 70
Instantaneous unit failure contingency, 70
Integer quadratic optimization
    distribution system coordination, 245–287
    algorithms, 268–276
    approximate integers
        voltage-constrained case, 262–265,
            274–276
        voltage-unconstrained case, 265–274
    approximate switched voltage-constrained
        problem, 260–262
    approximation formulas, 249–259
    bounding results, 266
    control variables, 260
    design problems and, 285–286
    losses, constraints and distribution
        automation, 246
    organization, 248–249
    solution strategies, 264–25
    switched capacitors and regulators,
        246–248
    test results, 276–285
Integer voltage-unconstrained distinguished
    problem, 270–274
Integration methods, transient stability,
    26–27
Interconnected systems, reliability assessment,
    198–199
Internal cost function, optimal operational
    planning, 298–300
International Standards Organization Open
    Systems Interconnection, 94
Intrainstruction parallelism, 10
iPSC
    parallel algorithm implementation problems,
        44
    hypercube, 39–43
    power flow analysis, 21
Iterated timing analysis (ITA), 29, 32

J

Jacobian matrices
  algebraic equations, 20–21
  optimal operational planning, 295
    linearization of power balance equation
      and, 300–301
    nonlinear power balance equation,
      297–298
    sensitivity equations, 303–305
  power flow analysis and, 13–14
  voltage stability
    modal analysis, 128–133
    static analysis, 126–127
Job level parallelism, algorithm efficiency and,
  10

K

Kirchhoff's laws, reliability modeling, 212
Kuhn-Tucker equations
  hydro-electric power systems, load flow
    scheduling efficiency, 452–459
  multichain power system optimality, critical
    period operation, 410–415
    algorithm solutions, 418–419
    optimal equations, 417
  optimal long-term scheduling, series
    reservoirs 386–391, 394–395
  specified monthly generation, 427–433
    optimal equations, 436–437

L

Lagrange multipliers
  hydro-electric power systems
    load flow scheduling efficiency, 452–459,
      458–459
    optimal long-term scheduling
      series reservoirs, 386–391
    optimal operational planning, 295
      Q-optimization module — linear form,
        307–308
    specified monthly generation, 427–433
      optimal equations, 433–437
Least-squares curve, specified monthly
  generation, 426
Linear constraints, optimal operational
  planning, 312
Linearized solutions, parallel processing, 1
Linear matrix equations, power flow analysis,
  25

Linear programming, dynamic programming
  and, 374–375
Line data, optimal operational planning
  6-bus system, 313, 316
  14-bus system, 323
  30-bus system, 329
LINES relations, in RELAY, 61–62
LINKNET data structure
  subsystem coordination, 82–83
  transmission protection, 64–65
Liquid propane gas, optimization of operations,
  362, 366–367
Load buses, voltage stability, 127–128
Load compensation, voltage collapse
  prevention, 156
Load dependability, hydro-electric power
  systems, 461
Load duration curve, generating system
  reliability assessment
    loss of load probability (LOLP), 193–194
    system reliability modeling, 189–190
Load flow methods
  reliability modeling, 210–212
  scheduling efficiency, 446–468
Load following optimization
  hydro-electric power systems, 461–468
  results, 461, 464, 467
Load increase, voltage stability, 145–146
Load modeling, voltage stability, 124, 154
Load point indices, transmission system
  reliability, 177
Load tripping, voltage stability, 146–149
Local control, distribution system coordination,
  277–278
Long-term hydro plant operation, 375–379
  pumped storage plants, 375–376
  run-of-river plants, 376
  storage plants, 376–379
Loop enumeration algorithm, transmission
  protection, 66
Loosely coupled multiprocessor architecture,
  7–8
Loss of load expectation (LOLE)
  generating system reliability assessment,
    195–196
  reliability indices and, 175
Loss of load probability (LOLP)
  generating system reliability assessment,
    192–195
    capacity outage probability table, 193–194
    cumulant method, 194–195
    interconnected systems, 198–199
    load duration curve, 193–194
    reliability indices and, 174–176

Loss technology
  distribution system coordination, 278–283
  integer quadratic optimization, 246
LU decomposition, transient stability, 27–28

M

Man-machine dialog
  distance relay coordination, 77–78
  overcurrent relay coordination, 73–74
Margin quantification, voltage stability, 123
Markov models
  bulk power system reliability assessment
    protection system malfunctions, 206–207
    state enumeration, 217–218
    weather conditions, 205–206
  system reliability modeling, 187–188
  state space representation, 181
Masking, reliability assessment, 220
Matrix differential equation, state reliability
    indices, 182–183
Max-flow/min-cut algorithm, reliability
    modeling, 212
Mean time spent in state, state reliability
    indices, 182–183
Medium load, distribution system coordination,
    278–281
Mega-watt hours (MWh)
  optimal long-term scheduling, 380–382
  hydro-electric power systems, 446–447, 450
Message passing machines
  parallel algorithm implementation problems,
    43–44
  pipelining and, 8
Mho distance relays, distance relay
    coordination, 76
Mid-term time frame
  generating systems, 237–239
  operational reliability, 231
MIMD (Multiple-Input-Multiple-Data)
    machine
  algorithm efficiency and, 10–11
  pipelining and, 7
Minimum discharge requirement violation,
    465–466
Minimum loss reactive flows, 252–253
Minimum norm formulation
  hydro-electric power systems, 446–468
    load flow scheduling efficiency, 452–459
    multichain power system optimality,
      401–402
    critical period operation, 410–415
    optimal long-term scheduling, 386–391

specified monthly generation, 427–433
Mismatch equations
  algebraic equations, 20–21
  power flow analysis and, 14, 16
Mixed gas and oxygen demand, optimization of
    operations, 363–364
Modal analysis, voltage stability, 128–133
Modified Newton, transient stability, 29, 37
Monte Carlo simulation
  bulk power system reliability assessment,
    221–226
    random number generation, 221–222
    sample size and accuracy, 224
    sequential sampling, 222–223
    snapshot sampling, 222
    variance reduction techniques, 225–226
  generating system reliability assessment
    energy limited units, 199
    interconnected systems, 199
  parallel processing, 3
Monthly inflow probabilities, series reservoirs,
    396, 398
Monthly reservoir inflows, specified monthly
    generation, 438
Multi-area generating system reliability
    assessment, 199
Multichain power systems, critical water
    conditions, 400–422
  cubic spline technique, 419–422
  minimum norm formulation, 410–415
  optimal equations, 415–417
  solution algorithms, 418–429
  water conservation equations, 404–410
Multicomputer (distributed processing)
    systems, 8–9
Multiple factorization, algebraic equations,
    23–24
Multiple matrix solutions, power flow analysis,
    17–18
Multipliers' vector, simplex method, 349–351
Multiprocessing, pipelining and, 4, 6–8
Multipurpose stream use requirements,
    specified monthly generation, 424–425
Multireservoir power systems, optimization
    techniques, 372
Multistage linear programming, optimal
    operational planning, 341–369
  algorithm, 347–349
  applications, 360–368
  basis factorization, 343–346
  simplex method, 347
    factorization updating, 356–358
    initialization, 359
    multipliers' vector, 349–351

Multistage linear programming (*continued*)
    optimality testing, 352
    pivoting, 352–356
    relative cost factor, 351–352
Multi-state generating unit model, 185–186

N

N-1 rule, transmission system reliability,
    177–178
Near-term power system reliability, 231–233
Near-term time frame, 231
Network flow model, 212
Neural networks, parallel processing and,
    47–48
Newton power flow algorithm
    algebraic equations, 20–21
    properties of, 13–19
    schematic, 15
Newton-Raphson method
    bulk power system reliability modeling
        full AC load flow, 210–211
    optimal operational planning
        linearization of power balance equation
            and, 300–301
        nonlinear power balance equation,
            297–298
    voltage stability, static analysis, 126–127
Newton's method, 1
    transient stability, 29–30
Neyer's algorithm, approximation formulas,
    254
Nonlinear power balance equations
    optimal operational planning, 297–298
    incremental cost function, 298–300
    power flow analysis and, 13–19
Nonuniformity, hydro-electric power systems,
    461, 465
Numerical testing, distribution coordination
    systems, 257–259

O

"One-at-a-time" stochastic optimization,
    373–374
One-step (SOR-Newton), transient stability,
    32–33
On-line relay coordination programs, 94–95
On-load tap changers, voltage stability and,
    117
Operational constraints, reservoir modeling,
    378–379

Operational reliability
    near-term power system, 231–233
        component modeling, 233–235
        generating system operating reserve,
            235–237
        short- and mid-term reliability, 237–239
        reliability-constrained unit commitment,
            239–240
        time horizons, 231
Operator action, voltage collapse prevention,
    158
Optimal boiler output, optimization of
    operations, 362, 365
Optimal continuous losses, distribution
    coordination systems, 248
Optimal equations
    critical period operation, 415–417
    hydro-electric power systems, load flow
        scheduling efficiency, 457–459
    series reservoirs, 392–395
    specified monthly generation, 433–437
Optimal holder level, multistage linear
    programming, 362, 365
Optimality testing, multistage linear
    programming, 352
Optimal monthly releases
    reservoir storage, 399–400
    series reservoirs, 396, 399
Optimal operational planning
    multistage linear programming, 341–369
        algorithm, 347–349
        applications, 360–368
        basis factorization, 343–346
        initialization, 359
        simplex method, 347
            factorization updating, 356–358
            multipliers' vector, 349–351
            optimality testing, 352
            pivoting, 352–356
            relative cost factor, 351–352
    real and reactive power dispatches, 293–338
        incremental cost function, 298–300
        P-Q decomposition, 300–311
            P-optimization module, 305–306
            power balance equation linearization,
                300–301
            Q-optimization module, 307–311
            sensitivity equations, 301–305
        simulation results, 318–338
            6–bus system, 313–319
            IEEE 14–bus system, 319–326
            IEEE 30–bus system, 326, 329–336
Optimal power flow (OPF), parallel processing,
    3

Optimal power generation policy, multistage
    linear programming, 362, 366
Optimal storage, specified monthly generation,
    438, 443–445
Optimal switched losses, distribution
    coordination systems, 248
Overcurrent distance, 92–93
Overcurrent relay coordination
    adaptive transmission protection, 89–90
    relay protection, 68–74
Overhead transmission lines, 203

P

Parallel algorithms
    power system analysis with, 9–45
        algebraic equations, 20–25
        BX algorithm, 17–18
        efficiency of, 10–12
        Gauss-Jacobi algorithms, 18–19
        Gauss-Seidel algorithms, 18–19
        implementation, 38–45
        nonlinear equation solution, 13–19
        transient stability simulation, 26–38
        XB algorithm, 16–17
Parallel architecture, 1
    classification of, 4–9
Parallel-in-time methods, transient stability,
    36–37
"Parallelism-in-space" algorithms, 35
Parallel processing
    adaptive transmission protection, 97–98
    power systems analysis and
        future research trends, 45–48
        need for, 1–3
        parallel algorithms, 9–45
            algebraic equations, 20–25
            efficiency of, 10–12
            implementation, 38–45
            Newton algorithms, 13–16
            nonlinear equation solution, 13–19
            quasi-Newton algorithms, 16–19
            transient stability simulation, 26–38
        parallel architecture, classification, 4–9
            multicomputer (distributed processing)
                systems, 8–9
            multiprocessor systems, 6–8
            processor arrays, 6
            vector (pipelined) computers, 5–6
Partitioning
    algorithm efficiency and, 12
    parallel processing for transient stability, 34
Peaking operations, system reliability

modeling, 186–188
Percentage required load, 438–443
Performance indices, reliability assessment,
    219–220
Permanent faults, adaptive transmission
    protection, 89–90
Permanent outages, reliability modeling, 203
Picard iteration, transient stability, 31–32
Pickup setting, adaptive transmission
    protection, 88–92
Pipelining
    parallel processing with, 4–6
    time-point-pipelining, 35–36
Pivoting
    multistage linear programming
        simplex method, 352–356
        PIVOT1 and PIVOT2, 358
PJM method, operational reliability, 235–236
Potential energy concept, optimal long-term
    scheduling, 384–386
Power balance equation, optimal operational
    planning, 300–301
Power factor constants, voltage stability,
    114–117
Power purchasing costs, multistage linear
    programming, 362, 364, 366
Power transfer limits, voltage stability, 123
P-Q optimization, operational planning,
    293–296, 300–301
    IEEE 30–bus system, 331, 333–334
    P-opimization module, 305–306
    power balance equation linearization,
        300–301
    Q-opimization module
        linear form, 307–308
        quadratic form, 309–311
        sensitivity equations, 301–305
        six-bus system, 315, 318–319
P (real power) optimization module
    operational planning, 305–306
        cost function, 296–297
        IEEE 30–bus system, 331, 333–334
        six-bus system, 315, 318
Precedence relationships, algebraic equations,
    23–25
Primary/backup (P/B) relay pairs, transmission
    protection, 64
    coordination, 67–68
Primary constraints, subsystem coordination,
    81–82
Probabilistic techniques
    reliability of power systems and, 166–167
        indices and criteria, 167–168
        unit commitment, 239–240

Probability bounds, reliability assessment,
    220–221
Probability distribution, reliability modeling,
    185–186
Probability methods, operational reliability
    generating system operating reserve,
        235–237
    short- and mid-term time frames, 238
Probability model, reliability modeling,
    188–189
Processor architecture, future research trends
    in, 46
Processor arrays, pipelining and, 4, 6
Profit realization, specified monthly generation,
    438–443
Protected component failure rates, 206
Protection/control coordination, voltage
    collapse, 156–157
Protection, of power systems
    adaptive transmission protection, 88–105
        benefits of, 88–92
        communication and control issues, 93–94
        definitions, 88
        distributed computing approach, 97,
            99–100
        future research trends, 101–105
            complexity levels, 101–103
        hardware components, 92–93
        human factors, 95–96
        relay coordination parallelism, 97–98
        software needs, 94–95
        supercomputer approach, 100
    artificial intelligence and, 106
    bulk power system reliability modeling,
        206–208
        event tree method, 207–208
        failure rate adjustment, 206
        Markov models, 206–207
        station failures, 208
        zone boundaries, 207–208
    combined transmission and distribution
        protection, 105–106
    historical aspects, 58–59
    overview, 57–58
    subsystem coordination, 78–88
        identification and coordination algorithms,
            83
        procedures for, 82–83
        relay response to system changes, 82
        terminology, 79–81
        test system and results, 84–88
    transmission protection software, 59–78
        database structure, 60–63
        distance relation coordination, 75–78
        overcurrent relay coordination, 68–74

topological analysis, 63–68
Protective device operation, voltage stability,
    154
Protective relay, 58
Pseudo-binary tree strategy, 43–44
Pseudo-rate variables, monthly generation,
    426
Pumped storage plants, long-term modeling,
    375–376

                      Q

Q-optimization module
    optimal operational planning
        IEEE 14–bus system, 319, 326
        IEEE 30–bus system, 331, 333–334
        linear form, 307–308
        quadratic form, 309–311
        six-bus system, 315, 318
Quadratic function
    optimal operational planning, 298–300
    multireservoir power systems optimization,
        375
Quasi-Newton algorithms, 13, 16–19
Query modes, overcurrent relay coordination,
    73–74
Q-V curves, voltage stability, 114–117
    criteria for, 140–142
    test systems, 119–122

                      R

Radial lines, overcurrent relay coordination,
    70–71
Randomized algorithm
    approximate integer voltage-unconstrained
        case, 268–269
    distribution system coordination, 278–280,
        282
Random numbers generation, reliability
    assessment, 221–222
Reactive power compensation, voltage collapse
    prevention, 156
Rectangular matrices, specified monthly
    generation, 431–433
Regulated compensators, voltage collapse
    prevention, 156
Regulators, distribution coordination systems,
    246–248
Relational Information Management (RIM),
    61–63
Relative cost factor, multistage linear
    programming, 351–352

Relative sequence vector (RSV), transmission protection,
   enumeration, 67
   topological analysis, 64–65
Relaxation techniques
   parallel algorithm implementation problems, 44
   parallel processing for transient stability, 28–29
   iteration, 32
Relay coordination
   adaptive transmission protection, 97–98
   overcurrent relay coordination, 69
   transmission protection, SSP, 67–68
RELAY programming
   adaptive transmission protection, 89
   DataBase Management System (DBMS) and, 60–63
   historical background on, 58–59
   transmission protection with, 59–78
Relay reliability, adaptive transmission protection, 91
Relay response, subsystem coordination, 82
Relay settings, subsystem coordination, 86–87
Release rates, series reservoirs, 396, 399
Reliability concepts, power systems, 163–171
   bulk power systems
      indices, state enumeration, 215–221
      modeling, 202–213
      reliability assessment, 215–229
   concepts, 163–165
   defined, 163–164
   deterministic and probabilistic approaches, 166–167
   generating systems
      assessment methods, 192–200
      indices and criteria, 174–176
      reliability modeling, 184–191
   large composite system studies, 170–171
   operational reliability, 231–240
      short- and mid-term time frames, 237–238
   parallel processing, 3
   probabilistic indices and criteria, 167–168
   severity of disturbances and, 178
   significance of, 163
   state-space representation, 180–183
   studies of, 165
   system failure conditions, 168–170
   transmission systems, 176–178
Reliability-constrained unit commitment, 239–240
Reservoir constants, 437
Reservoir modeling, long-term operations, 377–378
Reservoir outflow limits, 405–406

Reservoir storage
   critical period operation, 405–406
   optimal long-term scheduling, 399–400
Ring technique, parallel processing and, 8
Rounding algorithm
   approximate integer voltage-constraints, 275–276
   approximate integer voltage-unconstrained case, 267–269
Rounding algorithms, distribution system coordination, 278–280, 282
Row replacement rules, multistage linear programming, 357
Rule curves results, hydro-electric power systems, 461–463, 467
Runge-Kutta technique, transient stability, 26–27
Run-of-river plants, long-term operation modeling, 376

S

"Safety factor," reliability of power systems and, 166–167
Scalar partitioning, transient stability, 32–33
Scheduled outages, reliability modeling, 209–210
Security, reliability of power systems and, 164
   indices, 175
   operational reliability, 232–233
      generating system operating reserve, 236–237
   See also Protection, of power systems
Sensitivity equations, 301–305
Sequent Balance computer, 42
Sequential mode, overcurrent relay coordination, 73–74
Sequential sampling, reliability assessment, 222–223
Series capacitors, voltage collapse prevention, 156
Series reservoirs, optimal long-term scheduling, 379–400
   installation characteristics, 397
   mega-watt hours (MWh), 380–382
   minimum norm formulation, 386–391
   monthly inflows and probabilities, 396–398
   monthly releases, 396, 399
   monthly reservoir storage, 399–400
   multireservoir power system example, 380–381
   optimal equations, 392–395
   solution algorithm, 395–396
   system modeling, 383–386

Series reservoirs (*continued*)
  upper/lower variable limits, 383
  water conservation equation, 382
  water conversion factor, 380–382
Set covering problems, transmission protection,
    66–67
Set of sequential pairs (SSP)
  overcurrent relay coordination, 69
    time dial setting, 72–73
  subsystem coordination, 83
  transmission protection, 64
    enumeration, 66–67
    relay coordination, 67–68
Severity of disturbance, as reliability criteria,
    178
  state enumeration, 219–220
Shared memory architecture, 44–45
  pipelining and, 7–8
Short-term time frame
  generating systems, 237–239
  operational reliability, 231
Shunt capacitors, voltage stability, 116–117
  excessive dependence on, 155
SIMD processors, power flow analysis, 48
Simplex method, 342, 347
  algorithm outline, 347–349
  factorization updating, 356–358
  multipliers' vector, 349–351
  optimality testing, 352
  pivoting, 352–356
  relative cost factor, 351–352
Simultaneous implicit approach, transient
    stability, 27–28
Six-bus system, optimal operational planning,
    313–320
  initial loadflow, 315, 317
  line and generator data, 313, 316
  on-line diagram, 314
  P-optimization, 315
  P-Q optimization, 315, 319
  Q-optimization, 315
Skeleton distribution system
  approximation formulas, 255–256
  base-case loadflow, 251–253
Slow dynamic decay, voltage stability, 154
Snapshot sampling, reliability assessment, 222
Software components
  adaptive transmission protection, 94–95
  future research trends in, 46
Solution algorithms
  critical period operation, 418–419
  optimal long-term scheduling, 395–396
SOR-Newton algorithm
  fault length and, 46–47
  implementation problems, 39–43

Sparse inverse factors, algebraic equations,
    23–24
Sparse matrix
  parallel processing, 1
  transmission protection data scheme, 64–65
Specified monthly generation, power system
    optimization and, 422–446
Speedup
  algorithm efficiency and, 10
  parallel algorithm implementation problems,
    41
SPICE technique, transient stability, 27, 29
Spinning reserve, voltage collapse prevention,
    157
Stability margin, voltage collapse prevention,
    157
Staircase coefficient matrix
  basis factorization, 344–346
  multistage linear programming, 342
Standby component modeling, operational
    reliability, 234–235
Starting points, transmission protection, 64–65
State enumeration
  bulk power system reliability assessment,
    215–221
  operational reliability, near-term time frame,
    233
State evaluation, reliability modeling
  load flow methods, 210–212
  remedial measures, 212–213
State reliability indices, 182–183
State-space representation,
  reliability of power systems, 180–183
  truncation of, 218–219
Static contingency analysis
  parallel power analysis and, 47
  voltage stability, 125–135
    modal analysis, 128–133
    stability indicators, 133–135
    V-Q sensitivity, 126–128
Static VAR compensators (SVC), voltage
    stability, 125
Station equipment failure, reliability modeling,
    208
Steam demands, optimization of operations,
    361–362
Steam turbine generators
  multistage linear programming, 360–361
  operational range, 362–363
Stochastic dynamic programming (SDP),
    372–373
Stochasticity of river flows, 445–446
Storage plants, long-term operation modeling,
    376–379
  operational constraints, 378–379

reservoir models, 377–378
Structural disturbances, subsystem
    coordination, 79–80
Substation bus loads, reliability modeling, 204
Subsystem coordination
    protection with, 78–88
        algorithms for (SCA), 78–79, 83
        procedures, 82–83
        relay response to system changes, 82
        terminology, 79–82
        test results, 84–88
Subsystem database setup, 83
Subsystem identification algorithms, 83
Successive zone coordination algorithm
    (SZCA), 76–77
Supercomputing, adaptive transmission
    protection, 100
Surplus power uniformity, 447–448
Switched capacitors
    adaptive transmission protection, 103–105
    distribution system coordination, 246
        design issues and, 286
Switching schedule
    approximate integer voltage-constraints,
        275–276
    distribution system coordination, 247–248
    benchmark procedures, 280
Synchronization, algorithm efficiency and,
    11–12
System failure conditions, reliability indices
    and, 168–170
System integrity, 164
System load models
    bulk power system reliability modeling,
        204–205
    system reliability modeling, 189–191
System-minutes approach, reliability indices
    and, 175

                        T

Tailwater elevation
    hydro-electric power systems, 446
    multichain power system optimality, 407
Tap increments
    approximate integer voltage-constraints,
        275–276
    distribution coordination systems, 246
Task parallelism, algorithm efficiency and,
    10
Taylor expansions, distribution coordination
    systems, 249, 254–256
TDMC, distance relay coordination, 75–76
Temporary outages, reliability modeling, 203

Tightly coupled multiprocessor architecture,
    7–8
Time delay pickup tap setting, 70
Time-dial settings
    adaptive transmission protection, 88–92
    overcurrent relay coordination, 71–73
Time domain analysis, voltage stability, 135
Time horizons, operational reliability, 231
Time-point-pipelining (TPP)
    parallel processing for transient stability,
        35–36
Topological analysis, 63–68
Topological Analysis Program (TAP), 65, 68
Toroidal method, for transient stability, 38
Transient outages, reliability modeling, 203
Transient stability program, parallel processing,
    2, 26–38
Transmission components, reliability modeling,
    209–210
Transmission line loss, voltage stability, 153
Transmission protection
    data base structure and, 61–63
    distance relay coordination, 75–78
        man-machine dialog, 77–78
        models, 76
        successive zone coordination algorithm,
            76–77
        zone 1 setting, 76
    distribution combined with, 105–106
    overcurrent relay coordination, 68–74
        man-machine dialog, 73–74
        relay settings, 69–73
            instantaneous tap setting, 70
            instantaneous unit failure contingency,
                70
            models, 69
            radial line inclusion, 70–71
            time delay pickup tap setting, 70
            time-dial setting, 71–73
        system evaluation, 74
    reliability indices and criteria, 176–178
    software developments for, 59–78
    topological analysis and, 63–68
Transparency, implementation problems, 45
Transportation model, reliability modeling,
    212
Trapezoidal integration, transient stability,
    36–37
Travelling-wave based relays, 93
Travelling window technique, transient stability,
    38
Triangular factorization, power flow analysis,
    20–21
Triangular matrices, specified monthly
    generation, 429–433

Turbine optimal release, specified monthly
  generation, 438–443
Two-state generating unit model, reliability
  modeling, 184–185

U

Underload tap changer (UTLC)
  dynamic voltage analysis, 135
  voltage stability, 124
Undervoltage protection, 157
Updating factorization, simplex method,
  356–358
Upstream loads, distribution coordination
  systems, 253–254

V

Variance reduction, reliability assessment,
  225–226
Vector computers, pipelining and, 4–6
Vector matrix
  hydro-electric power systems, 455–459
  specified monthly generation, 429–433
Very dishonest Newton (VDHN)
  fault length and, 46–47
  implementation problems, 39–43
  parallel processing for transient stability,
    30–31
Vi/Ei criteria, voltage stability, 143
Voltage collapse
  prevention of, 156–159
    load compensation of generator AVRs, 156
    operating measures, 157–158
    protection and control coordination,
      156–157
    protection schemes, 158–159
    reactive power compensation, 156
    undervoltage protection, 157
  voltage stability and, 116
Voltage constraints
  approximate integer voltage-constrained
    case, 264
  approximate switched voltage-constrained
    case, 261–263
Voltage profile, distribution system
  coordination, 278–283
Voltage stability
  analytical techniques, 118–135
  current practice, 119–123

  dynamic analysis, 135
  static analysis, 125–135
    modal analysis, 128–133
    voltage stability indicator, 133–135
    V-Q sensitivity, 126–128
    system modelling, 124–125
  concepts and definitions, 114–117
  criteria for, 136–144
  incidents of, 145–153
    generation loss, 149–152
    load increase, 145–146
    load tripping, 146–149
    transmission line loss, 153
  overview of, 111–113
  preventive measures, 154–159
Voltage stability indicator, 133–135
Voltage-weak points, voltage stability, 123
V-P curves, voltage stability, 114–117
  criteria for, 138–139
  test systems, 123
V-Q sensitivity, voltage stability
  eigenvalues, 130–133
  static analysis, 126–128

W

Water conservation equation
  critical period operation, 404–410
  hydro-electric power systems, 448–449
  optimal long-term scheduling, 382
  specified monthly generation, 423–424
Water-to-energy conversion factor, 407–408
Waveform-Newton (WN) algorithm, 29, 34–35
Waveform relation (WR) method, 29, 33–34
Waveform-relaxation-Newton (WRN)
  algorithm, 35
Weather conditions, reliability modeling,
  205–206
Westinghouse KD relays, 76
Windows
  adaptive transmission protection, 89–90
  parallel processing for transient stability, 38
  subsystem coordination, 79–80
"Worst-case" conditions, reliability and,
  166–167

X

XB algorithm, power flow analysis, 17–18